PRENTICE HALL'S
ENVIRONMENTAL TECHNOLOGY SERIES

Volume 5

Waste Management Concepts

Prentice Hall's Environmental Technology Series

Available:

Planned:

PRENTICE HALL'S
ENVIRONMENTAL TECHNOLOGY SERIES

Volume 5

Waste Management Concepts

NEAL K. OSTLER, Editor
Salt Lake Community College

JOHN T. NIELSEN, Editor

Prentice Hall
Upper Saddle River, New Jersey   *Columbus, Ohio*

Library of Congress Cataloging-in-Publication Data

Waste management concepts / Neal K. Ostler, John T. Nielsen, editors.
      p.   cm. — (Prentice Hall's environmental technology series ;
  v. 5)
   Includes bibliographical references and index.
   ISBN 0-02-389545-4
   1. Hazardous wastes—Management.   I. Ostler, Neal K.
  II. Nielsen, John T.   III. Series.
  TD1030.W37   1998
  363.72'87—dc21
                                                     97-22365
                                                          CIP

Editor: Stephen Helba
Production Editor: Mary Harlan
Design Coordinator: Julia Zonneveld Van Hook
Text Designer: Custom Editorial Productions, Inc.
Cover Designer: Brian Deep
Production Manager: Pamela D. Bennett
Marketing Manager: Debbie Yarnell
Editorial/Production Supervision: Custom Editorial Productions, Inc.
Illustrations: Custom Editorial Productions, Inc.

This book was set in Utopia by Custom Editorial Productions, Inc., and was printed and bound by Quebecor Printing/Book Press.

 © 1998 by Prentice-Hall, Inc.
Simon & Schuster/A Viacom Company
Upper Saddle River, New Jersey 07458

Printed in the United States of America

10 9 8 7 6 5 4 3 2 1

ISBN 0-02-389545-4

Prentice-Hall International (UK) Limited, *London*
Prentice-Hall of Australia Pty. Limited, *Sydney*
Prentice-Hall Canada Inc., *Toronto*
Prentice-Hall Hispanoamericana, S. A., *Mexico*
Prentice-Hall of India Private Limited, *New Delhi*
Prentice-Hall of Japan, Inc., *Tokyo*
Simon & Schuster Asia Pte. Ltd., *Singapore*
Editora Prentice-Hall do Brasil, Ltda., *Rio de Janeiro*

Contents

Appendix E

Appendix F

Appendix G

Appendix H

Preface

Development of this Text

To prepare myself for the opportunity to teach a class primarily focused on the Resource Conservation and Recovery Act (RCRA, 1976), with no suitable text available at the time, I obtained a copy of Utah's "Rules of Hazardous Waste Management," from the Division of Solid and Hazardous Waste with the Utah Department of Environmental Quality (DEQ). Although these "Rules" were purely a regulatory manual, they provided the needed materials for a substantial portion of the curriculum in that first basic course. Subsequently I found it essential to add sections of study on both low-level radioactive waste and medical waste management concepts, and to supplement the course with discussions on waste minimization, pollution prevention, and remedial technologies. From this experience evolved the framework of a text on Waste Management Concepts.

Salt Lake Community College offers a two-year AS degree in this field and participates as a member of the Partnership for Environmental Technology and Education (PETE). The creation of PETE was the result of efforts (early 1990s) on the part of industry, the academic community, and government agencies (especially the Department of Energy) to establish an appropriate curriculum and provide industry with the ability to create and then to fill technician-level positions throughout the country. There are now (as this text goes to press) over 350 colleges and universities participating as active members of National PETE and implementing certificate and/or degree programs in fields directly related to PETE-developed curriculum guidelines and professional competencies for the environmental technician.

Government regulations of any sort are not easy to write about, and this includes those pertinent to the management of waste. Particular care has been taken to generalize the content of each chapter so that the material would be as nonperishable as possible. The reader is advised that this text is not a procedural document or a "how to" manual. This text is intended to provide the environmental technician with the ability to articulate waste management concepts that are basic to industry.

Goal of the Text

This text is the fifth in a planned six-volume series of Prentice Hall's Environmental Technology Series and was written primarily to serve as a course textbook for the environmental technician. Even though waste management is a complex subject, this text has been written with simplified terminology and language and is suitable as an aid to any person involved in a general study of the regulatory mechanisms for the management of different varieties of waste within the United States government.

The goal for the text was to include a full range of waste management concepts important to the hazardous materials or environmental technician and, therefore, includes the following subjects:

▶ Pollution prevention and its focus on the minimization of waste.
▶ Comparison of regulated solid (nonhazardous) wastes with those that are hazardous or exempt from EPA regulation.
▶ The hazardous waste generator with its related responsibility to identify and classify the waste it produces, as well as storage requirements for both small- and large-quantity generators.
▶ Transportation requirements that include "how to" guidelines in the preparation of the Hazardous Waste Manifest.
▶ Regulatory requirements for facilities involved in the treatment, storage, and disposal of hazardous waste (known as TSDFs).
▶ The permitting process for TSDFs.
▶ Enforcement mechanisms used by the EPA to obtain compliance with the regulations.
▶ Medical waste management concepts regulated by federal, state, and local health departments.
▶ Radioactive waste management concepts regulated by the U.S. Department of Energy and the Nuclear Regulatory Commission.
▶ Preparation of Phase I and Phase II Site Assessments necessary in the planning stages for hazardous waste site response and remediation.
▶ Modern hazardous waste remedial technologies and research.

In addition to these subjects, Chapter 6 includes additional information on Underground Storage Tank programs and Chapter 9, Medical Waste Management, includes discussion of OSHA's Bloodborne Pathogens Standard. The text also provides a substantial number of appendixes, such as blank copies of the Hazardous Waste Manifest accompanied by step-by-step directions for its completion, to help the student develop needed skills in understanding and handling of hazardous waste.

Acknowledgments

The idea for this text and for the entire series evolved from the lack of suitable training materials for use in the environmental technology classroom. To make all of the volumes in the Environmental Technology Series as authoritative and current as possible, I sought contributing authors to supplement the core manuscript with chapters related to their individual expertise and pertinent to the basic topic.

I would like to express my appreciation to Dan Smith and Ryan Dupont, with the Water Research Laboratory at Utah State University, and to Mike Williams of Kleinfelders, Inc., a national consulting firm with offices in Salt Lake City. Each of these chapter authors has made a significant contribution to this text, and I would encourage you to review *About the Authors* for a closer look at each of these professionals.

In the numerous years that have passed since the idea for the Environmental Technology Series was first conceived, I have received continued support from Salt Lake Community College. I offer my thanks to John Latkiewicz, my division director since the fall of 1993, for his fairness and objectivity as my supervisor, and to George Van DeWater, who is involved so closely with PETE.

I want to thank John T. Nielsen, my co-editor in this volume and in Volume 2, for the confidence and moral support he has given me throughout this project.

I wish to thank those who reviewed the manuscript: Rhonda L. Howard of Coconino Community College, Norm R. Sunderland of Utah State University, and Robert R. Treloar of Paradise Valley Community College.

Work on this text has involved a good deal of personal time, and I wish to acknowledge the support of my wife, Karen, and my children and stepchildren. They have come to understand what it means when Dad is "working on the texts," and I wish to express my love and appreciation for their concessions.

Finally, it has been my sincere pleasure to work with my project editor at Prentice Hall, Stephen Helba, with the assistants in his office, and with the dedicated staff at Custom Editorial Productions. Thank you all for helping to get the collective manuscripts finally ready for printing of this fifth volume.

Neal K. Ostler

About the Authors

Neal K. Ostler

Neal K. Ostler lectures on environmental technology at Salt Lake Community College and is founder of the Environmental Training Center where he coordinates and delivers a variety of noncredit workshops and seminars in the environmental health and safety arena. His background and experience include over twenty years in law enforcement, where he became adept at investigative reporting and emergency response operations. While employed at the Utah State Prison and assigned to maximum security, Mr. Ostler designed, developed, and delivered an in-cell Alcohol and Substance Abuse (biblio-therapy) Program for inmates housed in limited movement facilities that was the first of its kind in the U.S.

Mr. Ostler first began to develop his hazmat credentials while employed as a Motor Carrier Investigator with the Utah Department of Commerce. He is certified as a "Train-the-Trainer" in a variety of subject areas that include HAZWOPER, Confined Space, DOT HM-181, and Hazard Communications. His credentials include that of Certified Hazardous Materials Manager (CHMM) and he participates in his local chapter of the National Association of Environmental Professionals. In the fall of 1996 he obtained the credential of Designated Trainer for ISO 14000 Awareness from the Global Environmental Training Foundation. Mr. Ostler is a graduate of the University of Utah and has attended over 1,400 hours of related workshops and seminars in addition to completing the AS of Environmental Technology Degree at SLCC.

John T. Nielsen

John T. Nielsen is Senior Counsel and Director of Government Affairs for Intermountain Health Care (IHC), a large nonprofit integrated health care delivery system with corporate offices in Salt Lake City. He is also Counsel with the law firm of Van Cott,

Bagley, Cornwall, McCarthy and Associates. He has a broad range of experience as a public lawyer at all levels of government.

Mr. Nielsen served as Utah's Commissioner of Public Safety from 1985 to 1989, where he founded the Utah Hazardous Materials Institute and instituted a Hazardous Materials Response Program utilizing the Utah Highway Patrol and the state's comprehensive emergency planning operations. His law practice consists of civil and criminal litigation, administrative and regulatory law.

Daniel P. Smith

Dr. Daniel P. Smith has researched and applied biological treatment processes for over fifteen years. He received a Ph.D. from Stanford University, where he conducted fundamental research on substrate interactions in anaerobic processes. He has studied methanogenic processes extensively and has published research articles on hydrogenotrophic control in *Biotechnology and Bioengineering Science and Technology*. A key publication in the *Research Journal of the Water Pollution Control Federation* detailed a mathematical modeling approach that describes bioenergetic and kinetic interactions in non-steady-state bioreactors. Dr. Smith has worked as an environmental engineer for the U.S. Army Corps of Engineers and for state environmental agencies, and as a private consultant. He was a visiting research engineer at ENEA, a government-sponsored process research laboratory in Italy. There he developed control systems for hybrid upflow anaerobic sludge blanket reactors for treatment of agriculture wastewaters. Dr. Smith joined the faculty of the Civil and Environmental Engineering Department at Utah State University in September 1994. Prior to that he was Senior Environmental Process Engineer at Rhone Poulenc Inc.

R. Ryan Dupont

R. Ryan Dupont is an Associate Professor of Civil and Environmental Engineering and the Associate Director of the Utah Water Research Laboratory at Utah State University, where he teaches undergraduate and graduate students, carries out applied environmental research, and publishes in the areas of hazardous waste management and soil bioremediation. Dr. Dupont has a B.S. degree in Civil Engineering and M.S. and Ph.D. degrees in Environmental Health Engineering from the University of Kansas. He has been active in developing teaching materials in the hazardous waste and environmental management fields.

Michael A. Williams

Michael A. Williams earned his B.A. degree in geology and an M.S. degree in geology and water resources from Iowa State University. He is a registered Professional Geologist with the state of Wyoming. He has thirteen years of experience as a hydrogeologist in the petroleum, mining, and environmental industries. He also has eight years of experience conducting and supervising groundwater monitoring programs for RCRA, LUST, NPDES, and SDWA projects, and has prepared numerous sampling and analysis plans for sites contaminated with paint wastes, chlorinated solvents, petroleum hydrocarbons, and radionuclides. Mr. Williams is currently employed as a project hydrogeologist with Kleinfelder, a western geotechnical, environmental, and water resources engineering firm in Salt Lake City.

1

Overview of Pollution Prevention

Daniel Smith

Upon completion of this chapter, you will be able to meet the following objectives:

▶ Describe the basic concept of pollution prevention.

▶ Demonstrate the fundamental benefits of implementing a pollution prevention plan.

▶ Describe the various steps in the implementation of a pollution prevention program.

▶ Discuss the hierarchy of pollution prevention and its various levels of priority.

▶ Identify at least four different methods of separation technology for materials recovery.

▶ Differentiate between waste treatment, reduction of waste, and recycling materials.

OVERVIEW

Pollution prevention involves reducing waste materials generated from all aspects of a production process to the least amount possible, while still accomplishing the production goal. This is accomplished by a variety of methods including modifying the products themselves, changing the production methods to make them more efficient, recycling materials used in the process so there is less waste to be disposed of, lowering the amount of materials used where they are not really needed, changing the packaging, and even changing the product itself.

By reducing their generation at the source, lower quantities of wastes will have to be treated in waste treatment plants or disposed of in landfills. The reduction of pollution results because the more waste generation is reduced, the less waste will be released as pollutants into the environment. The goal of pollution prevention is to reduce the total "environmental footprint" of a production process, which means the overall impact of the production activity, both at the production site and at other locations. High-energy usage at a production plant, for example, results in greater power requirements from an energy generation plant that may be many miles away. Inefficient or wasteful energy usage at the production facility results in a higher emission of air pollutants from the generation station supplying the plant. Pollution prevention (P^2) is concerned with all the energy and raw materials used by a process, all the wastes generated, and all materials released into the environment in the form of solids, liquids (e.g., wastewaters), and emissions to the atmosphere. P^2 represents a "multimedia" approach to environmental impact, with the guiding principle being that it is better to prevent wastes from being generated in the first place than to have to treat or dispose of them later.

BENEFITS AND INCENTIVES OF POLLUTION PREVENTION

The benefits of reducing waste generation are summarized in the saying "pollution prevention pays." More efficient production methods can lower the costs of raw materials and energy, making an operation more profitable and competitive. More waste generation means higher costs per product unit for treating and disposing of waste materials. Waste management and disposal also create potential liability to the generator and can influence employee morale and public image. The ultimate benefit of a P^2 program is increased protection of public health and the natural environment. The benefits of a pollution prevention effort are summarized as follows:

▶ Lower costs for purchase of raw materials.
▶ Lower costs for waste treatment and disposal.
▶ Reduced liability for waste disposal.
▶ Lower energy consumption.
▶ More competitive production process.
▶ Reduce regulatory burdens.
▶ Higher employee morale and safety.
▶ More favorable public image.
▶ Protection of public health and the environment.

Pollution Prevention Hierarchy

Pollution prevention follows a hierarchical approach to management of wastes generated from production. As shown in Figure 1–1, source reduction is at the top of the hierarchy, followed by recycling, waste treatment, and final disposal. The lower in the hierarchy, the less desirable an option is. Source reduction receives the highest priority because it can reduce or even eliminate the need for the less desirable methods that are lower on the hierarchy. In Figure 1–1 example applications and activities are given for source reduction and the other waste management methods. The more effective any method in the hierarchy, the less need there is for the next lower priority item. For example, a new product design or more efficient production process can substantially reduce the need for recycling, while effective water recycling can substantially reduce the quantity of wastewater generated. Lower flow rates of wastewater would reduce the mass of biosolids that would have to be disposed of in a landfill.

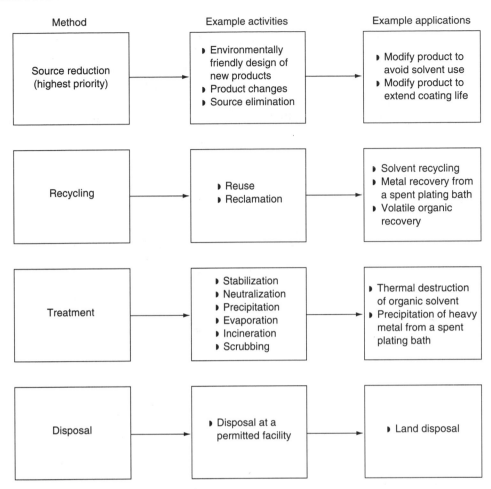

Method	Example activities	Example applications
Source reduction (highest priority)	• Environmentally friendly design of new products • Product changes • Source elimination	• Modify product to avoid solvent use • Modify product to extend coating life
Recycling	• Reuse • Reclamation	• Solvent recycling • Metal recovery from a spent plating bath • Volatile organic recovery
Treatment	• Stabilization • Neutralization • Precipitation • Evaporation • Incineration • Scrubbing	• Thermal destruction of organic solvent • Precipitation of heavy metal from a spent plating bath
Disposal	• Disposal at a permitted facility	• Land disposal

▶ **FIGURE 1–1**
Waste management prevention hierarchy (*Facility Pollution Prevention Guide,* U.S. EPA, EPA/600/R-92/088, 1992).

By starting at the top of the environmental management hierarchy and working down, waste generation and the overall environmental footprint of a production activity can be minimized. Waste management through a combination of source reduction, recycling, waste treatment, and final disposal constitutes a total *systems approach* to pollution prevention. The P² hierarchy is applied to all aspects of the production system to develop the combination of methods that will be most effective in reducing the total impact of waste generation. Pollution prevention is a "multimedia" approach to waste management, and is concerned with waste materials generated and entering the environment in all forms, from solid wastes, slurries, and sludges, to liquid wastes and suspensions and gaseous pollutants.

Source Reduction
Source reduction is the highest priority in P², and entails either modifying the product itself or modifying the process by which it is produced. *Product modification*

involves examining the purpose for making the product in the first place, and the possible modification of the product or its substitution with another more environmentally friendly product. In considering the product itself a *life cycle assessment* may be performed, in which the overall environmental impact of a product throughout its entire useful life is used as the basis for comparison with alternatives (Figure 1–2). A life cycle assessment tracks material and energy flows and transformations beginning from the acquisition of raw materials, through the use of the product for its intended purpose, to the final disposal of waste materials generated

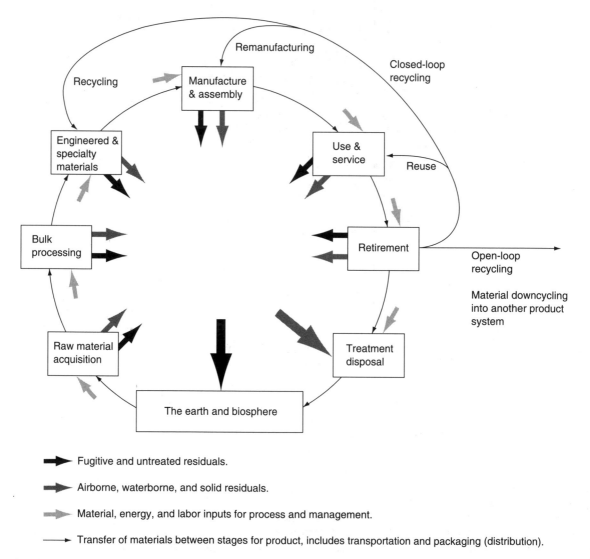

➧ **FIGURE 1–2**
Life cycle of a product (*Life Cycle Design Guidance Manual,* U.S. EPA, EPA/600/R-92/226, 1993).

from production and the product itself. The life cycle of a product consists of the following stages:

▶ Acquisition of raw materials.
▶ Processing of bulk materials.
▶ Production of engineered and specialty materials.
▶ Manufacture and assembly.
▶ Use of the product.
▶ Retirement.
▶ Disposal of the product and wastes generated from it.
▶ Disposal of the wastes generated from the product's manufacture and use.

Some parts of a product's life cycle are circular, others are not. Most products require direct or indirect utilization of numerous raw materials. The total life cycle for any individual product can be considerably more complex than the simplified diagram shown in Figure 1–2.

A life cycle assessment, therefore, represents a comprehensive, cradle-to-grave analysis of the environmental impact of a product and can be used to judge the relative merits of modifying the product or substituting it with an alternative. A life cycle assessment can lead to the redesign of products to minimize their environmental impact over their life cycle. Examples of environmentally responsible product design are as follows:

▶ Using recycled materials.
▶ Using renewable natural materials.
▶ Recycling product and wastes.
▶ Reducing use of toxic solvents.
▶ Replacing toxic solvents with alternative materials.
▶ Reusing scrap material.
▶ Reducing packaging requirements.
▶ Replacing solvent-based inks and coatings with water-based ones.
▶ Producing more durable products with longer lives.

Process modification involves changing the methods by which the product is made. Process changes can be accomplished by changing the input materials, improving operating practices, or changing the manufacturing technology used to produce the product. An example of changing input materials is using purer raw materials as feedstocks. This would reduce the amount of unwanted by-products generated per unit of production. Substituting less toxic materials as feedstocks would reduce the amount of hazardous materials in the generated wastes.

Many operating practices can be improved to reduce waste generation at the source. Improvements in operations and maintenance procedures, materials handling, production scheduling, and inventory control can all decrease waste generation, and all must be implemented through good management practices at the plant. The segregation of process and waste streams can allow more materials to be recycled and lower the overall quantity of wastes to be disposed. Examples of improved operating practices that can reduce waste generation include the following:

▶ Segregating hazardous and nonhazardous waste streams.
▶ Training operators.

- Covering solvent tanks when not in direct use.
- Having operators pay more careful attention to operating procedures.
- Improving maintenance scheduling and record keeping for higher efficiency.
- Optimizing purchasing and inventory procedures.
- Stopping leaks, drips, and miscellaneous spills.
- Using drip pans and splash guards.
- Minimizing spills and losses from transport of parts and materials.
- Conserving electrical energy through management of lighting and appliances.

Production technology changes might have done much to decrease waste generation, especially if they had been designed when waste generation was not as great a consideration as it is today. Still, new equipment and technology, more careful operating controls, and automation can reduce waste generation rates. Examples of technology changes for P^2 include the following:

- Redesigning equipment and piping to reduce volume.
- Redesigning to reduce losses during maintenance and cleaning.
- Changing to a powder coating system.
- Installing centrifuge to capture latex particles from wastewater for reuse.
- Using mechanical stripping and cleaning devices rather than solvent systems.
- Installing a vapor recovery system to capture and reuse vaporous emissions.
- Installing more efficient motors.
- Reducing energy consumption by automated controls.

With management support and trained plant personnel, some process modifications may be relatively simple and straightforward to implement. Other process modifications for waste reduction may require detailed studies before they can be implemented without compromising safety or product quality. The production of synthetic organic chemicals in high-temperature, high-pressure reactors is an example. Sophisticated computer simulation or experimental testing may be required to assess the feasibility of reducing wastes by changing the production methods.

Recycling

Recycling materials can take place either within a production process, between different processes, or after the product has reached a consumer market and been returned. Recycling involves reusing previously used material (which would formerly have been waste material) as a substitute for input material into either the process where it was originally used or into another process. *Recovery,* or *reclamation,* involves processing a waste material to remove or extract a component of the waste that is valuable or can be used again in the same or another process. Both reuse and recovery operations can be accomplished either at the site where the materials were first used or off-site (not at the point of original use). On-site reuse is preferable to off-site recycling or reuse because there are no risks or liabilities associated with transporting of the material. The use of a waste material as a fuel supplement, either on-site or off-site, recovers the energy value of the material and can be considered a form of recycling.

Many types of processing technologies can be employed to recover beneficial waste components for reuse. Most recycling technologies serve to separate or concentrate the desired material from the other materials with which it is mixed. Separation and concentration make it possible for the material to be reapplied for the same purpose or to be reused in another application. The following are some separation methods used for materials recovery:

1. Vapor-liquid separation, including distillation, evaporation, and gas absorption.
2. Solid-liquid separation, including filtration, centrifugation, and sedimentation.
3. Liquid-liquid separation, including liquid-liquid extraction, and decantation.
4. Solute recovery, including reverse osmosis, ultrafiltration, ion exchange, and precipitation.

The materials to be recovered can be present in different phases, as gases, liquids, or solids. It is useful to classify the separation of desired materials from waste mixtures in which they are contained according to the phases that are being separated. *Vapor-liquid separation* methods are commonly used to recover liquids for recycling. The most commonly used process for recovering liquids from process streams for recycling is *distillation*. It can be used to recover a material that has a lower boiling point than other components in the waste stream. Distillation is often used to recover organic compounds from water so they can be recycled within a process. By heating the waste stream, the organic compound with a lower boiling point will volatilize (change from a liquid to a gas) before the waste in which it is contained. The volatilized organic compound can then be captured, condensed, and recycled. An example of the distillation concept is to recover solvents from the manufacture of solvent-based paints or from carbon-adsorption systems.

Evaporation processes can be applied to sludges, slurries, process streams containing dissolved or suspended solids, and liquid wastes containing nonvolatile dissolved liquid. *Evaporation* is accomplished by heating the mixture and concentrating the desired nonvolatile component in a concentrated solution or thick liquor. The desired material for recovery is the thick liquor remaining after the other materials have evaporated; the vapor is not collected and condensed unless it contains organic compounds. Another process is *gas absorption*, which is used to separate chemical vapors from gas streams. In gas absorption, one or more components of a gas mixture are dissolved into a liquid. The vapor to be recovered must be readily soluble in the liquid. Gas absorption is usually performed in a contacting tower, where liquid and gas flow countercurrent to each other. An example of the gas absorption process is the removal of ammonia vapors from a stream of air by absorption into water.

Solid-liquid separation processes are used to remove suspended solids from a liquid stream. In *filtration*, the liquid is passed through a porous medium, which retains the solids as the liquid passes through. The porous medium may be a bed of sand or other granular material, a screen, or a fibrous fabric, paper, or cloth. For large particles, a thick bed of sand may be used in a process called *bed filtration granular media filtration*. A filter cloth may be used to filter finer materials. Precoating the medium with materials such as diatomaceous earth or ground cellulose may aid in removing small particles. Filters usually are cleaned by passing water through the filter in a direction opposite the flow of the waste stream. Granular media filters often are used after a gravity separation process for additional removal of suspended solids. An example of the use of filtration is the removal of dirt particles from organic solvent streams after they have been used in a process. The spent solvent can be reused in the process after the contaminating particles have been filtered out.

Centrifugation uses the differences in densities between the carrier liquid and suspended particles to separate the particles from the liquid by centrifugal force. The centrifugal force induced by rapid spinning causes the higher density solid particles to separate from the liquid. An example is the separation of latex particles from an aqueous process stream; removal of the particles reduces the wastewater organic content and the separated particles can be recovered for product use or recycled as raw materials for the production process. *Sedimentation,* used primarily for aqueous streams, removes particles from a liquid stream by gravity forces under quiescent, nonmixed conditions. It is accomplished in sedimentation tanks where the water is held long enough for the particles to settle. After the particles have settled out, the clarified liquid is decanted. Sedimentation is most applicable for waste streams containing low solid concentrations.

Liquid-liquid separation is accomplished through extraction and decantation. Liquid-liquid extraction is used to recover organic substances that are dissolved in water or organic solvent in concentrations ranging from a few hundred parts per million to a few percent. In liquid-liquid extraction, the waste stream is contacted with another liquid stream in which the waste stream solvent is immiscible. The organic solutes in the waste stream dissolve into the extracting solvent and are removed with the extracting solvent flow. The organic solutes can then be recovered by stripping or distillation and then can be recycled.

Decantation can be used to separate immiscible fluids of different densities. The liquid waste flows slowly from one end to the other of a separator tank, where two distinct layers are formed. The layers are withdrawn separately. Continuous flow, gravity decanters, and batch centrifuges all are employed for liquid-liquid extraction.

Solute recovery operations separate and recover dissolved organic and inorganic components from waste streams. Solute recovery methods include membrane separation, ion exchange, and precipitation. Two common *membrane separation* processes are reverse osmosis and ultrafiltration. *Reverse osmosis* uses a membrane that has selective permeability to solvent and solute. The solvent (water, for example) can pass through the membrane, but the dissolved organic or inorganic material cannot. The impurities concentrate on one side of the membrane while the purified solvent accumulates on the other. Reverse osmosis can be used to separate ions and small molecules in true solution in the solvent. One application for reverse osmosis is treating wastewater from metal machining operations, where oil-water emulsions are used for lubricating and cleaning tools. During this process application, the oil-water emulsion is contaminated with heavy metals, which are toxic materials. Reverse osmosis can be used to separate oil from water, yielding a water that can be discharged and an oil that can be further refined and reused.

Ultrafiltration uses a special membrane that allows the solvent to pass through under pressure while the larger solutes or colloids are retained. Particles as small as 0.001 microns can be removed and recycled. A common use for ultrafiltration in recycling is the recovery of pigments and resins from electrophoretic paint application processes. The small solutes retained by the membrane are returned to the electropaint tank for reuse rather than becoming a waste material.

In *ion exchange,* a solid material called an ion exchanger is used to collect and accumulate specific ions that are present in an aqueous waste stream. As ions from the process stream are collected on the ion exchange resin, other ions from the exchange resin are released. Ion exchange is accomplished in ion exchange beds that contain the resin. The liquid stream is passed through the bed until the capacity of the solid resin is exhausted. The resin is then regenerated by passing another liquid stream through the bed to restore the original ion exchange capacity. Ion exchange is used to remove diluted concentrations of metallic anions and cations, inorganic anions, and organic acids and amines from aqueous waste streams. One common use is the recovery of hexavalent chromium from wastes at metal plating operations, where the purified water stream can be reintroduced to the plating bath solution.

Precipitation uses the addition of chemicals to a waste stream to produce insoluble materials that precipitate from the liquid. These materials are then removed from the solvent by solid-liquid separation processes such as sedimentation or granular media filtration. One example is the addition of lime to raise the pH of an aqueous process stream. As the pH increases, the solubility of metal ions decreases because they are converted to hydroxide forms. The metal ions precipitate out of solution as hydroxides.

Waste Treatment

Waste treatment changes the form or composition of a waste stream through controlled physical, chemical, or biological processes that reduce the amount of waste material. Although it is a part of the waste management hierarchy, waste treatment is not a pollution prevention measure. In fact, one goal of pollution prevention is to reduce the dependency on waste treatment as a method for dealing with waste materials. The more successful that product modification and recycling are in reducing waste generation, the fewer waste treatment processes will be needed. In the past, the design of products and production processes often did not consider the amount of waste materials generated. Waste treatment technology was applied (or not applied) to treat the generated wastes. An example is a large synthetic-organic chemical plant, where liquid waste streams from numerous individual production processes were commonly collected and then conveyed to large "end-of-pipe" wastewater treatment plants. The philosophy of pollution prevention is to minimize waste generation as much as is feasible. Success in at-source waste reduction and materials recycling should reduce the need for end-of-pipe wastewater treatment plants. However, water pollution control plants have contributed greatly to environmental protection, and end-of-pipe treatment will continue to be a very significant waste management tool well into the foreseeable future.

Waste treatment processes are quite numerous and are often classified into physical, chemical, and biological processes. Some of the separation processes discussed here as recycling methods are also applied as waste treatment methods. The separation processes are considered waste treatment if they are used to remove contaminants from wastewaters before discharge, without the purpose of recycling and recovering materials. Several references at the end of this chapter contain detailed descriptions about many types of waste treatment processes.

Disposal

Disposal, the lowest item on the waste management hierarchy, is the least desirable. Disposal refers to the ultimate disposition of waste materials from a production process: transporting to a permitted hazardous waste landfill, incinerating, or disposing in a nonhazardous waste landfill. In a sense, the need for ultimate waste disposal is a measure of the inability of P^2 efforts to suitably eliminate the generation of waste materials at the source or to recycle waste materials. Although the idealized aim of a P^2 effort is total materials recycling and complete elimination of waste materials through the entire product life cycle, the inability to reach this goal in practice does not mean that P^2 efforts are not beneficial. As with waste treatment, the dependency on waste disposal should decrease as pollution prevention efforts are implemented.

IMPLEMENTING A POLLUTION PREVENTION PROGRAM

A general program for implementing pollution prevention at a facility is shown in Figure 1–3. The first series of steps involves setting up a P^2 program, organizing the program, setting goals, and doing a preliminary assessment of P^2 opportunities. Effective P^2 requires the strong support of management and all personnel involved in the production process. Often the implementation of P^2 measures will challenge long-established attitudes and require that existing methods and procedures be changed. In this aspect, implementing effective P^2 can be as much a process of cultural or institutional evolution as a technical challenge.

A preliminary assessment is conducted to identify potential steps to reduce waste generation and energy usage and to prioritize the areas on which attention should be focused. Data can be collected from numerous sources such as plant purchasing records, regulatory reporting forms, process flow data, and production and inventory data. The purpose of examining this data is to assess individual sources of waste generation and prioritize the sources for further action. Typical considerations for prioritizing waste generation sources for further study are as follows:

▶ Mass of wastes generated.
▶ Hazardous properties of the waste.
▶ Potential for success in waste reduction.
▶ Recovery of valuable by-products.
▶ Reduction of energy usage.
▶ Safety hazards to employees.
▶ Costs of waste treatment.
▶ Compliance with regulations.
▶ Liability of waste generation.

Once priorities have been established, an overall program plan can be developed to conduct more detailed investigations and assessments.

▶ **FIGURE 1–3**
Pollution prevention pro-
gram overview (*Facility Pollution Prevention Guide,* U.S.
EPA, EPA/600/R-92/088,
1992).

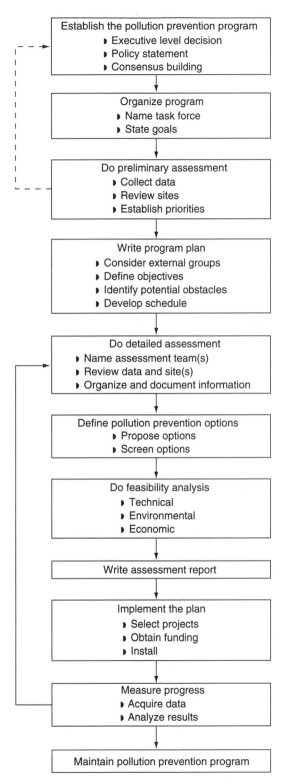

Pollution Prevention Assessment

Pollution prevention assessment refers to examining a process in detail, quantifying the amount of waste generated, and determining feasible methods to reduce the waste material generated from the process. The P^2 assessment includes a detailed examination of the physical configuration of the process; preparation of process flow diagrams (PFDs); identification and characterization of operating procedures; site visits and interviewing operators; quantification of the flow of materials into and out of process units in solid, liquid, and gas phases; calculation of mass balances on raw materials; calculation of energy usage; and developing unit waste generation factors and waste ratios. The term *process mapping* describes the combination of developing the comprehensive process flow diagram and defining the complete sequence of operating steps and procedures used in production.

The PFD for a chemical process shows the flow of materials into the process, through individual operations within the process, and the outflows of the product and waste streams. An example of PFD is shown in Figure 1–4. For the overall process or any unit operation within the process, a mass balance on an individual chemical can be written as

$$\text{Mass out} = \text{Mass in} + \text{Generation} - \text{Consumption}$$

The purpose of the process is to produce a desired product that is generated in the process. *Generation* is the formation of the product and other by-products within the process; *consumption* is the use of the materials put into the process (feedstocks) to produce the product. Both unused feedstocks and by-products are waste materials unless they can be recycled or recovered on-site or off-site. Therefore, it is desirable to use feedstocks as efficiently as possible and to generate as little by-product as possible.

The mass balance must include inflows and outflows of materials in all phases (gas, liquid, and solid) through all methods of entry and exit (e.g., pipes, emissions valves, bottom tars, and slurries). The mass balance can be very useful in quantifying losses of chemicals. Volatile solvents that do not participate in a chemical reaction, for example, have a generation and consumption of zero, so the mass balance should indicate that the mass out equals the mass in. If mass out is less than mass in, some of the volatile chemical is unaccounted for. It may have been lost through emissions to the atmosphere. In this case, the mass balance is used as evidence of a possible unknown loss of chemical by fugitive emissions. On the other hand, if a chemical is consumed in the reaction, it can be either incorporated into the product or exit the reactor as a waste material. In this case, mass out for the individual chemical will be less than mass in. The difference between mass in and mass out must be accounted for by some other form of the original chemical.

Although material and energy balances should be as complete as possible, the limitations in making balances should be understood. Often, PFDs are incomplete, so any calculated balances will not contain part of the materials flows. Most processes have numerous streams, many of which can exchange chemicals to other phases. The exact composition of many process streams is unknown and, thus, cannot be analyzed. Many plant operations or product mixes change frequently, and individual measurements cannot quantify average mass-flow rates of components.

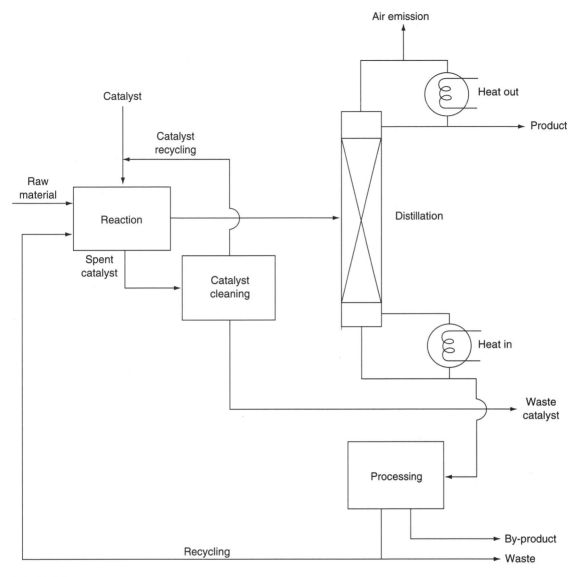

▶ FIGURE 1–4
Example process flow diagram (*Facility Pollution Prevention Guide,* U.S. EPA, EPA/600/R-92/088, 1992).

Nevertheless, every effort should be made to make the PFD and material balances as complete and accurate as possible.

Even if an accurate material balance cannot be generated, it may be possible to calculate unit waste-generation factors for specific pollutants. A unit waste-generation factor is the mass of waste material generated per unit of product formed:

$$f = \frac{\text{Mass of waste formed}}{\text{Unit of product}}$$

The unit waste-generation factor is useful because it normalizes waste generation to units of production. With a unit waste-generation factor, the amount of wastes generated at one time or under a given operating condition can be logically compared with the waste production rate under a different condition. Another useful method of quantifying waste generation relative to production is a waste ratio:

$$R = \frac{\text{Mass of waste formed}}{(\text{Mass of waste} + \text{Mass of product})}$$

Any loss of weight during a production process is considered waste material.

Once a production process has been described and waste generation sources have been determined to the extent feasible, alternatives for source reduction and recycling can be developed. A feasibility analysis must be conducted on all alternatives to determine which, if any, waste-reducing alternatives should be implemented. The feasibility of waste reduction alternatives should be evaluated from technical, reliability, operational, safety, economic, and other factors, with input from all personnel involved in the production process. The data, investigations, findings, and recommendations of the feasibility analyses should be completely documented and summarized in a written report for review by all concerned parties before a decision is made to proceed with implementation.

Once a process has been modified, it is important to continue to monitor the process to ensure that the projected waste generation is realized. This will allow the process to be fine-tuned for waste-generation reduction, and further waste generation steps may become apparent. Unit waste-generation factors represent one parameter that would be expected to decline as a result of implementing a pollution prevention project.

POLLUTION PREVENTION RESOURCES

Resource centers for pollution prevention information and technology transfer are available from industry trade groups, governments, and universities. Many if not all states have established pollution prevention offices that can be valuable sources of information on pollution prevention methodologies, educational programs, applications to specific industrial categories, and specific case studies of pollution prevention implementation. See Appendix A for a list of state pollution prevention offices.

QUESTIONS FOR REVIEW

1. What is the hierarchy of methods used in waste management? What is the order of priority and why?
2. What are the benefits of pollution prevention for an industry?
3. What is life cycle assessment?
4. What is the purpose of a materials balance approach? What are some of its limitations in practical application?

5. What is a unit waste-generation factor and why is it useful?
6. A vapor degreaser at a machining plant is used to clean oil from machined parts. The amount of oil on the machined parts going into the degreaser is 90 gallons per year. A solvent, 1,1,1-trichloroethane (TCA), is used to remove the grease. The quantity of fresh solvent added to the process is 2,500 gallons per year. Effluent

▶ TABLE 1–1

	Product A	Product B
Energy for raw material purchase and disposal of product (Joules / unit)	350	600
Energy for product manufacture and use (Joules / unit)	900	1,600
Emissions from raw material purchase and disposal of product (grams / unit)	0.90	2.8
Emissions from product manufacture and use (grams / unit)	1.8	3.0

from the degreaser is 25 percent oil and 75 percent TCA. Draw a process flow diagram. What are the annual flow of liquid waste and the annual liquid volume of TCA that evaporates into the atmosphere? (Assume that all oil from the machined parts is removed in the degreaser.)

7. A life cycle assessment is being performed on two possible products for manufacture, Product A and Product B. Table 1–1 contains data on energy requirements and atmospheric emissions from two parts of the product's life: (1) raw material acquisition and product disposal and (2) product manufacture and use. Which product is preferable from an energy usage perspective and from an atmospheric emissions perspective? Which is preferable overall?

8. An absorption tower is used to recover ammonia from a gas mixture that contains 15 percent ammonia and 85 percent nitrogen by volume. In the absorption tower, 80 percent of the ammonia in the gas is captured in the water. The mass fraction of ammonia in the water leaving the column is 0.03, and none of the nitrogen in the gas stream enters the water stream. What volume of water is required to recover ammonia from 200 moles of the gas mixture? (The specific gravity of the effluent water/ammonia solution is 0.99.)

ACTIVITIES

1. How has your state implemented the concepts of pollution prevention? Contact your state, county, or local government agency and develop a written profile of what policies, regulations, and programs are currently in place and what efforts are being made to give incentives or otherwise promote the implementation of such programs in industry.

2. Identify a small or medium size business entity in your community that generates hazardous waste. Make contact with the environmental health and safety officer to request a tour and talk with him or her regarding the facility's pollution prevention and recycling programs.

3. Using the Internet, provide a list of pollution prevention home pages and projects that are currently being used or implemented by industry, academic institutions, and government, both large and small.

4. Contact your local community recycling coordinator to determine what products and materials are currently being diverted from the landfill or incinerator and are now being recycled. List specific examples of recycled materials and develop a rough model of the potential savings that can be expected from the implementation of a recycling program for a business, an individual, or a community.

REFERENCES

Hazardous Waste Treatment Processes Including Environmental Audits and Waste Reduction Manual of Practice FD-18
Water Environment Federation
Alexandria, VA

Pollution Prevention Case Studies Compendium
U.S. EPA/600/R-92/046
Risk Reduction Engineering Laboratory
Cincinnati, OH 45268

Pollution Prevention in California (July 1992)
California Department of Toxic Substances Control
Alternative Technology Division

Waste Minimization Opportunity Assessment Manual
U.S. EPA/625/7-88/003
Hazardous Waste Engineering Research Laboratory
Cincinnati, OH 45268

Industrial Pollution Prevention Opportunities for the 1990s
U.S. EPA/600/8-91/052
Office of Research and Development
Washington, D.C. 20460

A Primer for Financial Analysis of Pollution Prevention Projects
U.S. EPA/600/R-93/059
Office of Research and Development
Washington, D.C. 20460

Proven Profits from Pollution Prevention Case Studies in Resource Conservation and Waste Reduction
Institute for Local Self-Reliance
Washington, D.C. 20009

Life Cycle Design Guidance Manual
U.S. EPA/600/R-92/226
Office of Research and Development
Washington, D.C. 20460

User's Guide: Strategic Waste Minimization Initiative Version 2.0
U.S. EPA/625/11-91/004
Office of Research and Development
Washington, D.C. 20460

Facility Pollution Prevention Guide
U.S. EPA/600/R-92/088
Office of Research and Development
Risk Reduction Engineering Laboratory
Cincinnati, OH 45268

Guides to Pollution Prevention
Selected Hospital Waste Streams
U.S. EPA/625/7-90/009
Risk Reduction Engineering Laboratory
Cincinnati, OH 45268

Guides to Pollution Prevention
The Paint Manufacturing Industry
U.S. EPA/625/7-90/005
Risk Reduction Engineering Laboratory
Cincinnati, OH 45268

Guides to Pollution Prevention
Municipal Pretreatment Programs
U.S. EPA/625/R-93/006
Risk Reduction Engineering Laboratory
Cincinnati, OH 45268

2

Basic EPA Concepts of Hazardous Waste

Neal K. Ostler

Upon completion of this chapter, you will be able to meet the following objectives:

▶ Discuss the generic EPA definition of hazardous waste as it incorporates the definition and concept of solid waste.

▶ Identify the different EPA classifications of hazardous waste according to the physical characteristics exhibited.

▶ Outline the basic organization of wastes into lists designated as D, F, K, P, and U wastes.

▶ Identify the various wastes that are excluded from the EPA's definition of hazardous waste.

▶ Identify the role of state hazardous waste programs and the relationship of the state agencies with the federal EPA.

OVERVIEW

It would be nearly impossible to assess the entire level of environmental destruction that has occurred as the result of mismanagement of hazardous waste. Improper disposal of hazardous waste has polluted streams, rivers, lakes, and other surface waters, killing aquatic life, destroying wildlife, and stripping areas of vegetation. In other cases, respiratory illnesses, skin diseases (including skin cancer), and elevated levels of toxic materials in the blood and tissues of humans and domestic livestock have been traced to the vaporization of volatile organic materials from wastes that were disposed of improperly. In still other cases, the mismanagement of hazardous waste has resulted in fires, explosions, or the generation of toxic gases that have killed or seriously injured workers and firefighters.

The improper management of hazardous waste is probably one of the most serious environmental problems in the United States. As late as 1990, the Environmental Protection Agency (EPA) estimated that only 40 percent of all hazardous waste was managed in an environmentally sound manner, which meant that the remainder had been transported, treated, stored, or disposed of in a way that potentially threatened human health and the environment. The amount of hazardous waste produced has steadily increased and will continue to do so with increased population and production of more and more "high-tech" toys, equipment, and communications systems to satisfy the public demand.

The most serious consequence of waste mismanagement, particularly hazardous waste, is probably the threat of groundwater pollution. Groundwater is the source of drinking water for about half the nation's population, and much of it is threatened with pollution from the open dumping of wastes or from improperly operated landfills and surface impoundments. In some areas of the United States, residents obtain drinking water from other sources, such as by trucking it in because local groundwater supplies are badly contaminated with toxic or cancer-causing chemicals and heavy metals.

Subtitle C of the Resource Conservation and Recovery Act (RCRA)

The purpose of the Subtitle C program under RCRA (§3001–3019 of the act) is to create a federal "cradle-to-grave" system to manage hazardous waste and prevent continuation of the mismanagement of hazardous wastes. This program includes provisions for cleaning accidental releases of hazardous materials.

"Cradle-to-Grave" Management

Although the scope of the EPA has now broadened to include the concepts of pollution prevention and waste minimization discussed in other chapters in this text, the primary thrust of RCRA is to manage waste from the point it is first created ("cradle") throughout its entire lifetime, to the place or point that it is disposed of either in a landfill or an incinerator ("grave").

RCRA is a plan for managing wastes, particularly hazardous waste, that includes the following that will be discussed in different chapters of this textbook:

▶ Identification of waste that is hazardous.
▶ Generators of hazardous waste.
▶ Transporters of hazardous waste.
▶ Owners and operators of facilities that treat, store, or dispose of hazardous wastes.
▶ Issuance of operating permits to owners or operators of treatment, storage, and disposal facilities, and provision for corrective action for hazardous waste releases.
▶ Enforcement of the Subtitle C program.
▶ Transfer of the responsibilities of the Subtitle C program from the federal government to the states.

The statutory and regulatory requirements entailed for each of these items also will be discussed and elaborated on in separate chapters.

Only those wastes that are found to be hazardous are subject to Subtitle C regulations under RCRA. Determining what is a hazardous waste is a complex task because wastes are potentially hazardous for different reasons and the universe of potential hazardous wastes is large and diverse. Hazardous waste may consist of specific chemical substances and their mixtures, certain generic waste streams, and many specific products.

The difficult task of identifying hazardous waste was addressed (under Section 3001 of RCRA) by a congressionally directed EPA study to establish criteria for identifying the characteristics of hazardous waste, to develop lists of particular wastes as hazardous, and to promulgate the criteria and lists.

All solid waste generators must determine whether their solid waste is hazardous and, if so, is subject to regulation under RCRA Subtitle C. These include not only the largest manufacturers but also the smallest dry cleaner on the corner; all are responsible for identifying their solid and hazardous waste streams.

RCRA's Subtitle C program is based on this identification and listing of wastes. The purpose of this chapter is to define the term *hazardous waste,* describe how the EPA determines whether a solid waste is hazardous, and discuss those wastes specifically excluded from Subtitle C regulation.

The federal waste identification framework sets a baseline standard with which all states must comply. Many states have chosen, however, to go beyond that standard by adapting and applying more stringent requirements. For example, RCRA does not list waste oil as hazardous, but many states regulate it as such. A state program that has been delegated the Subtitle C program for that state must be followed. Therefore, understanding whether federal or state requirements apply in a specific situation is important.

DEFINITION AND CLASSIFICATIONS OF HAZARDOUS WASTE

Congress defined the term *hazardous waste* in Section 1004(5) of RCRA as the following:

> *Solid waste, or combination of solid wastes, which because of its quantity, concentration, or physical, chemical, or infectious characteristics may: (1) cause or significantly contribute to an increase in mortality or an increase in serious irreversible, or incapacitating reversible, illness or (2) pose a substantial present or potential hazard to human health or the environment when improperly treated, stored, transported, disposed of, or otherwise managed.*

A hazardous waste is a solid waste that may be listed as a hazardous waste, may exhibit characteristics that make it hazardous, or is not exempted specifically and managed by regulations under some agency other than the EPA or another department of the U.S. government. Hazardous waste is classified further into the subcategories of solid waste, characteristic waste, listed waste, mixed and derivative wastes, exempted waste, and recycled nonsolid waste materials.

Solid Waste

As noted, RCRA defines *hazardous wastes* in terms of properties of a solid waste. Therefore, if a waste is not a solid waste, it cannot be a hazardous waste. Accordingly, it is important to understand what constitutes a solid waste under RCRA. The reader is advised that separate states have some latitude to clarify for themselves what constitutes such waste.

For the purpose of RCRA, a *solid waste* means any discarded material including a solid, liquid, semisolid, or contained gaseous material resulting from an industrial, commercial, mining, or agricultural operation and community activities. Examples of solid waste include any garbage, refuse, or sludge (including sludge from a waste treatment plant, water supply treatment facility, or air pollution control device or facility). This definition does not include solid or dissolved material in domestic sewage that is governed by the Federal Water Pollution Control Act and related regulations.

Solid wastes, therefore, include almost any discarded material that is not specifically excluded, including materials that are abandoned, recycled, or inherently waste-like. **Abandoned waste** includes materials that have been (1) disposed of, (2) burned or incinerated, or (3) in lieu of either (1) or (2), accumulated, stored, or treated. This category does not include recycled materials. **Recycled materials** are used, reused, or reclaimed. If a material is recycled in any of the following manners, it is a solid waste:

1. **Used in a manner constituting disposal.** Certain materials are solid waste if they have been either recycled or accumulated, stored, or treated beforehand in a manner that would actually constitute disposal. Examples are materials that have been (1) applied to or placed on the land, (2) used to produce products that are applied to or placed on the land, or (3) contained in products that are applied to or placed on the land. Usually, this does not include products that are designed or produced with the intent of being applied to the land as their ordinary manner of use.

2. **Burned for energy recovery.** Certain materials are solid waste if they are identified particularly and are to be used to produce a fuel, are contained in fuels (in which case the fuel itself is solid waste), or they are to be burned for recovery of energy. An example is household waste burned in municipal incinerators to create steam-generated electricity. Commercial chemicals, such as butane or propane, produced for the purpose of being a fuel are not solid wastes.

3. **Reclaimed.** Certain materials are solid waste when they are to be reclaimed. A material is reclaimed if it is processed to recover a usable product or if it is to be regenerated. An example is lead recovered from spent batteries or regenerated spent solvents.

4. **Accumulated speculatively.** Certain materials are solid waste when accumulated speculatively, such as when someone stockpiles the items in a yard or warehouse in the hope of finding a buyer or discovering another use, but does not yet intend the items to be discarded.

5. **Recycled nonsolid waste.** Certain materials are not solid waste when recycled if it can be shown that they are to be recycled by being (a) used or reused as ingredients in an industrial process to make a product, provided the materials are not being reclaimed; (b) used or reused as effective substitutes for commercial products; or (c) returned to the original process from which they are generated

without first being reclaimed and are thus a substitute for a raw material feedstock in a process using raw materials as principal feedstock.

Inherently wastelike materials include certain materials that are solid waste because of their inherent nature when they are to be recycled in any manner. Examples include materials listed as F022, F023, F026, and F028 (listed hazardous waste is covered later in this chapter). The EPA has developed certain criteria to add wastes to the F list that include (1) materials with toxic constituents not ordinarily found in their raw materials, and (2) products for which the materials are intended to substitute and are not used or reused during the recycling process. Either type of material is included on the F list because it may pose a substantial hazard to human health and the environment when recycled.

Characteristic Waste

Characteristic waste refers to a solid waste that is classified as hazardous due to its inherent characteristics. To be classified as characteristic waste, an item must meet these criteria: (1) cause or contribute to an increase in mortality or an increase in serious, irreversible, or incapacitating illness; (2) pose a substantial present or potential hazard to human health or the environment when stored, treated, transported, disposed of, or otherwise mismanaged; (3) be measured by an easily accessible and affordable standardized test; or (4) be easily detected by the generator or some other person knowledgeable of his or her solid waste stream.

A solid waste is therefore a hazardous waste if it exhibits any of the following characteristics as determined by using one of the testing protocols identified by the EPA:

1. **Ignitability.** A solid waste that exhibits any of the following properties is considered a hazardous waste due to its ignitability:
   a. A liquid, except an aqueous solution, that contains less than 24 percent alcohol and has a flash point less than 60°C (140°F).
   b. A nonliquid capable, under normal conditions, of spontaneous and sustained combustion.
   c. An ignitable compressed gas per Department of Transportation (DOT) regulations.
   d. An oxidizer, per DOT regulation.

   The EPA's reason for including ignitability as a characteristic is to identify wastes that can cause fires during transportation, storage, or disposal. Examples of ignitable wastes include waste oils and used solvents.

   NOTE: A solid waste that exhibits the characteristic of ignitability but is not listed as a hazardous waste has the EPA hazardous waste number D001.

2. **Corrosivity.** A solid waste that exhibits any of the following properties is considered a hazardous waste due to its corrosivity:
   a. An aqueous material with pH less than or equal to 2 or greater than or equal to 12.5.
   b. A liquid that corrodes steel at a rate of more than .25 inch per year at a temperature of 55°C (130°F).

The EPA chose pH as an indicator of corrosivity because wastes with high or low pH can react dangerously with other wastes or cause toxic contaminants to migrate from certain wastes. It chose steel corrosion because wastes capable of corroding steel can escape from their containers and liberate other wastes. Examples of corrosive wastes include acidic wastes and used pickle liquor (employed to clean steel during its manufacture).

> **NOTE: A solid waste that exhibits the characteristic of corrosivity but is not listed as a hazardous waste has the EPA hazardous waste number D002.**

3. **Reactivity.** A solid waste that exhibits any of the following properties is considered a hazardous waste due to its reactivity:
   a. Is normally unstable and reacts violently without detonating.
   b. Reacts violently with water.
   c. Forms an explosive mixture with water.
   d. Generates toxic gases, vapors, or fumes when mixed with water.
   e. Contains cyanide or sulfide and generates toxic gases, vapors, or fumes at a pH of between 2 and 12.5.
   f. Is capable of detonation if heated under confinement or subjected to strong initiating source.
   g. Is capable of detonation at standard temperature and pressure.
   h. Is listed by DOT as Class I or Division 1 or 2 explosive. See 49 Code of Federal Regulations (CFR) 173.50—Explosives.

   Reactivity was chosen as a characteristic to identify unstable wastes that can pose a problem, such as an explosion, at any stage of the waste management cycle. Examples of reactive wastes include water from TNT operations and used cyanide solvents.

> **NOTE: A solid waste that exhibits the characteristic of reactivity but is not listed as a hazardous waste has the EPA hazardous waste number D003.**

4. **Toxicity.** The term *EP toxicity* often confuses newcomers to the topic because, in addition to referring to a characteristic of a waste, it also refers to a test. The original test, called the *extraction procedure* (EP), was designed to identify wastes likely to leach hazardous concentrations of particular toxic constituents into the groundwater as a result of improper management. During the procedure, constituents are extracted from the waste in a manner designed to simulate the leaching actions that occur in landfills. The extract is then analyzed to determine whether it possesses any of the toxic contaminants listed in Table 2–1. If the concentrations of the toxic constituent exceed the levels shown in the table, the waste is classified as hazardous.

   The EP toxicity test has been criticized because it fails to adequately simulate the flow of toxic contaminants to drinking water. Under the Hazardous and Solid Waste Amendments of 1984 (HSWA), Congress directed EPA to examine the EP toxicity test to determine whether modifications or additions could improve it as a diagnostic tool.

   The EPA has now replaced the EP test with the *toxicity characteristic leaching procedure* (TCLP). This test involves a sample preparation procedure designed to determine the mobility of both organic and inorganic constituents present in

▶ **TABLE 2-1**
Toxicity characteristic constituents—numerical.

EPA HW[1] Number	Constituent	CTRL Basis	CTRLs (mg/l) ×	DAF of 100 =	Regulatory Level (mg/l)
D004*	Arsenic	MCL	0.05		5.0
D005*	Barium	MCL	1.0		100.0
D006*	Cadmium	MCL	0.01		1.0
D007*	Chromium	MCL	0.05		5.0
D008*	Lead	MCL	0.05		5.0
D009*	Mercury	MCL	0.002		0.2
D010*	Selenium	MCL	0.01		1.0
D011*	Silver	MCL	0.05		5.0
D012*	Endrin	MCL	0.0002		0.02
D013*	Lindane	MCL	0.004		0.4
D014*	Methoxychlor	MCL	0.1		10.0
D015*	Toxaphene	MCL	0.005		0.5
D016*	2-4-D	MCL	0.1		10.0
D017*	2,4,5-TP (Silvex)	MCL	0.01		1.0
D018	Benzene	MCL	0.005		0.5
D019	Carbon tetrachloride	MCL	0.003		0.03
D020	Chlordane	RSD	0.003		0.03
D021	Chlorobenzene	RID	1.0		100.0
D022	Chloroform	RSD	0.06		6.0
D023	0-Cresol	RID	2.0		200.0[2]
D024	m-Cresol	RID	2.0		200.0[2]
D025	p-Cresol	RID	2.0		200.0[2]
D026	Cresol	RID	2.0		200.0[2]
D027	1.4-Dichlorobenzene	MCL	0.075		7.5
D028	1.2-Dichloroethane	MCL	0.005		0.5
D029	1,1-Dichloroethylene	MCL	0.007		0.7
D030	2,4-Dinitrotoluene	RSD	0.0005		0.1[3]
D031	Heptachlor (and its epoxide)	RSD	0.00008		0.008
D032	Hexachlorobenzene	RSD	0.0002		0.1[3]
D033	Hexachloro-1,3-butadiene	RSD	0.005		0.5
D034	Hexachloroethane	RSD	0.03		3.0
D035	Methyl ethyl ketone	RID	2.0		200.0
D036	Nitrobenzene	RID	0.02		2.0
D037	Pentachlorophenol	RID	1.0		100.0
D038	Pyridine	RID	0.04		5.0[3]
D039	Tetrachloroethylene	RSD	0.007		0.7
D040	Trichloroethylene	MCL	0.005		0.5
D041	2,3,5-Trichlorophenol	RID	4.0		400.0
D042	2,4,6-Trichlorophenol	RSD	0.02		2.0
D043	Vinyl chloride	MCL	0.002		0.02

* Fourteen original constituents based on drinking water standards.

1 Hazardous waste number.

2 In o-, m-, and p-Cresol concentrations cannot be differentiated; the total Cresol concentration is used. Total Cresol regulatory level: 200 mg/l.

3 If quantitation limit is greater than the calculated regulatory level, quantitation limit becomes regulatory level.

Key: CTRL, Chronic Toxicity Reference Level; DAF, Dilution Attentuation Factor; MCL, Maximum Contaminant Level; RFD, Reference Dose; RSD, Risk Specific Dose.

▶ **TABLE 2-2**
Comparison of the extraction procedure (EP) and the toxicity characteristic leaching procedure (TCLP).

Item	EP	TCLP
Leaching media	0.5 M acetic acid added to distilled deionized water to a pH of 5 with 400-ml maximum addition continual pH adjustment	0.1 M pH 2.9 acetic acid solution for moderate to high alkaline wastes and 0.1 M pH 4.9 acetate buffer for other wastes
Liquid/solid separation	0.45-µm filtration to 75 psi in 10 psi increments unspecified filter type	0.6 to 0.8-µm glass-fiber filter filtration to 50 psi
Monolithic material/particle size reduction	use of structural integrity procedure or grinding and milling	grinding or milling only structural integrity procedure not used
Extraction vessels	unspecified design blade/ stirrer vessel acceptable	zero-headspace vessel required for volatiles bottles used for nonvolatiles blade stirrer vessel not used
Agitation	prose definition or acceptable agitation	rotary agitation only in an end-over-end fashion at 30 ± 2 rpm
Extraction time	24 h	18 h
Quality control requirements	standard additions required one blank per sample batch	standard additions required in some cases one blank per ten extractions and every new batch of abstract analysis specific to analyte

liquid, solid, and multiphasic wastes that simulate landfill leachate. Certain solid wastes that did not fail the EP test may fail the TCLP and, therefore, are classified and regulated as hazardous waste. Table 2-2 compares the old extraction procedure with the newly adopted TCLP procedure. The final TCLP rule (March 1990; found at 40 CFR 261, Appendix A, Method 1311) added twenty-five new constituents to the existing fourteen EP constituents and represents a wide range of organic compounds with varied chemical and physical properties and industrial uses.

> **NOTE: A solid waste that exhibits the characteristics of toxicity but is not listed as a hazardous waste has the EPA hazardous waste number specified in the list found in Table 2-1. These numbers are D004–043.**

The toxicity characteristic (TC) regulatory levels are based on a codisposal scenario of hazardous waste disposed of with solid waste in an actively decomposing landfill overlying an aquifer.

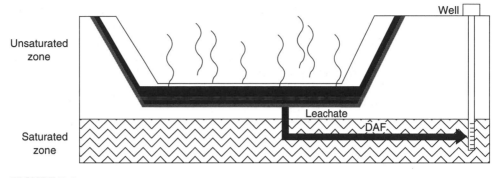

FIGURE 2–1
TCLP scenario.

In this scenario, shown in Figure 2–1, leachate from the waste travels through the unsaturated zone to the saturated zone, where it is transported to a drinking water well. The TCLP rule is more fully discussed in Appendix B of this book, which also includes a full list of the TC constituents.

The TCLP is now used for *land disposal restriction* compliance purposes and fully replaced the extraction procedure (EP) for both large and small generators as of March 1991.

Summary

Any solid waste that does not exhibit any of the characteristics of hazardous waste discussed in this section or is not present as a constituent of the waste in sufficient detectable quantity is not a hazardous waste.

The responsibility for determining whether a particular solid waste is hazardous falls on its generators. If a solid waste is neither listed nor excluded, as discussed in the following pages, the generator must either test the waste using standard methods (specified in 40 CFR 261) or have sufficient knowledge about it to assess whether it exhibits any of the hazardous waste characteristics. If the waste does exhibit a characteristic, it is considered hazardous and must be handled accordingly. Tests must be applied to each individual waste and cannot be used to assess a type of waste (other than to define the waste generically as hazardous). This provision was established to prevent a national company from making one waste determination and using the results nationwide, masking potential regional variations. The tests must also be run on representative samples to obtain results that adequately characterize the nature of the waste.

Listed Waste

A solid waste may be included in the list developed by the EPA or more restrictively by the states under EPA guidelines. The following criteria are used in designating wastes to be placed on lists:

1. It exhibits any of the previously discussed characteristics of hazardous waste.
2. It has been found to be fatal to humans in small doses (see the definition of acute hazardous waste in the Glossary).
3. It contains any of the toxic constituents listed in Table 2–1.

The EPA also includes protocols that are acceptable as criteria for detemining that certain solid and other wastes shall not be placed on any of the previously mentioned lists if it can be shown that the waste is not capable of posing the necessary substantial present or potential hazard to human health or the environment. These criteria include the nature of the toxicity presented by the constituent, the concentration of the waste, the degradation potential of the waste, the persistence of the constituent, the degree to which the waste bioaccumulates, the quantity of the waste generated, the nature and severity of the human health and environmental damage that has already been measured as a result of accidental releases, and other appropriate factors.

A solid waste therefore is hazardous if it is named on one of the following lists developed by the EPA. If it is on one of these lists, it has been assigned an EPA hazardous waste number that precedes its name.

1. **Nonspecific source wastes.** According to 40 CFR 261.31, these are generic wastes, commonly produced by manufacturing and industrial processes. Examples of hazardous, nonspecific source wastes include the following:
   F001 Spent halogenated solvents containing at least 10 percent of chemicals such as trichloroethylene and carbon tetrachloride.
   F003 Spent nonhalogenated solvents such as acetone, ethyl ether, and n-butyl alcohol.
   F006 Wastewater treatment sludges from electroplating operations.
   F024 Wastes from distillation residues, heavy ends, tars, and reactor clean-out wastes from production of chlorinated aliphatic hydrocarbons.
   F999 Residues from the demilitarization, treatment, and testing of nerve, military, and chemical agents such as CX, GA, GB, and HN-1.
2. **Specific source wastes.** According to 40 CFR 261.32, this list consists of wastes from specifically identified industries such as wood preserving, petroleum refining, and organic chemical manufacturing. Examples include the following:
   K001 Bottom sediment sludge and wastewaters from the production of wood preservatives.
   K009 Distillation bottoms from the production of acetaldehyde from ethylene.
   K016 Wastewater treatment sludge from the mercury cell process in chlorine production.
   K047 Pink/red water from TNT operations.
   K060 Ammonia-still lime sludge from coking operations.
3. **Commercial chemical products.** According to 40 CFR 261.33(e), this list consists of specific commercial chemical products or manufacturing chemical intermediates that are considered to be acutely hazardous wastes. Examples include the following:
   P010 Arsenic acid
   P021 Calcium cyanide
   P031 Cyanogen
   P056 Fluorine
   P081 Nitroglycerine
   P095 Phosgene
   P108 Strychnine and salts

4. **Commercial chemical products subject to small quantity exclusion.** According to 40 CFR 261.33(f), this list consists of specific commercial chemical products, manufacturing chemical intermediates, and off-specification commercial products that are considered to be toxic wastes. Examples include the following:
   U001 Acetaldehyde
   U002 Acetone
   U019 Benzene
   U211 Carbon tetrachloride
   U061 DDT
   U122 Formaldehyde
   U142 Kepone
   U151 Mercury
   U220 Toluene
   U243 Vinyl chloride

> NOTE: In most cases, the characteristic of the waste caused the EPA to list it. In addition to categorizing the wastes as a waste, the EPA also assigns it one or more of the following hazard codes:

I—Ignitable waste
C—Corrosive waste
R—Reactive waste
E—EP toxic waste
H—Acutely hazardous waste
T—Toxic waste

Delisting Hazardous Waste

The EPA recognized that the criteria for designation as a hazardous waste might not be applicable in all cases. To make allowances or special provisions for these cases, EPA created a *delisting* process, which allows any person (e.g., waste handler or a member of the general public) to petition the EPA to drop a listed waste from regulation under Subtitle C of RCRA. The following are the principal methods to demonstrate that a waste should be removed from the EPA lists:

1. **Facility-specific variations.** Petitioners trying to have their wastes delisted under this process must prove to the EPA that the wastes are not hazardous because of facility-specific variations in raw materials, processes, or other factors.
2. **Other constituents in the waste.** When evaluating a delisting petition, the EPA must consider the factors that caused the waste to be classified as hazardous, such as the fact that the waste has constituents in addition to those for which the waste was listed.
3. **On-site treatment of the waste.** Facilities that treat listed wastes often want to show that the treated listed wastes are no longer hazardous. For example, treating some hazardous waste by incineration may be cost effective only if the resulting ash is not considered a hazardous waste. If this can be demonstrated successfully, the disposal of the wastes do not have to follow Subtitle C regulations. The owner or operator must conduct studies to show that, once treated, the listed hazardous waste is no longer hazardous.

4. **Ceasing to exhibit characteristics.** Wastes automatically cease to be considered hazardous when they no longer exhibit any characteristic of hazardous waste. It is also worth noting that the general mixture-derived-from rule does not apply to characteristic wastes. (Of course, if mixtures or residues themselves exhibit a characteristic, they are considered hazardous wastes.)

Should EPA determine, upon completion of their evaluation, that the waste is not hazardous due to conditions at the facility, that waste is removed from Subtitle C's regulatory jurisdiction. Delisting is done on a case-by-case basis. Therefore, if a waste is delisted at one facility, it is not automatically delisted at other facilities.

Mixed and Derivative Wastes

When the EPA was setting the conditions for identifying hazardous wastes, it had to answer many questions regarding hazardous waste management. Two questions revolved around the Department of Energy and the Nuclear Regulatory Commission, which were already managing radioactive wastes, and the Department of Agriculture, which had the responsibility to manage many pesticides and fertilizers. One of the principal issues was how to classify a waste mixture that contains both a listed hazardous waste and a nonhazardous solid waste. The general rules for mixed waste, waste residues, and certain exceptions are discussed in the following sections.

The General Mixture Rule

The EPA decided that any waste mixture containing a listed hazardous waste is considered a hazardous waste and must be managed accordingly. This applies regardless of the percentage of the waste mixture that is composed of listed hazardous wastes. Without such a regulation, generators could evade Subtitle C requirements simply by commingling listed wastes with nonhazardous solid waste. Most of these waste mixtures would not be included in the four Subtitle C characteristics (ignitability, corrosivity, reactivity, and toxicity) because they would contain wastes that were listed for reasons other than exhibiting those characteristics; for example, being acutely toxic. Allowing this situation would leave a major loophole in the Subtitle C management system and create inconsistencies in how wastes are managed under that system.

Exceptions to the General Mixture Rule A few exceptions to the general mixture rule exist.

1. The resultant mixture of a wastewater discharge subject to regulation by the Clean Water Act with low concentrations of a listed waste, as specified in 40 CFR 261.3, is not considered a listed hazardous waste. Of course, if such a mixture exhibits one of the characteristics, it is deemed hazardous.
2. A mixture of nonhazardous wastes and listed wastes that are listed because they exhibit a characteristic is not considered hazardous if it no longer exhibits any of the characteristics identified in Subpart C of 40 CFR Part 261.
3. A mixture of nonhazardous wastes and characteristic hazardous wastes is not considered hazardous if it no longer exhibits any of the characteristics set forth in Subpart C of 40 CFR Part 261.

4. Certain concentrations of spent solvents and laboratory wastewater that are discharged in low concentrations and do not pose a threat to human health or the environment are excepted.
5. *De minimis* losses of discarded commercial chemical products or intermediaries used as raw materials in manufacturing or produced as by-products are excepted. These include minor losses from spills and transfer of materials, process leaks, and similar incidental discharges.
6. The heat exchanger bundle cleaning sludge in the petroleum refining industry is excluded from the general mixture rule.

The Residue Rule

The EPA also had to determine how to classify residues from the treatment, storage, or disposal of a listed hazardous waste. It decided that any residue derived from a listed hazardous waste is therefore a hazardous waste. The EPA established this "derived from" rule because wastes derived from hazardous wastes can be reasonably assumed to be hazardous themselves unless it can be shown that they do not pose a threat to human health and the environment.

EXCLUDED WASTE

Early in the development of RCRA, Congress decided that certain types of solid waste should not be considered hazardous under Subtitle C. Excluded wastes include a number of common solid wastes that do not present a significant threat to human health or the environment, or are currently managed under other programs in a way that minimizes any threat. They include the following:

1. **Household wastes** are left to state and local governments to manage. Recent developments have indicated a need for the EPA to establish some type of guidelines and incentives for state and local governments to follow in regulating these wastes, especially those with hazardous characteristics.
2. **Domestic sewage** and any mixture that passes through a sewer system to a publicly owned treatment works (POTW) and untreated sanitary wastes are regulated under the Clean Water Act.
3. **Industrial wastewater and other point source discharges** from the actual point source are regulated under the Clean Water Act. This exclusion does not include such wastewaters while they are being collected, stored, or treated before discharge, nor does it exclude sludges that are generated by such treatment.
4. **Irrigation return flows** may contain a variety of diluted pesticides. They are regulated by the Department of Agriculture and are monitored by state-operated programs under the federal Insecticide, Fungicide, and Rodenticide Act under 40 CFR 152.707.
5. **Radioactive materials** as defined by the Atomic Energy Act are managed by the Department of Energy and the Nuclear Regulatory Act. Some mixed (radioactive and hazardous) wastes come under RCRA management guidelines.
6. **Mixed waste** (radioactive and hazardous) represents one of the most difficult challenges facing the states and the EPA. Because mixed waste is considered

hazardous under RCRA and radioactive under the Atomic Energy Act, the NRC and the EPA work together to address their management. Generators of mixed waste must comply not only with the minimum technical requirements of RCRA, but also with NRC regulations. Thus, because the hazardous and radioactive waste components cannot be readily separated, the design of facilities, drafting of operating requirements for RCRA permits or NRC licenses, and the development of clean-up solutions must be done in a manner that adequately addresses the hazards posed by both the radioactive and hazardous components of the waste.

7. **Wastes recovered from municipal facilities** in most instances are not considered to be hazardous wastes if such facilities receive and burn only household wastes and/or solid nonhazardous waste from commercial or industrial sources and do not accept hazardous waste.

8. **Agricultural wastes,** those wastes produced by growing and harvesting crops or raising of animals, including manure, are excluded from hazardous wastes.

9. **Mining overburden** that is returned to the mine site is excluded waste. Other mining and processing wastes, such as solid wastes from the extraction, beneficiation, and processing of ores and minerals including coal, are still being studied and are provisionally exempt from Subtitle C.

10. **Coal combustion wastes** including fly ash wastes, bottom ash wastes, slag wastes, and flue gas emission control wastes are excluded.

11. **Cement kiln dust waste** is excluded.

12. **Certain chromium-containing wastes** are excluded.

13. **Laboratory samples** of water, soil, or air that are collected for the sole purpose of testing to determine their chemical characteristics and composition are excluded. To qualify as an exemption, the sample collector must follow strict guidelines of the Department of Transportation and the U.S. Postal Service regarding proper packaging, labeling, and so on.

14. **Residues of hazardous waste in empty containers** also are excluded from regulation under Subtitle C. Specifically, any hazardous waste remaining in a liner or empty container is exempt if all of the waste has been removed by methods such as pumping or pouring. Additionally, no more than 2.5 centimeters of waste can remain in the bottom of the container (less than 3 percent by weight if the container is under 110 gallons, less than 0.3 percent if over 110 gallons). Special requirements apply if the waste is acutely hazardous. The liner or container must be triple rinsed or cleaned by an equally effective method. The rinseate then must be handled as a hazardous waste.

NOTE: Language excluding these wastes from RCRA regulation is contained in 40 CFR 261.4(a).

15. **Wastes from oil and gas production** are exempted from consideration as hazardous waste. Congress provisionally exempted a few of these waste streams including wastes from oil and gas production, mining and the combustion of coal or other fossil fuels, as well as cement kiln dust waste. To determine whether those wastes should be regulated at all, Congress, under Section 8002 of RCRA, directed the EPA to conduct studies and report on whether the results indicate

that the wastes should be regulated under Subtitle C. The EPA completed several of these large-volume studies. As a result, it recently determined that regulation of wastes from the development and production of crude oil, natural gas, and geothermal energy is not warranted under Subtitle C. Instead, the EPA has chosen to expand existing regulatory programs under Subtitle D, the Clean Water Act, and the Safe Drinking Water Act. The agency has also determined that regulation under Subtitle C of mining wastes (referred to as *Bevill wastes*) from extraction and beneficiation currently is not warranted. Instead, the EPA plans to develop a program to regulate these mining wastes under Subtitle D of RCRA. These wastes are detailed in 40 CFR 261.4(b).

Recycled Material

Consistent with RCRA's mandate to foster resource recovery, the regulations concerning hazardous waste exempt some secondary materials from RCRA when they are recycled using specified procedures. The exemption is based on the type of secondary material being recycled and the recycling process used. Table 1 of 40 CFR 261.2 identifies classes of hazardous waste secondary materials, including the following:

- **Spent materials** have been used and can no longer serve their original purpose without reprocessing.
- **Sludges** are residues from treatment of air or wastewater (pollution control operations).
- **By-products** are residual materials from industrial, commercial, mining, and agricultural operations.
- **Commercial chemical products** include chemical products and intermediates.
- **Scrap metal** refers to bits and pieces of metal from metal-processing operations or consumer use.

Recycled Nonsolid Waste Materials

When recycled by either reclamation or specutative accumulation, some secondary materials are exempted from RCRA regulation. Similarly, RCRA (40 CFR 261.6) exempts or reduces (40 CFR 266) the regulatory requirements for certain hazardous waste recyclable materials or recycling procedures, which were addressed previously.

RCRA RELATIONSHIPS OF STATES AND THE EPA

The original mission of the U.S. Constitution was to give the separate states autonomy and independence with a united government to protect individual rights and afford a strong national defense. In keeping with this concept, Congress passed RCRA with the intent that the states assume responsibility for implementing it with oversight from the federal government. State responsibility for Subtitle C of RCRA involves developing a state hazardous waste program and having it approved by EPA.

In 1976 a state could choose either interim or final authorization. All interim authorizations expired on January 1, 1993. For a state program to receive final authorization, its program must afford regulations and a mechanism to implement them that are fully equivalent (no less stringent than) and consistent with the federal program. However, the states may impose requirements that are more stringent or broader in scope than the federal requirements. A state's program must also provide for public availability of information, an adequate enforcement authority to carry out its provisions, and public notice and hearing in the permitting process in substantially the same manner and to the same degree as the federal program.

The purpose of this section is to explain the process that is involved in a state obtaining EPA authorization and how it must operate to continue under EPA oversight.

The Program Application Process (40 CFR 271.5)

Any state seeking final authorization for its hazardous waste program must submit an application to the EPA administrator. This application must contain the following elements:

1. **A letter from the governor** requesting program approval, which is self-explanatory but necessary.
2. **Copies of all applicable state statutes and regulations,** including those governing state administrative procedures. This item needs no further explanation.
3. **Documentation of public participation activities** that have occurred or are planned to allow full democratic process in establishing the program and in promulgating state regulations for its support, compliance, and enforcement.
4. **A description of the state program** that tells how the state intends to administer the program in place of the federal program and includes, in narrative form, the following:
   a. Scope, structure, coverage, and processes of the state program.
   b. State agency or agencies responsible for running the program.
   c. State-level staff who will carry out the program.
   d. The state's compliance tracking and enforcement ability.
   e. The state's manifest system.
   f. Applicable state procedures, including permitting procedures and any state administrative or judicial review procedures.
   g. Any forms used to administer the program under state law.
   h. A proposed budget for running the program and an itemization of the sources and amounts of funding available to support the program's operation.
   i. An estimate of the number of generators, transporters, and on-site and off-site disposal facilities.
   j. The annual amounts (if available) of hazardous wastes generated, transported into and out of the state, and stored, treated, or disposed of within the state.
5. **An attorney general's statement** identifying the legal authorities (statutes, regulations, and case law) on which the state is relying to demonstrate equivalence with the federal program and that must be fully effective at the time the program is authorized. The statement must be signed by the attorney general or an authorized designate. (See 40 CFR 271.7.)

6. **A memorandum of agreement (MOA)** that outlines the nature of the responsibilities for compliance and enforcement by the state with oversight from the EPA with different levels of coordination between the two in implementing the program. MOAs contain both state-specific provisions and several other general provisions including the following:
   a. Specifications of the reports and the frequency with which the state must submit them to the EPA.
   b. Coordination of compliance monitoring and other enforcement activities between the state and the EPA.
   c. Conduct of an EPA overview of program administration and enforcement.
   d. Joint processing of permits for those facilities that require a permit from both the state and the EPA.
   e. Specification of the types of permit applications that will be sent to the regional administrator for review and comment.
   f. Transfer of permitting responsibilities upon authorization.
      (See 40 CFR 271.8.)

Should the state choose to develop a program more stringent and/or broader in scope than required by RCRA, the program description should address those parts of the program that go beyond what is required under Subtitle C. (See 40 CFR 271.6.)

Review of the Proposed State Program

The first review of the proposed state program must be carefully documented. It must be reviewed even before the application is submitted to the EPA for approval. The state must inform the public of its intent to seek program approval by issuing a public notice and holding a public hearing if sufficient interest is expressed.

The regional EPA administrator determines whether the state's program should be authorized. The following schedule must be adhered to for the administrator to make the determination.

1. **Tentative determination.** A completed application must be either tentatively approved or disapproved within ninety days from its receipt by the regional administrator and published in the *Federal Register.*
2. **Public input.** A thirty-day period for public comment and review of the tentative decision must be allowed and, if sufficient interest is expressed, a public hearing must be held.
3. **Final determination.** Within ninety days of publication in the *Federal Register* of the tentative determination, the regional administrator, taking into account any comments submitted, must decide whether to approve the state's program. This final determination must also be published in the *Federal Register.*

Revision of State Programs

The State program must be revised as often and as expeditiously as changes are made in the regulatory authority at the federal level. If the state initiates a statutory or regulatory change to be consistent with federal changes, it submits a copy of the amended statute or rule, a modified program description, revised MOA, attorney general's statement, and any other pertinent documents to the EPA. The agency then reviews the modifications using the same standards by which it reviewed the state's initial program application. Any revisions become effective upon EPA approval and notice in the *Federal Register.*

All state programs must incorporate and implement any required modifications within certain time frames: States must modify their programs by July 1 each year to reflect all changes to the federal program occurring during the twelve months (referred to as *clusters*) preceding the previous July 1. If a state must amend or adopt a new statute or if its rule-making procedures preclude it from meeting these cluster time frames, the dates can be extended. (See 40 CFR 271.21.)

It is important to note that rules promulgated pursuant to RCRA take effect only in nonauthorized states. An authorized state must modify its program, submit an application, and obtain approval from the EPA before an RCRA rule may be implemented. Conversely, rules promulgated pursuant to HSWA are effective in both authorized and nonauthorized states. EPA implements and enforces HSWA rules until states modify their programs, submit applications, and receive approval.

Withdrawal of Approval of Programs (40 CFR 271.23)

State RCRA programs are continually subject to review. Should the EPA administrator determine that a state's program no longer complies with the appropriate regulatory requirements, and the state fails to take corrective action, authorization of the state's program may be withdrawn. Such circumstances include a failure to do any of the following:

1. Issue permits that conform to the regulatory requirements.
2. Inspect and monitor activities subject to regulation.
3. Take appropriate enforcement action.
4. Comply with the terms of the MOA.

If program approval is withdrawn, responsibility for administering the program reverts to the *federal* government.

Transfer of Responsibility Back to the EPA

For a number of reasons, states with approved programs may voluntarily transfer the program back to the EPA. The state must give the administrator a 180-day notice and submit a plan for the orderly transfer of all relevant program information necessary for the agency to administer the program. A possible reason for the state to make such a transfer would be its failure to obtain the needed financial support from the state's executive appropriations committee, an arm of the state legislature.

Grants and Oversight

Because the states are the primary implementers of RCRA, they may receive financial assistance from the federal government under RCRA Section 3011. Both authorized and nonauthorized states are eligible to participate in the RCRA grant program. This money is intended as an incentive to help the states develop and implement their own RCRA hazardous waste programs. States and regions negotiate the specific work to be accomplished with these grant funds, which are awarded annually.

Such financial grants give the EPA additional power to oversee state programs through a condition of the grant. This allows the EPA to ensure that the program implemented adequately protects human health and the environment.

Setting Priorities

RIP Flexibility

The *RCRA Implementation Plan (RIP)* is a concept created by the EPA to acknowledge that regions and states have unique environmental problems and allow them to substitute activities necessary to address environmentally significant problems for national priorities. This RIP "flex" requires EPA headquarters' approval.

Agency Operating Guidance

The *Agency Operating Guidance* is a document published annually that outlines the EPA's goals and priority program activities, and identifies the national direction and priorities for implementing each EPA program, including RCRA. The RCRA priorities in this document form the basis for regional and state workload negotiations for the coming year.

State Grants

The Agency Operating Guidance also establishes the formula used to determine RCRA grant allotments. This formula is based on population, the hazardous waste generated in the region, and other factors. States submit proposed work plans that outline planned activities in the upcoming year, including permitting, enforcement, and program management. Each EPA regional office receives an allotment based on factors contained in the grant allocation formula and then negotiates with each state during the spring and summer. The grant award is made in October.

NOTE: States that receive RCRA grant funds must provide a 25 percent match.

State Oversight

An important role of the EPA's regional staff is to provide ongoing oversight of the entire state RCRA program. The purpose of oversight is to (1) promote a consistent national RCRA implementation, (2) encourage the coordination and agreement between the EPA and the states on technical and management issues, (3) ensure proper enforcement by the state, and (4) ensure appropriate expenditure of federal grant funds. Several agency guidance documents are available to assist regions in state oversight. These include the *National Criteria for a Quality Hazardous Waste Management Program* (OSWER Directive 9545.00-1), which contains standards and requirements for planning and overseeing an adequate RCRA program. This document can also help the EPA assess a state's progress and identify areas in which the state requires assistance.

NOTE: EPA guidance stipulates an annual mid- and end-of-year review of the RCRA program.

Information Management

Extensive RCRA program reporting requirements are made of both the EPA and the states. Many apply to the regulated community as well. The objectives of RCRA reporting requirements are to ensure that the program is adequately managed at the headquarters, regional, and state levels, and to provide accurate, up-to-date information to Congress and the public. The EPA uses several systems to track these reporting requirements:

1. **The RCRA Information System (RCRIS)** is a national database in which the EPA maintains RCRA program information. The original system, replaced by RCRIS in 1991, was the *Hazardous Waste Data Management System (HWDMS)*. The agency tracks accomplishments through RCRIS reports and its accountability system, the *Strategic Planning and Management System (SPMS)*. Examples of information tracked include permits issued/denied, inspection and violation data, and monitoring activities for all generators, transporters, and TSDFs. The RCRIS is maintained by regions and states that submit monthly updates to the database.

2. **The Strategic Planning and Management System (SPMS)** is an accountability system that facilitates integrated planning, tracking, and reporting of major activities within each of the EPA's high priority programs. The Agency Operating Guidance documents and SPMS measures are developed concurrently and finalized by March 1 of each year. Regions and states negotiate appropriate targets for many of the programs identified by the SPMS system. Progress is monitored on a quarterly basis with data stored in the RCRIS.

3. **The Biennial Reports** are required by RCRA Sections 3002 and 3005. As discussed earlier in this section, they establish requirements for generators and TSDFs submitting detailed activity reports. These reports must be submitted to the EPA on March 1 of each even-numbered year for the previous year's hazardous waste activity. Many states require that this reporting be done annually. States compile these reports and submit information to the EPA regions by September of the even-numbered year. These data, known as the biennial reports, are entered directly into the Biennial Report Data System (BIRDS), which provides information on the status of the RCRA program.

SUMMARY

▶ All generators of solid waste must determine if their waste is hazardous or nonhazardous. If it is nonhazardous, they may still be subject to many local solid and hazardous waste regulations under state and local government, but they are not subject to RCRA. If their waste is hazardous, then they are subject to regulation under Subtitle C of RCRA. The Subtitle C regulations specify that a solid waste is hazardous if it is not excluded and meets one of four conditions:

    a. Exhibits any of four characteristics:
        1. Ignitability
        2. Corrosivity
        3. Reactivity
        4. EP Toxicity
    b. Is listed
    c. Is a mixture
    d. Is derived from the treatment, storage, or disposal of a listed waste.

▶ Through delisting, any person may petition EPA to exclude listed waste from regulation under Subtitle C.

▶ The recycling of hazardous waste may, under certain circumstances, exempt it from Subtitle C regulation.

▶ Each state with an EPA-approved program assumes responsibility for implementing RCRA and accepts oversight from the federal government. A state seeking this final authorization for its program must submit an application to the regional administrator containing the following elements:

1. A letter from the governor requesting program approval.
2. Copies of all applicable state statutes and regulations.
3. Documentation of public participation activities.
4. A program description.
5. An attorney general's statement.
6. A memorandum of agreement.

▶ Before approving an application, EPA must be satisfied that the state program (1) is equivalent to, no less stringent than, and consistent with the federal program (state requirements may be more stringent or broader in scope, however.); (2) provides adequate enforcement authority; (3) provides for public notice and hearing prior to the issuance of a permit; and (4) provides for public availability of information in substantially the same manner and to the same degree as the federal program.

▶ Approved state programs are subject to revision, withdrawal of approval, and transfer of program responsibilities back to the EPA.

▶ States are the primary implementers of RCRA and may receive annual grants from EPA under RCRA Section 3011. States negotiate annual work plans with the EPA regions, and their progress is monitored primarily by regional staff, with headquarters' assistance.

▶ Implementers of RCRA are subject to extensive reporting requirements that apply also to the regulated community. These requirements include quarterly SPMS reporting and biennial reports. Most RCRA program data are tracked on the EPA national database, RCRIS.

QUESTIONS FOR REVIEW

1. What is the title of the act of Congress that governs the management of solid and hazardous waste?
2. What is a one-paragraph definition of *hazardous waste*?
3. Does the definition of a "solid waste" include gaseous wastes but only those in a liquid or solid state? Explain your answer.
4. What materials are specifically excluded from the definition of solid and hazardous waste? Provide a list and a brief description of each material.
5. What are the four categories of physical characteristics exhibited by hazardous waste that

the EPA uses for their classification? Provide a short definition of each.
6. What types of wastes are listed and identified by the following designations?
   a. D wastes
   b. F wastes
   c. K wastes
   d. P wastes
   e. U wastes
7. What are the principal methods of having a waste "delisted" by the EPA?
8. What is the general mixture rule? Write a one-sentence description.
9. What types of programs are the best to imple-

ment to prevent a waste from becoming a "solid and hazardous waste"?

10. Under which EPA act or legislation are the provisions for the regulation of polychlorinated biphenyls (PCBs) found?

11. Summarize the relationships between the EPA and the state agencies involved with implemen-

tation of a state's hazardous waste management program. Be sure to describe the application process, management of information, and the reporting process involved in this relationship.

ACTIVITIES

1. Recontact the small or medium business entity that generates hazardous waste in your community. Prepare a list of their solid and hazardous waste(s). From that list, determine the following for each waste (set up this information in a table format).

   a. Classification by characteristic.

   b. The EPA hazardous waste number for each waste if one has been assigned.

   c. The name of the primary ingredient or constituent of a mixed waste that makes it hazardous.

   d. Whether the waste is hazardous, acutely hazardous, or merely a solid waste.

2. Contact the agency in your state responsible for regulating agricultural wastes. Obtain program information and other brochures or pamphlets, and summarize their programs and regulatory provisions in manner suitable to be used for a five- to ten-minute presentation.

3

Generators of Hazardous Waste

Neal K. Ostler

Upon completion of this chapter, you will be able to meet the following objectives:

▶ Define the EPA term *generator.*

▶ Identify the steps in applying for an EPA identification number.

▶ List the categories of generators and describe the parameters of each in terms of quantity of hazardous waste generated.

▶ Describe the regulatory framework of generator requirements.

WHAT IS A GENERATOR?

A *generator* is any person, by site, whose act or process produces hazardous waste identified or listed by the EPA or whose act first causes a hazardous waste to become subject to its regulation. *Generation* is the act or process of producing hazardous waste.

A generator can be a large manufacturing or metal refining company with very large daily quantities of hazardous waste or a small auto body repair shop with very small annual quantities. The person who first makes the waste subject to Subtitle C regulations may include someone who imports a hazardous waste, initiates a shipment of a hazardous waste from a treatment, storage, or disposal facility (TSDF), or mixes hazardous wastes of different Department of Transportation (DOT) shipping descriptions by placing them into a single container.

39

The hazardous waste generator is the first link in the cradle-to-grave plan to manage hazardous waste that was established under the Resource Conservation and Recovery Act (RCRA).

Who Must Comply?

Any generator of more than 100 kilograms of hazardous waste or 1 kilogram of acutely hazardous waste per month, with a few exceptions, must comply with all of the generator regulations developed under Subtitle C of RCRA (40 CFR Part 262). Subtitle C requires generators to identify and fully document that the waste they produce is hazardous waste along with storage times. They must also identify that what they produce is properly transported to a TSDF approved by RCRA.

The subset of solid waste generators that must comply with Subtitle C regulations is, simply stated, made up of those whose solid waste is determined to be hazardous waste either by character or by EPA lists.

The procedures that solid waste generators use in determining whether they are subject to Subtitle C were outlined in Chapter 2. Once a generator determines that all or part of the waste produced is hazardous, he or she must comply with the regulatory requirements of Subtitle C.

CATEGORIES OF GENERATORS

The criteria used for determining which category suits a particular generator depend mainly on the quantity of hazardous waste produced annually. Under RCRA, there are three categories of generators:

▶ Large-quantity generators.
▶ Small-quantity generators.
▶ Conditionally exempt small-quantity generators.

The Large-Quantity Generator

In the early stages of the RCRA program, the EPA recognized that the regulations for hazardous waste would impose a substantial burden on the regulated community. Thus, in issuing its waste regulations, the agency first focused on those generators who produce the most volume of hazardous waste. *Large-quantity generators* (LQGs) are those facilities that generate more than 1,000 kilograms per month of hazardous waste or more than 1 kilogram of acutely hazardous waste per month. By nature of the size of its business, a large-quantity generator produces the greatest volume of hazardous waste in the United States. The EPA's 1985 biennial survey of generators estimated that 274 of the 275 million metric tons of hazardous waste came from large-quantity generators.

The Small-Quantity Generator

Small-quantity generators (SQGs) are those facilities that generate more than 100 but less than 1,000 kilograms of hazardous waste at a site per month (and accumulate less than 6,000 kilograms at any one time) or less than 1 kilogram of acutely hazardous waste per month (and accumulate less than 1 kilogram at any one time).

When the EPA regulations were first published on May 19, 1980, they exempted SQGs from most of the hazardous waste requirements. An SQG now must comply with most of the requirements but is afforded different accumulation periods. Congress amended the definition of SQGs in the Hazardous and Solid Waste Amendments of 1984 (HSWA), reducing the cut-off point from 1,000 kilograms to 100 kilograms because of concern that hazardous wastes exempted from regulation due to the SQG exclusion could be causing environmental harm.

Congress gave the EPA authority to vary the regulatory requirements applicable to SQGs from those applied to large generators. Congress was concerned that full regulation of SQGs, which produce only a small portion of the nation's hazardous waste, might be economically burdensome and inappropriate. But it required the EPA to provide a close watch and ensure that the relaxed requirements still protected human health and the environment.

> **NOTE: SQGs that store more than 6,000 kilograms of their waste on site must meet all large-quantity generator requirements.**

The Conditionally Exempt Small-Quantity Generator

Under current regulations, a facility that does not generate enough hazardous waste to qualify as a small-quantity generator is given the category of conditionally exempt SQG. A *conditionally exempt small-quantity generator* is defined as a facility that generates less than 100 kilograms per month of hazardous waste or less than 1 kilogram per month of acutely hazardous waste. A generator that qualifies as conditionally exempt is exempt from full regulation under Subtitle C. The conditionally exempt SQG, however, must still do the following:

▶ Identify the waste to determine whether it is hazardous.
▶ Not accumulate more than 1,000 kilograms of hazardous waste at any time.
▶ Properly treat or dispose of the waste on-site or ensure that the waste is sent to a permitted or interim status TSDF, a permitted municipal or industrial solid waste facility, or a recycling facility.

As soon as a conditionally exempt SQG generates more than 100 kilograms per month of hazardous waste, it becomes fully regulated as an SQG. In addition, any facility that generates in excess of 1 kilogram per month of acutely hazardous waste is regulated as an LQG.

> **NOTE: It is important to note that state classification of generators may be different from those outlined here. Some states regulate all generators of hazardous waste (i.e., have no exempt category) while some classify generators by waste type rather than by volume.**

REGULATORY REQUIREMENTS

All large- and small-quantity generators are subject to the regulations contained in 40 CFR 262. Such regulations require the following:

▶ Obtaining an EPA identification number.
▶ Manifesting of hazardous waste.
▶ Proper recordkeeping and reporting.

Each requirement is discussed here with different requirements for large- and small-quantity generators noted where appropriate.

The EPA Identification Number

One way RCRA requirements monitor and track generators is to assign each generator a unique identification number. Without this number, the generator is barred from treating, storing, disposing of, transporting, or offering for transportation any hazardous waste. Furthermore, the generator is forbidden from offering the hazardous waste to any transporter or TSDF that does not have an EPA ID number.

Obtaining an EPA Identification Number

Once a generator has identified waste as hazardous, he or she must obtain an EPA identification number before offering the waste for transport, treatment, storage, or disposal. (This means that prior to producing any waste, the generator must identify it. An EPA ID number is, in essence, a license or permit to produce a hazardous waste. It is illegal for a generator to transport hazardous waste off-site or to treat the waste without obtaining proper EPA permits and identification numbers. It is unlawful for a generator to offer hazardous waste to a transporter or to transport it himself or herself to a TSDF that does not have an EPA identification number.

To obtain a number, the generator must notify the EPA, under guidelines established in Section 3010 of RCRA. The EPA then assigns a twelve-digit identification number to the applicant. (The generator completes EPA Form 8700-12, Notification of Regulated Waste Activity, which is found, with instructions, in Appendix C.)

In most states, the EPA representative state agency, which coordinates with one of the ten regional EPA offices, will assist in obtaining an ID number.

> **NOTE: A conditionally exempt small-quantity generator is not required under Subpart C to obtain an EPA identification number but may be required by the state to do so.**

Waste Accumulation Guidelines

Prior to transportation, in addition to adopting the following DOT regulations, the EPA also developed regulations that cover the accumulation of waste at the site where it was generated. The DOT states that a large-quantity generator may accumulate hazardous waste on-site for ninety days or less as long as the following requirements are met:

- **Proper storage.** The waste must be properly stored in containers or tanks marked with the words *hazardous waste* and the date on which accumulation began.
- **Emergency plan.** A contingency plan and procedures to use in an emergency must be developed. Large-quantity generators are required to have a written contingency plan, but small-quantity generators are not.
- **Personnel training.** Facility personnel must be trained to handle hazardous waste properly. Large-quantity generators are required to have an established training program. Small-quantity generators are currently exempt from this requirement but must ensure that employees handling waste are familiar with proper procedures.

The ninety-day period allows a generator to collect enough waste to make transportation more cost effective; that is, instead of paying to haul several small shipments of waste, the generator can accumulate waste until there is enough for one big shipment.

**NOTE: A generator accumulating hazardous waste on-site
for more than ninety days is considered an operator of a
storage facility and must comply with the Subtitle C requirements for such facilities. Under temporary, unforeseen, and
uncontrollable circumstances, the ninety-day period may
be extended for up to thirty days by the regional administrator
on a case-by-case basis.**

Small-quantity generators can store waste on-site for up to 180 days, providing the following criteria are met:

▸ The on-site quantity of waste cannot exceed 6,000 kilograms at any time.
▸ The facility must have basic safety information including, at a minimum, the telephone number of the fire department and a coordinator for emergency activities.
▸ The generator must also ensure that personnel are familiar with emergency procedures that must be followed during spills and accidents.

For more information on safety requirements, see 40 CFR 262.34(d).

**NOTE: Small-quantity generators that must transport waste
for 200 miles or more for off-site treatment, storage, or
disposal are allowed to accumulate waste for up to 270 days.
The generator must still comply with the basic safety
requirements outlined.**

Pretransportation Requirements

Pretransportation requirements and regulations are designed to ensure safe transportation of a hazardous waste from its origin to its ultimate disposal. In developing these regulations, the EPA adopted those used by the DOT for transporting hazardous materials (49 CFR parts 172, 173, 178, and 179). These DOT regulations require the following:

▸ Proper packaging to prevent leakage of hazardous waste during both normal transport conditions and potentially dangerous situations. An example is a drum that falls out of a truck.
▸ Labeling, marking, and placarding the packaged waste to identify the characteristics and dangers associated with transporting wastes. For each container holding less than 110 gallons, the label or marking shall include the following language:

HAZARDOUS WASTE
Federal law prohibits improper disposal. If found, contact the nearest police or public safety authority or the U.S. Environmental Protection Agency.

Generator's name and address

Manifest document number

See Figure 3–1 for an example of a hazardous waste label. The requirements for proper transportation of hazardous waste will be discussed later.

**NOTE: These pretransport requirements and regulations
apply only to generators shipping waste off-site and
must be only to an RCRA TSDF.**

HAZARDOUS WASTE

FEDERAL LAW PROHIBITS IMPROPER DISPOSAL

IF FOUND, CONTACT THE NEAREST POLICE, OR
PUBLIC SAFETY AUTHORITY, OR THE
U.S. ENVIRONMENTAL PROTECTION AGENCY

PROPER D.O.T.
SHIPPING NAME _____ UN OR NA# _____

GENERATOR INFORMATION:

NAME _____

ADDRESS _____

CITY _____ STATE _____ ZIP _____

EPA   EPA
ID NO. _____ WASTE NO. _____

ACCUMULATION   MANIFEST
START DATE _____ DOCUMENT NO. _____

HANDLE WITH CARE!

CONTAINS HAZARDOUS OR TOXIC WASTES

▶ **FIGURE 3-1**
Hazardous waste label.

The Manifest

Consistent with the objectives of Subtitle C, designed to manage hazardous waste from cradle to grave, the EPA developed a special shipping requirement called the *Uniform Hazardous Waste Manifest* (the manifest), which is the key to accomplishing this objective.

The manifest provides generators with a mechanism through which they can track the movement of hazardous waste from the point of generation to the point of ultimate treatment, storage, or disposal. Briefly, the manifest must include the following minimum information:

▶ Name and EPA identification number of the generator, the transporter(s), and the facility where the waste is to be treated, stored, or disposed.
▶ DOT description of the waste being transported.
▶ Quantities of the waste being transported.
▶ Address of the treatment, storage, or disposal facility to which the generator is sending the waste (called the *designated facility*).
▶ A certification that the generator has in place a program to reduce the volume and toxicity of the waste to the degree economically practical, as determined by the generator, and that the treatment, storage, or disposal method chosen by the generator is the most practical method currently available that minimizes the risk to human health and the environment.

The manifest is the critical key to a controlled tracking system. Each time the waste is transferred, such as from a transporter to the designated facility or from one transporter to another, the manifest must be signed to acknowledge receipt of the waste. A copy is retained by each link in the transportation chain. Once the waste arrives at the designated TSDF, the owner or operator of that facility must send a copy of the manifest back to the generator. This system provides documentation that the generator's hazardous waste reached its ultimate destination.

The manifesting system has provisions should the generator not receive back a copy of the manifest: If thirty-five days pass from the date on which the waste was accepted by the initial transporter and the generator has not received a copy of the manifest from the designated facility, the generator must contact the transporter and/or the designated facility to determine the whereabouts of the waste. If forty-five days pass and the manifest still has not been received, the generator must submit an exception report (described later).

NOTE: Appendix D is an example, with complete instructions, of a Uniform Hazardous Waste Manifest.

Recordkeeping and Reporting

Recordkeeping and reporting requirements for generators provide the EPA and the states with a method to track the quantities of waste generated and the movement of hazardous wastes.

Regulations for generators are found in 40 CFR 262. They provide three primary recordkeeping and reporting requirements: *biennial* reporting, *exception* reporting, and *three-year retention* of reports, manifests, and test results.

Biennial Reporting

Large-quantity generators that transport hazardous waste off-site must submit a biennial report to the regional administrator by March 1 of each even-numbered year. Most small-quantity generators are exempt from this requirement. The 1995 Hazardous Waste Report (found in Appendix E) details the generator's activities during the previous calendar year and includes the following:

- The EPA identification number, name, and address of the generator.
- The calendar year covered by the report.
- The EPA identification number and name of each transporter used throughout the year.
- The EPA identification number, name, and address of each off-site treatment, storage, or disposal facility to which waste was sent during the year.
- A description, EPA hazardous waste number, DOT hazard class, and quantity of each hazardous waste shipped off-site.
- Quantities and nature of the hazardous waste generated.
- Efforts made to reduce the volume and toxicity of the wastes generated consistent with the Waste Minimization Program.
- Changes in volume or toxicity actually achieved, compared with those achieved in previous years.
- The certification signed by the generator or authorized representative.

> **NOTE: Generators that treat, store, or dispose of their hazardous waste on-site also must submit a biennial report that contains a description of the type and quantity of hazardous waste the facility handled during the year and the method(s) of treatment, storage, or disposal used.**

Exception Reports

Two time frames are provided for missing or tardy manifest copies:

1. **Thirty-five day rule.** A generator that has shipped hazardous waste off-site and does not receive a copy of the manifest with the handwritten signature of the owner or operator of the designated TSDF within thirty-five days of the date the waste was accepted by the initial transporter shall contact the transporter and/or the owner or operator of the designated TSDF to determine the status of the hazardous waste. This means that they must write a letter or in some way document the contact.

2. **Forty-five day rule.** A generator shall submit an exception report to the EPA, or the state authorized agency, if he or she has not received a signed copy of the manifest from the owner or operator of the designated facility within forty-five days of the date the waste was accepted by the initial transporter.

The exception report consists of (1) a legible copy of the manifest for which the generator does not have confirmation of delivery and (2) a cover letter signed by the generator or authorized representative explaining the efforts taken by the generator to locate the manifest and/or the hazardous waste and the results of those efforts. This is a time when nonrequired but recommended recordkeeping concepts should be adopted and good logbook entries and other records should be kept by the generator.

Three-Year Retention of Reports, Manifests, and Test Records

The generator is required to keep a copy of each biennial report and any exception reports for a period of at least three years from the date the report was submitted. The generator also is required to keep a copy of all manifests for three years or until he or she receives a copy of the manifest signed and dated from the owner or operator of the designated facility. The manifest from the facility must then be kept for at least three years from the date on which the hazardous waste was accepted by the initial transporter. Finally, the records of the waste analyses and determinations undertaken by the generator must be kept for at least three years from the date that the waste was last sent to an on-site or off-site TSDF.

> **NOTE: The periods of retention mentioned here can be extended automatically during the course of any unresolved enforcement action regarding the regulated activity or as requested by the administrator.**

ADDITIONAL GENERATOR REQUIREMENTS AND EXCLUSIONS

Additional generator requirements apply to persons who export their wastes or submit their waste for land disposal. In addition, farmers have been excluded from complying with generator requirements under certain circumstances. These requirements are also discussed here.

International Shipments

Additional notification requirements prohibit the export of hazardous waste unless the exporter obtains prior written consent of the receiving country. When importing hazardous waste, a person must meet all manifest requirements.

Before a generator can export a hazardous waste, the administrator must first be notified by the exporter sixty days prior to when the waste is scheduled to leave the United States. This notification must be completed only for the first shipment in any twelve-month period, unless basic information about the nature and frequency of the shipments changes. If the importing country agrees to accept the hazardous waste, the EPA sends an acknowledgement of consent form to the exporter, who may then export the waste to the importing country. Recordkeeping and reporting requirements are similar to those for domestic shipments of hazardous waste. These notices shall be sent to:

Office of International Activities (A106)
United States Environmental Protection Agency
Washington, D.C. 20460

NOTE: The statement of acceptance or written consent by the receiving country must be attached to the manifest accompanying each waste shipment and the generator shall require that the foreign consignee confirm the delivery. The generator must file an exception report if confirmation is not received according to the preceding guidelines.

Farmer Exclusion

Farmers who generate hazardous waste need not comply with the Subtitle C regulations when the wastes being disposed of are pesticides used only by them and the empty pesticide containers are triple rinsed and pesticide residues are disposed on the farm following the instructions on the pesticide label.

SUMMARY

- Generators regulated under RCRA fall into three categories:
  1. Large-quantity generators.
  2. Small-quantity generators.
  3. Conditionally exempt small-quantity generators.
- Regulations under Subtitle C for large- and small-quantity generators include the following:
  a. Obtaining an EPA identification number.
  b. Handling hazardous waste properly before transporting it.
  c. Adhering to time frames for accumulating waste.
  d. Completing the Uniform Hazardous Waste Manifest.
  e. Meeting recordkeeping and reporting requirements.
- The regulations included in HWSA require generators to evaluate their waste to determine whether it must be treated prior to land disposal. HWSA requires the EPA to establish waste-specific treatment standards in accordance with specified schedules.
- Depending on whether a treatment standard has been established, recordkeeping requirements for generators differ.

▶ Exempted from some or all of the generator requirements under certain circumstances are:

    a. Generators who treat, store, or dispose of their waste on-site.

    b. Generators of less than 1,000 kilograms per month who ship their waste offsite.

QUESTIONS FOR REVIEW

1. What does the term *generator* mean?
2. What EPA term describes the production of hazardous waste?
3. Who must comply with RCRA's requirements for generators?
4. How does a company obtain an EPA identification number?
5. What are the differences among a large-quantity generator, a small-quantity generator, and a conditionally exempt small-quantity generator?
6. What basic regulatory requirements apply to any size of generator of hazardous waste?
7. For each size of generator, what are the guidelines for accumulating hazardous waste?
8. Do any pretransportation requirements exist for generators of hazardous waste? If so, describe each.
9. What is the purpose of the Uniform Hazardous Waste Manifest?
10. Why are there five minimum requirements of the Uniform Hazardous Waste Manifest?
11. How long must a generator keep the records related to its hazardous waste activity, including its copies of the Uniform Hazardous Waste Manifest?
12. What are the record-keeping requirements for generators? Outline them.

ACTIVITIES

1. Contact your state agency for management of solid and hazardous waste and determine the overall number of small-quantity and large-quantity generators of hazardous waste in your state.
2. After consulting the requirements and obtaining the proper forms, prepare an application and write (but do not mail) a letter requesting a twelve-digit EPA identification. Address the letter to the designated EPA office for your state.
3. Contact a hazardous waste generator in your community and obtain permission to have a supervised visit to examine its completion and filing procedures of the following record-keeping activities:

    a. Manifesting.

    b. Accumulating wastes.

    c. Biennial reporting.

4. Identify the contact agency and person(s) in your state and at the EPA with whom a small-quantity generator would correspond to obtain exceptions for its waste streams that would make the generator a "conditionally exempt small-quantity generator."

4

Transportation of Hazardous Waste

Neal K. Ostler

Upon completion of this chapter, you will be able to meet the following objectives:

▶ Define transportation, transporter, transport vehicle, and transfer facility.

▶ Identify the requirements a generator must meet before transporting hazardous waste.

▶ Outline the steps required in preparing a load of hazardous waste for shipment.

▶ Identify the requirements a transporter must meet before transporting hazardous waste.

▶ Recognize and complete a Hazardous Waste Manifest.

▶ Identify the different manifest requirements of the generator and the transporter.

▶ Outline the immediate action, notification, and reporting requirements that must be taken in the event of release or spill of hazardous waste during transportation.

OVERVIEW

A *transporter* is anyone engaged in the off-site movement of hazardous waste by air, rail, roadway, or water if such hazardous waste is required to be shipped on a *Uniform Hazardous Waste Manifest. Transportation* means the movement of hazardous waste by any of the same modes. A *transport vehicle* means each separate cargo-carrying trailer, railroad freight car, or motor vehicle by any of the same modes. A *transfer facility* is any transportation-related facility including loading docks, parking areas, storage areas, and other similar areas where shipments of hazardous waste are held during the normal course of transportation.

Department of Transportation (DOT) regulations require that any hazardous material (hazmat) employee be trained in DOT regulations pertinent to their job by the hazmat employer. Such an employer may be anyone who transports, causes to transport, or is involved with the decisions regarding containers used in the transportation of hazardous materials in commerce. The hazmat employee is anyone who loads, unloads or handles hazardous materials or containers for hazmat transportation or is responsible for hazmat preparation, safety, or actually operates a vehicle used for such transportation. The training is required in three areas of the facility: awareness and familiarization for workers, safety and response to emergency spills or releases of hazmat, and function-specific training applicable to the job the employee performs, such as preparation of shipping papers, package selection, or marking, labeling and placarding responsibilities.

> **NOTE: The purpose of this chapter is to provide the reader a basic orientation regarding specific requirements of DOT/EPA for the transportation of hazardous waste. However, the orientation cannot appropriately be used to substitute for proper DOT HM-126/181 training requirements.**

Transportation of hazardous waste is a critical link between the generator and the ultimate off-site hazardous waste treatment, storage, or disposal facility (TSDF). The regulations pertinent to the transporter were developed jointly by the EPA and the DOT to avoid contradictory requirements between the two agencies, and although the regulations are integrated, they are not contained in the same part of the Code of Federal Regulations (CFR). Pertinent transporter regulations are found in both of the following:

- 49 CFR 171-179 (the Hazardous Materials Transportation Act)
- 40 CFR 263 (Subtitle C of the Resource Conservation and Recovery Act—RCRA)

The remainder of this chapter summarizes the Subtitle C regulations applicable to transporters. Readers interested in obtaining a more complete picture of transporter regulations should also review the DOT regulations.

THE GENERATOR: PRETRANSPORTATION REQUIREMENTS

The pretransportation requirements of the generator were briefly discussed in Chapter 3; however, more in-depth discussion is included in this chapter regarding the requirements of the DOT. These regulations were developed to help ensure the protection of the public and the environment when hazardous waste is being transported and include proper packaging; labeling, marking, and placarding; identification and use of shipping names used in preparation of the Uniform Hazardous Waste Manifest; certification by the generator; and emergency response information. Examples of DOT placards are found in Figure 4–1.

▶ **FIGURE 4–1**
Department of Transportation hazardous waste placards.

Packaging: §173

DOT regulations regarding packages were developed to help ensure that no leakage would occur during normal transportation, causing potentially dangerous situations. It is the responsibility of the generator to make a proper selection of packaging for his or her particular wastes. Important definitions for the variety of quantities shipped in a single package as determined by the size of the package are as follows:

Bulk packaging. A package with no intermediate form of containerization or containment with a maximum capacity of more than 110 gallons (450 L), a net mass of more than 882 pounds and a maximum of 119 gallons, and a water capacity of more than 1,000 pounds.

Nonbulk packaging. Any package smaller than a bulk package but with the total weight being the sum of both the package and the contents.

Limited quantity. The maximum amount of a hazardous material/waste for which there is a specific labeling or packaging exception (meaning there may be no such requirements).

Reportable quantity (RQ). This amount must be shown as a reportable quantity on the shipping papers and reported if released or potentially released by accident or otherwise into the environment.

Residue. A transport container must remain placarded as residue, and shipping papers must be prepared to indicate such until the container has been triple-rinsed and certified. A container is not considered empty until then.

Column-8 of the Hazardous Materials Table identifies the packaging selection(s) for a particular shipment. These selections and their references in 49 CFR 173 follow:

§173.155. Limited quantities of miscellaneous hazardous materials (Class 9) are **excepted** from labeling and specification packaging requirements. This provides for the transportation of samples that are still subject to the requirements of shipping papers.

§173.203. This reference provides packaging information for nonbulk liquid, Class 9 hazardous materials.

§173.213. Reference for nonbulk solids.

§173.240. Bulk packaging for hazmat other than Class 1 and Class 7. Hazardous waste is a Class 9 material.

§173.241. Bulk packaging for liquid and solid materials with low-risk hazards.

Labeling and Marking: §172.300 & 400

DOT requires that every package, freight container, or transport vehicle containing hazardous material be labeled and marked appropriately. Different standards apply to bulk packaging as opposed to nonbulk.

Nonbulk Packaging

Packages considered less than "bulk" must be marked with a shipping name (see Figure 4–2, a sample of a shipping paper entry) and must have a Class 9 label. Each container less than 110 gallons must include the following specific language on its label or marking:

HAZARDOUS WASTE

Federal law prohibits improper disposal. If found, contact the nearest police or public safety authority or the U.S. Environmental Protection Agency.

Generator's name and address

Manifest document number

Bulk Packaging

Packages larger than 110 gallons or weighing more than 882 pounds or otherwise meeting the definition of bulk packaging must be marked in one of the following three ways:

1. With a four-digit number in the white diamond placard.
2. With an orange rectangle containing the four-digit number.
3. On a Class 9 placard.

SHIPPING PAPERS

Shipper's name:			To:			24-hr. phone
Qty.	Container	Name	Haz/Class	ID#	Pkg. Grp.	Gross Wt.
50	55 lb. bags	HAZARDOUS WASTE, n.o.s.	9	NA3082	III	2,750 lbs.
1	1 qt. plastic	Fouracetic Acid	6.1	UN2642	I	-

▶ **FIGURE 4–2**
A properly prepared shipping label.

> **NOTE: If a load is placarded the driver must comply with commercial driver license requirements.**

Using Proper Shipping Names and Descriptions: §172.200

Whether you are using an invoice, a docket sheet, a bill of lading, or some other method of receipt for the movement of hazardous materials, it will serve as the required shipping paper if it is prepared properly. The regulation requires that a hazardous material and waste be described according to the guidelines on the shipping paper in the Hazardous Materials Table found at 49 CFR 101.

If the shipment is a hazardous waste being shipped to a permitted TSDF, it must be accompanied by a manifest (described later in this chapter) and follow regulations described in §172.205. The manifest is a special type of shipping paper but must contain the proper DOT shipping name and description.

The shipping paper or manifest must have the hazardous materials entered first, highlighted, or entered in a column designated as hazardous materials (HM). The shipping paper must be legible, printed in English, and contain no codes or abbreviations other than those authorized.

> **NOTE: The word waste must proceed the proper shipping name if it is an EPA listed waste.**

The shipping paper (see Figure 4–2) must have a proper name, hazard class, the UN identification number, the packaging group in roman numerals, and the total quantity of the material in units of measurement and types of packaging. If the material is a hazardous substance, it must include the initials "RQ" for reportable quantity. If the package is empty, the shipping paper may include the word *residue*.

> **NOTE: Shipping papers must include a 24-hour emergency telephone number where someone responsible can be contacted in the event of an incident involving the shipment. (See 49 CFR 172.604.)**

In addition to the information that identifies the hazardous material, the regulations may also require certain additional entries such as

▶ DOT exemption
▶ Limited quantity
▶ Hazardous substance
▶ Dangerous when wet

▶ N.O.S. constituents
▶ Poison constituents
▶ Poison-inhalation hazard

Certifying: §172.204

Any person who offers a hazardous material or waste for transportation must provide precise language and sign the shipping paper certification. The regulation provides two choices of language, essentially saying the shipper (or the generator) certifies that the materials are fully and properly classified, described, packaged, marked, and labeled, and are in proper condition for transportation according to applicable DOT regulations.

Shipping papers may be signed by mechanical means, but the Uniform Hazardous Waste Manifest must be by hand signature of the generator or a principal representative so authorized.

Providing Emergency Response Information: §172.600

Any person who offers for transportation a hazardous material or waste that is subject to the shipping paper and manifesting requirements must provide emergency response information that is immediately available for use at all times the hazardous material or waste is present. The required information is as follows:

▶ The basic name and description, as required on the shipping paper.
▶ Immediate hazards to health.
▶ Risks of fire or explosion.
▶ Immediate precautions to be taken in the event of an accident.
▶ Immediate methods for handling fires.
▶ Preliminary first-aid measures.

This information must be printed legibly in English and be kept handy according to specific directions of the regulations.

Often a shipper includes a material safety data sheet (MSDS), but it is recommended that the information on the MSDS be summarized for the driver so that he or she will be able to obtain the information quickly in the event of an emergency or accidental release of the hazardous materials.

THE REGULATED TRANSPORTER

If the cargo of waste is identified as RCRA hazardous waste and requires a manifest under 40 CFR 262, the transporter is required to comply with the requirements under Subtitle C of RCRA relating to transporters. The definition covers transport by air, rail, highway, or water. The transporter regulations apply neither to the on-site transportation of hazardous waste by generators who have their own treatment, storage, and disposal facilities or their own on-site hazardous waste management facility (HWMF), nor to TSDFs transporting wastes within their own facilities. In other words, the generator or transporter can move the hazardous waste from place to place within or upon his or her own facility, but prior to transporting it anywhere else, the transporter must comply with the regulatory requirements for transporters of hazardous waste.

A transporter who only stores manifested shipments of hazardous waste in appropriate containers at a transfer facility for a period of ten days or less is not subject to regulations with respect to storage of those wastes, because the regulations consider this short-term storage to be continuing transportation. However, if the storage time exceeds ten days, the transporter is considered a storage facility and must comply with the regulations for such a facility.

A transporter often becomes subject to RCRA regulations. An example is a transporter who brings hazardous waste into the United States or mixes hazardous wastes of different DOT shipping descriptions in the same container. Either action classifies the transporter as a generator, who then must comply with the regulations applicable to generators, as outlined in Chapter 3.

Regulatory Requirements for Transporters

A transporter is subject to a number of regulations that will be discussed in this section. These regulations include the following:

▶ Obtaining an EPA ID number.
▶ Complying with the manifest system.
▶ Handling and reporting discharges of hazardous waste.

The EPA Identification Number

Prior to transporting hazardous waste, a transporter must obtain a unique twelve-digit ID number from the EPA. This number is assigned on receipt of the transporter's notification of hazardous activity under Section 3010 of RCRA. Without this number for identification, the transporter is forbidden to handle any shipments of hazardous waste. Furthermore, a transporter may not accept waste from a generator unless that generator also has an EPA ID number. In most states, this number can be acquired by contacting the corresponding state agency for the EPA or by contacting the EPA regional office.

A more complete discussion regarding the EPA identification number is found in Chapter 2.

The Uniform Hazardous Waste Manifest

The manifest is the shipping document identified by the EPA as Form 8700-22 and, if necessary, EPA Form 8700-22A (for extra pages). DOT regulations for the manifest are found in 49 CFR §172.205.

> **NOTE: A copy of the manifest with instructions for completion
> is included as Appendix D of this book.**

The manifest document number refers to the twelve-digit identification number the EPA assigned the generator, plus a unique five-digit document number assigned to the manifest itself by the generator for in-house recording and reporting purposes.

The manifest is a special shipping paper that provides a key role in tracking hazardous waste from cradle to grave. This is accomplished by requiring signatures and providing copies to the generator, the transporter, a possible second transporter, and the owner or operator of the designated facility. Another copy is returned to the generator with all signatures attached.

This shipping document was developed cooperatively between the U.S. Department of Transportation and the EPA. DOT regulations provide that the Uniform Hazardous Waste Manifest, containing all of the information required, may serve as the shipping paper required. The DOT reference in DOT §172.205 is as follows:

> *No person may offer, transport, transfer or deliver a hazardous waste unless accompanied by the EPA Form 8700-22.* (See Appendix D.)

The manifest requirements for the generator, transporter, off-site TSDF, and manifest discrepancies are discussed here.

Generator Manifest Procedures: 40 CFR §262   The generator is responsible for originating the manifest, selecting a transporter that has an EPA identification number, and selecting a designated and alternate facility for delivery of the hazardous waste shipment by the transporter. Omissions, false coding, or illegible entries on the manifest are considered EPA violations. The manifest must be prepared strictly in accordance with 40 CFR 262.20. The manifest must include the following information:

- Generator's EPA identification number.
- Lines for generator's name, mailing address, and 24-hour phone number.
- Transporter name and EPA ID number.
- Designated facility name, EPA number, address, etc.
- DOT shipping name, hazard class, and UN ID number.
- Type of container, quantity, and unit of measure.
- Special handling instructions from the generator.
- Generator certification.

The requirements for manifesting do not apply to small-quantity generators (greater than 100 kilograms but less than 1,000 kilograms of hazardous waste per month) when the waste is to be reclaimed under a contractual agreement that specifies the type of waste and frequency of shipments. The generator is responsible for keeping a copy of the reclamation agreement for three years after the agreement ends.

The generator must read, sign by hand, and date the certification statement and keep the original and fully signed copies of the manifest for a period of three years from the date the hazardous waste was accepted by the initial transporter (see Appendix D).

Transporter Manifest Procedures   A transporter must comply with the following manifest procedures:

- Not accepting hazardous waste from a generator unless it is accompanied by a manifest signed by the generator.
- Hand signing and dating the manifest, acknowledging acceptance of the hazardous waste from the generator.
- Accepting responsibility for the full quantity of hazardous waste designated on the manifest. Therefore, the transporter should weigh, calculate, or personally make a count of the cargo before acknowledging receipt of the shipment by signing the manifest.
- Not transporting improperly labeled hazardous waste or hazardous waste containers that are leaking or appear to be damaged, because those packages become the responsibility of the transporter during transport.

▶ Signing the manifest and returning a signed copy of it to the generator before leaving the generator's property.

▶ Ensuring that at all times a copy of the manifest accompanies the hazardous waste shipment.

▶ Complying with all pertinent laws and regulations regarding the cleanup and reporting of a discharge of hazardous waste. (The responsibilities of this requirement will be discussed later in this chapter at greater length.)

▶ Obtaining the signature of a transporter or facility owner/operator to whom the original transporter delivers the shipment. The original transporter must retain one copy of the manifest for his or her own records and give the remaining copies of the manifest to the accepting transporter or designated facility.

▶ Delivering an entire quantity of waste accepted from either the generator or another transporter to the designated facility listed on the manifest. If the waste cannot be delivered as the manifest directs, the transporter must inform the generator and receive further instructions (i.e., return the waste or take it to another facility).

▶ Contacting the generator if the transporter is unable to deliver the hazardous waste to the designated or alternate facility named on the manifest. The generator shall either designate another facility or instruct the transporter to return the waste.

▶ When transporting hazardous waste out of the United States, indicating on the manifest the date that the shipment left the United States, signing the manifest, retaining one copy, and returning a signed copy to the generator.

▶ Retaining a copy of the manifest for three years from the date the hazardous waste is accepted by the initial transporter.

These manifest requirements do not apply to hazardous waste produced by generators of more than 100 kilograms but less than 1,000 kilograms of hazardous waste in a calendar month when the waste is reclaimed under a contractual reclamation agreement and the vehicle used for transport is owned and operated by the reclaimer. Nor do these manifest requirements apply to samples of hazardous waste taken for the express purpose of determining, through sampling and analyses protocols, the characteristics of the hazardous waste. The manifest requirements also do not apply to transporters who handle certain reclaimed wastes from small-quantity generators who are able to exempt their waste. However, the vehicle used to transport the waste from the recycling facility must be owned and operated by the reclaimer of the waste.

> NOTE: When a transporter is exempt from the manifest requirements, it is because the generator is exempted; however, each party must still comply with all parts of the shipping paper, in lieu of the manifest, and all other DOT transportation requirements.

Most of the manifest regulations relate to transportation by motor carrier or trucks. When the transportation is to be by water (bulk shipment) or by rail, many of the requirements are modified or not applicable. Both the generator and the transporter must examine the regulations to be sure that pertinent DOT procedures are followed. These procedures are found in 49 CFR 171–179.

Off-site TSDF Manifest Procedures Before a generator can designate a facility for a shipment of hazardous waste, and before a transporter can effect the delivery to the facility, the facility must have its own EPA identification number. This number goes on the manifest. The owner or operator of the TSDF, upon receiving a shipment of hazardous waste, must do the following:

- Ensure that the shipment of hazardous waste is accompanied by a manifest (or, in some cases of rail and water shipment, a shipping paper).
- Confirm that the contents and quantity of the shipment are consistent with that shown on the manifest (or shipping paper).
- Sign and date each copy of the manifest (or shipping paper) to certify that the hazardous waste described on the manifest was received.
- Note any discrepancies on each copy of the manifest as discussed in the following section.
- Immediately provide at least one signed copy of the manifest to the transporter.
- Send a copy containing all of the signatures and dates to the generator within thirty days after delivery or, if the manifest is not received within thirty days after delivery, a signed and dated copy of the shipping paper.
- Retain a copy of the signed and dated manifest and/or shipping paper for at least three years from the date of delivery.

Manifest Discrepancies Occasionally, there may be a disparity between the quantity or type of hazardous waste described on the manifest (or shipping paper) and that actually received by the TSDF owner or operator. This disparity is known as a *manifest discrepancy* and may be significant. Discrepancies in quantity are variations in piece count or variations of more than 10 percent in weight. Discrepancies in the type of waste can be discovered by visual observation or by waste sampling and analysis protocols. Examples of type differences are a waste solvent substituted for a waste acid, or toxic constituents in the waste not reported on the manifest.

Upon discovering a discrepancy that the TSDF considers significant, the owner or operator will attempt to reconcile the discrepancy with the generator or transporter as expeditiously as possible (usually by telephone conversation). If unable to resolve the discrepancy, the TSDF may decline to receive the waste. If not resolved within fifteen days after receiving the waste, the TSDF owner or operator must submit a letter describing the discrepancy and efforts to reconcile it, and a copy of the manifest (or shipping paper) to the state agency issuing the permit to the TSDF. Unreconciled discrepancies discovered during analyses performed at a later date must also be reported.

> **NOTE: A TSDF initiating a shipment from his or her facility shall comply with all parts of the RCRA requirements for generators and transporters.**

Transportation Discharges of Hazardous Waste

Although both the generator and transporter of hazardous waste may comply with all appropriate regulations, the business of transporting hazardous waste can still be very dangerous. There is always the possibility that an accident will occur. To respond to this possibility, the regulations require transporters to take **immediate action** to protect health and the environment if a release occurs, to **notify** certain federal, state, and local agencies, and to make certain follow-up **reporting requirements**.

Immediate Action In the event of a hazardous material spill that becomes hazardous waste, the person responsible for the material at the time of the spill shall take immediate action. In most cases this means diking off the discharge area to keep the spill from spreading and notifying the local authorities.

Notification When a serious accident or spill occurs, the transporter must notify the EPA and the DOT and, if it involves an etiologic agent, the Centers for Disease Control (404-633-5313). This can be done simultaneously by notifying the National Response Center (NRC)at 800-424-8802 or 202-267-2675. Specifically the NRC must be notified when any of the following incidents occurs:

1. A person is killed or seriously injured.
2. Estimated damage exceeds $50,000.
3. A general evacuation of the area occurs lasting one hour or more.
4. The spill involves disease-causing agents or radioactive material.
5. The spill exceeds a SuperFund reportable quantity.
6. A life-threatening situation exists. A general rule to follow can be when the following spill quantities are exceeded:
   a. One kilogram (2.204 lbs) of any acutely hazardous waste listed by the EPA as a P-Waste, or a smaller quantity if there is a potential threat to human health or the environment.
   b. One hundred kilograms (220.4 lbs.) of any hazardous material that, when spilled, becomes hazardous waste other than an acutely hazardous waste, or a smaller quantity if there is a potential threat to human health or the environment.

> **NOTE: The person reporting the spill should remain on the line**
> **until the other person has completed his or her inquiry,**
> **obtained all of the information needed, and given further**
> **directions. The manifest or shipping paper and any log book**
> **or incident notes should be available and the reporting party**
> **should be prepared to make as full a disclosure of the**
> **incident as possible.**

Reporting Requirements Because a transportation spill involves both the DOT (the mode of travel) and the EPA (the general environment), reporting requirements must be made to each agency.

> **DOT:** Within 30 days of the incident, the transporter must file a report with the DOT as required by 49 CFR 171.16. The basic guidelines of the regulations regarding hazardous wastes involved in a spill are that the transporter must supply a copy of the hazardous waste manifest, give an estimate of the quantity of waste removed from the site, identify where the waste was taken, and dispose of any unremoved waste. Consumer commodities, batteries, and small containers of paint are exempt from the thirty-day notification requirement unless they are being transported on a manifest as a hazardous waste. These reports are to be made in writing on specific forms required by the DOT and sent to The Director, Office of

Hazardous Materials Regulations, Materials Transportation Bureau, Department of Transportation, Washington, D.C. 20590.

NOTE: A water transporter who has discharged hazardous waste must give the same notice as required in 33 CFR 153.203 for oil and hazardous substances.

EPA: The EPA reporting guidelines require that within fifteen days of an incident involving any spill of a hazardous material (or a hazardous material that became a hazardous waste when spilled), the person responsible for the material at the time of the spill must submit a written report to the responsible state EPA counterpart. The written report shall contain the following minimum information:

a. The person's name, address, and telephone number.
b. The date, time, location, and nature of the incident.
c. The name, identity, and quantity of the material(s) involved.
d. The extent of any possible injuries.
e. An assessment of the actual and/or potential hazards to human health or the environment.
f. The estimated quantity and disposition of recovered materials that resulted from incident. This includes arrangements made for the treatment, storage, and disposal of all waste materials and contaminated equipment, tools, personal protective equipment (PPE), and such.

RCRA gives certain federal, state and/or local officials special authority to handle transportation accidents. Should any of those persons, with appropriate authority, determine that immediate removal of the waste is necessary to protect human health or the environment, the official can authorize an immediate waste removal by a transporter who lacks an EPA identification number and do so without the immediate need of a manifest. However, a log book entry of a shipping paper is still required.

STATE REGULATIONS OF TRANSPORTATION

In most cases, transporters will travel through many states and, thus, in addition to meeting federal requirements, must also comply with the requirements of each state they travel through. Some state programs are significantly more stringent than the federal program and, indeed, the most substantive requirements come from more than thirty states with transportation regulations that differ from the federal program. Many of these states have registration programs for each individual transportation company and require transporters to post bonds in case of an accident involving hazardous materials and waste. These state-to-state differences make it important that transporters consult the regulations of each state they travel through in addition to the federal requirements.

It is recommended that transporters check with the state counterpart to the EPA, the public service commission, the state department of transportation, and the state department of public safety or highway patrol.

SUMMARY

▶ A generator has a full complement of pretransportation requirements under regulations of both the EPA and DOT. These mainly involve properly packaging, labeling, marking, placarding, and preparing the hazardous materials and waste before shipment.

▶ A transporter of hazardous waste must comply with both DOT and EPA regulations. The RCRA Subtitle C regulations require a transporter to obtain an EPA identification number, comply with the manifest system, and properly handle and report discharges of hazardous waste.

▶ Under certain circumstances, a transporter of hazardous waste also may be subject to Subtitle C generator and/or storage facility requirements.

▶ Individual state hazardous waste transportation programs often add requirements beyond those of the federal program. Transporters should consult the regulations of states they travel through to ensure that they understand fully the hazardous waste transport requirements.

QUESTIONS FOR REVIEW

1. What is the definition of *transportation* of hazardous waste?

2. Where in the Code of Federal Regulations are the most pertinent directives related to the transportation of hazardous waste found?

3. What are the various amounts for each of the following quantity types that determine whether packaging and other requirements apply to the hazardous waste?
   a. Bulk
   b. Nonbulk
   c. Limited quantity
   d. Reportable quantity
   e. Residue

4. What type of labeling is required on nonbulk packages? On bulk packages?

5. Is the order of different description information on the shipping paper or the Hazardous Waste Manifest important? If so, in what order must the basic description occur?

6. What type of certification does the shipper or generator of hazardous materials and waste make when transporting those materials?

7. What requirements must a trucking company meet before it can legally transport hazardous waste?

8. What is the purpose and/or function of the Hazardous Waste Manifest? Why are there five copies included?

9. Who is responsible for placarding of a load of hazardous waste? Describe what the placard looks like and where it must be attached to the container.

10. What agencies of the federal and state government must be notified in the event of an accidental spill or release of hazardous materials and waste?

ACTIVITIES

1. Using the Yellow Pages, compile an alphabetical list of companies in your area that advertise as haulers or transporters of hazardous waste. Contact a dozen of them and attempt to obtain their EPA identification numbers. Find out whether they also have authority to transport hazardous waste under the U.S. Department of Commerce regulations.

2. Identify a hazardous material or waste that you are somewhat familiar with or that you've learned about from previous activities. On an actual shipping paper or a facsimile, complete the full description and other information needed to accompany a shipment of that material.

3. Contact a transporter in your community. Schedule a visit to or tour of the facility and examine the vehicles and containers it uses in the shipment of hazardous waste.

4. Determine whether your state or community has a program to monitor the movement of hazardous waste. If it does, describe the regulatory mechanism that has been implemented to monitor it.

5

Treatment, Storage, and Disposal of Hazardous Waste

Neal K. Ostler

Upon completion of this chapter, you will be able to meet the following objectives:

▶ Define the EPA term *treatment, storage, and disposal facility* (TSDF).

▶ Describe the following: *treatment, storage,* and *disposal,* as they apply to the management of hazardous waste.

▶ Identify the operations that are subject to the TSDF regulatory requirements.

▶ Define *totally enclosed treatment facility.*

▶ Outline the basic standards required of a TSDF.

OVERVIEW

The last link in the cradle-to-grave hazardous waste management system is the treatment, storage, and disposal facility (TSDF). Because treatment, storage, and disposal involve many different types of units (e.g., a landfill or an incinerator), the regulations governing TSDFs are far more extensive than those that have been described for generators and transporters. As a result, this chapter provides only a summary of the TSDF requirements.

Details of the TSDF regulations are found in Title 40 of the Code of Federal Regulations (CFR) Parts 264 and 265; the reader is encouraged also to examine the *Federal Register.* These regulations establish facility design and operating criteria as well as performance standards that owners and operators must meet to protect human health and the environment.

63

All TSDFs that handle hazardous waste are required to obtain an operating permit and abide by the treatment, storage, and disposal (TSD) regulations. (The process of permitting is described in Chapter 7.)

This chapter describes the general requirements of all facilities that treat, store, or dispose of hazardous waste and provides the administrative and nontechnical requirements that apply to both the interim-status TSDF and the permitted TSDF, which are nearly identical.

This chapter also discusses and describes the disposal restrictions mandated by the Hazardous and Solid Waste Amendments of 1984 (HSWA), curtailing the placement of untreated hazardous waste into or upon the land for disposal.

DEFINITION OF A TSDF

According to 40 CFR 260.10, a TSDF encompasses the different functions of treatment, storage, and disposal. These functions are defined as follows:

Treatment. Any method, technique, or process, including neutralization, designed to change the physical, chemical, or biological character or composition of any hazardous waste. This may mean any or all of the following:
  ▶ To neutralize such waste.
  ▶ To recover energy.
  ▶ To recover material resources.
  ▶ To render such waste nonhazardous or less hazardous.
  ▶ To make the waste safer to transport, store, or dispose of.
  ▶ To make the waste amenable for storage or volume reduction.

Storage. Holding hazardous waste for a temporary period, at the end of which the hazardous waste is treated, disposed of, or stored elsewhere.

Storage facility. A facility or part of a facility designated for holding hazardous waste for a temporary period, at the end of which the hazardous waste is treated, disposed of, or stored elsewhere (except for generators who accumulate or store hazardous waste on-site less than 90 days for subsequent transportation off-site).

Disposal. The discharging, depositing, injecting, dumping, spilling, leaking, incineration, or placing of any solid waste or hazardous waste into or on any land or water so that such waste or any constituent thereof may enter the environment or be emitted into the air or discharged into any waters, including groundwaters.

Disposal facility. A facility or part of a facility where a hazardous waste is intentionally placed, or on any land or any water at which the waste will remain after closure of the facility.

The regulations provide that no person shall own, construct, modify, or operate any facility for the purpose of treating, storing, or disposing of hazardous waste without first submitting a hazardous waste operation plan and receiving approval of the plan from the EPA or its state counterpart agency.

TWO SETS OF REGULATIONS

The Resource Conservation and Recovery Act (RCRA) required that the EPA develop regulations for all TSDFs. Although only one set was required, the EPA developed two sets: one for interim-status TSDFs, the other for permitted TSDFs.

An appropriate method for presentation of the general provisions, for some purposes, would be in the same order as they are found in the Code of Federal Regulations. However, the length of this material prohibits doing so in one chapter. Therefore, this chapter will present the administrative and nontechnical requirements governing both interim-status and permitted TSDFs, because the requirements for both are nearly identical. Chapter 6 will cover the technical requirements of both the interim-status and permitted TSDFs, allowing a more thorough comparison of the requirements for each and discussion of their differences.

REGULATORY REQUIREMENTS

The purpose of the **administrative and nontechnical** requirements governing TSDFs is to ensure that their owners and operators establish the necessary procedures and plans to run such facilities properly and to handle any emergencies or accidents. These regulations are in Subparts A–E of 40 CFR Parts 264 and 265, which are applicable to permitted and interim-status facilities, respectively.

Coverage and Exceptions (Subpart A)

Subpart A identifies who is subject to TSDF regulations and any circumstances under which a person is excluded or subject to only limited requirements.

In general, all owners or operators of facilities treating, storing, or disposing of hazardous waste must meet the appropriate treatment, storage, and disposal regulations. The exceptions to this include the following:

1. A person disposing of waste in an underground injection well and subject to a permit issued under a program promulgated and approved under the Safe Drinking Water Act.
2. The owner or operator of a public-owned treatment works (POTW) permitted under the Clean Water Act.
3. A generator of hazardous waste operating only within the allowable accumulation guidelines.
4. A farmer disposing of pesticides from his or her own use.
5. The owner or operator of a totally enclosed treatment facility. A *totally enclosed treatment facility* is a facility that treats hazardous waste that is connected directly to an industrial production process and is constructed and operated in such a manner to prevent the release of any hazardous waste or its constituents into the environment during treatment. (An example is a pipe in which acid is neutralized.)
6. The owner or operator of an elementary neutralization unit or a wastewater treatment unit.
7. A person cleaning up an emergency spill or hazardous waste discharge.

8. Facilities that reuse, recycle, or reclaim hazardous waste (except persons who produce, burn, and distribute hazardous waste–derived fuel and used oil recyclers).
9. A transporter storing manifested shipments for less than ten days while in the normal course of transportation.
10. A facility regulated by an authorized state program.

In addition, owners or operators of facilities regulated by other environmental laws under a permit-by-rule (discussed in Chapter 7) need to meet only minimum TSD requirements. This includes the POTWs, underground injection wells, and others discussed previously.

> NOTE: A small- or large-quantity generator that engages in operations that constitute the treatment, storage (other than accumulation), or disposal of hazardous waste is subject to all provisions of a TSDF. However, certain activities, such as compaction, bulking, and lab packing of the wastes, may qualify for special exception. A hazardous waste-handling facility operated by a generator must submit an application for approval and receive the approval to engage in such operations without first being permitted as a POTW.

General Facility Standards (Subpart B)

Before handling any hazardous wastes, every facility owner or operator must take the following steps:

1. Apply to the EPA for an identification number.
2. Ensure that their wastes are properly identified and handled.
3. Have waste analyses performed prior to treatment, storage, and disposal to ensure that the owners or operators possess sufficient information on the properties of the wastes they manage to be able to treat, store, or dispose of them in a manner that will not pose a threat to human health or the environment. The regulations also require owners or operators to perform detailed chemical and physical analyses of their wastes, to develop and follow a written waste analysis plan that specifies tests and test frequencies, and to test any incoming wastes.
4. Properly manage ignitable, reactive, or incompatible wastes. In general, all ignitable or reactive wastes must be protected from sources of ignition or reaction or treated to remove the cause of concern. Owners or operators also must ensure that treatment, storage, or disposal of ignitable, reactive, or incompatible waste does not result in damage to the containment structure (container, tank, surface impoundment, landfill cell, or pit) and/or threaten human health and the environment. Incompatible wastes must not be placed in the same containment structure if there is the potential for reaction.
5. Install security measures. Security requirements were developed to minimize the potential for the unauthorized entry of people or animals onto the active portions of facilities. To meet these security objectives, a barrier surrounding the active portion of the facility with controlled entry systems or 24-hour surveillance must be installed and warning signs posted. Owners or operators also must take precautions to avoid fires, explosions, generation of toxic gases, and any other events that would threaten human health, safety, and the environment. Owners and operators are exempt from these requirements if unauthorized or unknowing entry

will not result in injury and if the disturbance of waste or equipment will not result in environmental damage.

6. Conduct inspections. The regulations require an owner or operator to develop and follow a written inspection schedule to assess the status of the facility and detect potential problem areas. Any observations made during the inspections are recorded in the facility's operating log and kept on file for three years. All problems identified must be remedied.

7. Conduct training (40 CFR 265.16). The regulations explain in detail what needs to be accomplished in the training programs. The program must teach employees to perform their duties in a manner that ensures compliance with the regulations and with a focus on health and safety. The program must be directed by someone trained in hazardous waste management procedures and must teach relevant site-specific handling procedures. The training must familiarize employees with the company's hazardous waste contingency plan. The purpose of the training requirement is to reduce the potential for mistakes that might threaten human health and the environment. This is accomplished by ensuring that facility personnel acquire expertise in the areas to which they are assigned. Some of the basic provisions of the training requirements are as follows:

   a. Facility personnel must be trained within six months after beginning a job. Employees are not allowed to work unsupervised until the required training is completed.

   b. Records must be maintained for three years for employees who have left the company and until closure of the facility for current employees.

   c. The initial training received by the employees must be updated annually. This annual refresher is required to be a review of the initial training.

   d. Both on-the-job training and in-house training programs may be used to meet the training requirements.

   e. OSHA now requires TSDFs to provide a minimum of twenty-four hours of safety training (this may be waived if the employee has had equivalent training or work experience) and implement the following written training plans:
      ▸ A hazard communication plan
      ▸ A medical surveillance program
      ▸ A respiratory protection program
      ▸ A health and safety plan for employee decontamination procedures

   Figure 5-1 is an example of a generic hazardous waste management training plan.

8. Comply with location standards. Current standards prohibit locating a new facility where flood or seismic events could affect a waste management unit. Bulk liquid wastes are prohibited from placement in salt domes, salt beds, or underground mines or caves. Provisions of the HSWA required the agency to further strengthen these location criteria. The EPA is currently revising the regulations to reflect these new statutory directives.

Contingency Planning and Emergency Procedures (Subparts C and D—40 CFR 264.51, 56)

Known as *preparedness and prevention requirements,* the first steps of the contingency planning and emergency procedures (Subparts C and D) are very explicit:

Sponsor:	A–1 Waste Facility
Dates:	Anytime 1999
Site:	Salt Lake Community College
	Millcreek Center
	1621 E. 3900 So.
	SLC, UT 84106
Location:	Room #306
Trainer:	Neal K. Ostler and Company

DAY ONE:

Hrs		
1/2	07:00 – 07:25	Greetings & Orientation
11/2	07:30 – 08:45	Regulations Overview
2	08:50 – 10:30	Chemistry of Haz/Mat
1	10:35 – 11:25	Health Effects of Haz/Mat
		-LUNCH-
1	12:10 – 1:00	Health Effects (continued)
1	1:10 – 2:00	Medical Surveillance
1	2:10 – 3:00	Respiratory Protection
8		

DAY TWO:

1	07:00 – 08:20	Safety & Health Considerations
11/2	08:30 – 09:20	PPE Clothing
1	09:30 – 10:20	Decontamination
1	10:30 – 11:20	Heat Stress & Hearing Protection
		-LUNCH-
1	12:00 – 12:50	Atmospheric Monitoring
21/2	12:55 – 3:00	Level B Familiarization
8		

DAY THREE:

2	07:00 – 08:50	Emergency Response:
		"Awareness/Operations/
		Technician/Specialist"
11/2	09:00 – 10:15	Site Safety & Control
1	10:25 – 11:15	Confined Space Awareness
		-LUNCH-
21/2	11:50 – 2:05	Full Site Scenario
1	2:10 – 3:00	Quiz & Certificates
8		

<24 Hrs>

▶ **FIGURE 5–1**
Generic hazardous waste training plan.

install fire protection equipment and alarms, and arrange for coordination with the local authorities in emergency situations.

The next logical step is developing the contingency plan. The employer must describe the actions that facility personnel will take in response to fires, explosions, or any unplanned (sudden or nonsudden) release of a hazardous waste or its constituents into the air, on soil, or in surface water at the facility.

These two subparts, originally grouped as one, were developed to prepare for emergencies and may either be separate documents or part of the hazardous waste contingency plans. At a minimum, these plans must be in writing and provide the following basic elements:

1. Emergency escape procedures and routes.
2. Procedures to be followed by employees who remain to operate critical plant operations before they evacuate.
3. Procedures to account for all employees after the emergency evacuation has been completed.
4. Rescue and medical duties for those employees who are to perform them.
5. The preferred means to report fires, explosions, and other emergencies.
6. Names and means of contacting persons who can provide further information or explanation of duties under the plan.

Manifest System, Recordkeeping, and Reporting (Subpart E)

These requirements were detailed in Chapter 4 under the subject of manifesting, but in short, the manifest must be signed by the TSDF owner/operator and returned to the generator, thus completing the manifest loop established in the manifest regulations (40 CFR 262).

The owner/operator is responsible for ensuring that the waste described on the manifest is the same as the waste on the transport vehicle. This ensures that there are no significant discrepancies in the amount (e.g., an extra drum) or type of waste (e.g., acid waste instead of paint sludge) that was sent by the generator. If a significant discrepancy is discovered, the TSDF must reconcile the difference with the generator or transporter. If this is impossible, the EPA must be notified about the problem within thirty days of the incident.

Subpart E also includes requirements for recordkeeping and reporting. This includes operating records, biennial reports, unmanifested waste reports, and reports on releases, groundwater monitoring, and closure. These records and reports provide the regulating authority with information used to assess compliance with the hazardous waste regulations.

As part of the Community Right-to-Know portions of the SuperFund Act, the regulations also provide facility owners/operators and local authorities with information that may be used in responding to emergencies. Appendix F details the provisions necessary in a contingency plan and an emergency reponse procedure.

THE HAZARDOUS AND SOLID WASTE AMENDMENTS OF 1984

After RCRA's passage in 1976, many hazardous waste generators seemed to view the act as being merely new directives on storing and disposing of their hazardous waste. As a result, landfills became the primary disposal site for nearly every chemical waste because generators seldom looked for alternatives for their wastes. By 1984 it became obvious that a ban on the land disposal of certain hazardous wastes was necessary. Collectively, this *land ban* comprises the following four major sets of regulations that collectively are referred to as *land disposal restrictions* (*LDR*):

▶ Restriction on the land disposal of spend solvents and dioxins, which were regulated in 1986.
▶ Restriction of California wastes on land; these were regulated in 1987.
▶ Assessment of all EPA-listed wastes and appropriate LDR.
▶ EPA evaluation of underground injection of spend solvents, dioxins, and California list wastes according to the provisions of the Safe Drinking Water Act.

The LDR program's main purpose is to make unlawful and thereby discourage the placement of untreated wastes in or on the land when a better alternative, such as immobilizing and/or treating these wastes to eliminate or reduce their chemical hazard, may exist.

The primary focus of companies that generate or manage hazardous wastes today is the concern about the regulations that the EPA has generated since the passage of the LDR as they attempt to implement the HSWA requirements. As this book is written, twelve major final rules have been published regarding the LDR for managing different hazardous waste activities. Taken together, these rules compose the major sections of Title 40 of the Code of Federal Regulations. Part 148 deals with underground injection, and Part 268 deals with the restrictions applicable to all other types of land disposal.

Fundamentals of Land Disposal Restrictions (LDR)

These restrictions require the EPA to establish treatment standards for each hazardous waste to protect human health and the environment when the waste is disposed of on land. Each hazardous waste facility is required to use one or more specified treatment technologies or treat the waste to reduce the concentration of hazardous constituents to meet certain limits. The LDR assumes that the EPA requires the use of the best demonstrated available technology (BDAT) or the standard in the treatment of any wastes to which concentration limits are applied. An example is waste that does not require incineration, but whose treatment must reduce the concentrations of its hazardous constituents to at least the same level that incineration would provide.

The Compliance Strategy

Compliance with LDR involves taking the following six fundamental steps:

1. Determining whether the waste is an RCRA hazardous waste; if not, it is not subject to LDR. This process is identical to the steps taken by a generator in identifying its waste. Treatment standards for many newly defined and newly listed wastes have not been established. Although they still must be treated as hazardous wastes, they can be land disposed without meeting any treatment standards. Eventually, LDR will apply to all wastes.
2. Determining whether the waste is a waste water or a nonwaste water. *Wastewater* is defined as a waste containing less than 1 percent total organic carbon (TOC) and less than 1 percent total suspended solids (TSS). Material not meeting the definition of wastewater is a nonwaste water.

3. Determining the appropriate codes that apply to the wastes. Again, the process is identical to the steps a generator takes in classifying its wastes.
4. Identifying the EPA treatment standards for the applicable waste codes.
5. Determining whether the waste meets the treatment standards applying to it. If it does, it can be land disposed without further treatment. If not, further treatment is necessary until the waste meets the minimum standards.
6. Preparing and retaining all paperwork related to compliance.

The LDR program not only establishes standards of treatment before hazardous wastes can be disposed of on land, but also applies different guidelines and requirements to different waste management activities such as storage limits, accumulation times, methods of storage, and other regulations concerning the dilution and/or aggregation of hazardous wastes.

A serious study of LDR is necessary to come into full compliance with RCRA. The use of a hazardous waste consultant's services is often necessary for generators and may be the most reliable means of determining the proper methods to treat, store, and dispose of their wastes.

NOTE: For a more complete study of this subject, you may wish to obtain a subscription to *The Hazardous Waste Consultant* published by Elsevier Science, Inc.

SUMMARY

▶ Treatment, storage, and disposal facilities (TSDFs) are the last link in the cradle-to-grave hazardous waste management system; TSDFs must obtain a permit and abide by TSD regulations if they handle hazardous wastes.

▶ TSDFs fall into two categories: interim status facilities and permitted facilities. (Congress developed *interim status* to allow certain owners and operators of facilities in existence on November 19, 1980—or brought under Subtitle C regulation after this date by amendment—to continue operating as if they have a permit until their permit application is issued or denied.)

▶ There are two sets of TSD regulations:
1. Interim-status standards. These are "good-housekeeping" requirements, such as using proper tanks, found in 40 CFR 265.
2. Permit standards. These are facility-specific performance standards and "design and operating" requirements incorporated into the permit (e.g., tanks storing hazardous waste must be designed to specifications found in 40 CFR 264). The permit standard language in the regulations is general and serves as a guideline for permit writers in setting the specific design and operating requirements through best engineering judgment.

▶ Both TSDF regulations are composed of administrative and nontechnical requirements that ensure that owners or operators of TSDFs establish the necessary procedures and plans to run their facilities properly and to handle any emergencies or accidents. They cover the following:
1. Who is subject to the regulations.
2. General facility standards.
3. Preparedness and prevention.
4. Contingency plans and emergency procedures.

   5. Manifest system, recordkeeping, and reporting.
 ▶ The regulations also include technical requirements to ensure that owners or operators run their TSDFs in a way that minimizes the potential for threat to human health and the environment. (The technical requirements for both interim-status and permitted TSDFs will be discussed in Chapter 6.)
 ▶ HSWA Section 3004 requires the administrator to examine all listed hazardous wastes and some others to determine whether any should be banned from land disposal. Those wastes with concentrations of toxic constituents that threaten human health and the environment must be treated before they can be land disposed.

QUESTIONS FOR REVIEW

1. What are the definitions of the following terms?
   a. Treatment
   b. Storage
   c. Disposal
   d. TSDF
   e. Totally enclosed treatment facility
2. Where in the Code of Federal Regulations are the most pertinent directives for TSDF operations found?
3. What are the purposes of the administrative and on-technical requirements of a TSDF? What operations do these requirements cover?

4. What type of waste operation is an exception to the requirements of a TSDF?
5. What are the basic general standards required of a TSDF?
6. What are the basic elements that must be found in writing within a TSDF's contingency and emergency procedures plan?
7. What does the term *land disposal restrictions* mean?
8. What are *California wastes*?

ACTIVITIES

1. Using the Yellow Pages, compile an alphabetical list of companies in your area that advertise as facilities that treat, store, or accept hazardous waste; this includes most landfills and recycling centers.
2. Obtain the names of contact persons with your state or local environmental agency and identify the method of current compliance mechanism for enforcement of the LDR.

3. Contact a TSDF in your community. Schedule a visit to tour the facility.
4. Obtain a contingency plan or emergency procedures document and compare its written provisions (do an audit) with those procedures outlined in this chapter. Determine whether these plans or emergency procedures could be improved. Submit any suggestions to the TSDF in writing.

6

Technical Requirements of a TSDF

Neal K. Ostler

Upon completion of this chapter, you will be able to meet the following objectives:

▶ Describe the interim-status program for treatment, storage, and disposal facilities (TSDFs).

▶ Outline the general and specific standards required of interim-status facilities, including financial requirements, to assure the proper handling of hazardous materials and waste during the interim period.

▶ Outline the general and specific standards required of permitted TSDFs.

▶ Outline what materials or waste processes are exempt from Resource Conservation and Recovery Act (RCRA) compliance when recycled.

▶ Describe the RCRA program for management of underground storage tanks (USTs).

OVERVIEW

As you have already learned, the treatment, storage, and disposal facility (TSDF) is the last link in the EPA's hazardous waste management cradle-to-grave system. The previous chapter discussed the administrative and nontechnical requirements of both interim-status and permitted TSDFs. This chapter provides a separate overview of the technical requirements for both types of TSDFs. Though many of these requirements are nearly identical, they are reviewed here to give you an opportunity to compare them.

This chapter will also discuss requirements for facilities that govern procedures for the handling and recycling of materials.

> **NOTE: The details of the regulations governing the operation of a TSDF are very extensive, and you are reminded that this text cannot even remotely claim to be a guidance manual. You are encouraged to search Title 40 of the Code of Federal Regulations (CFR) Parts 264 and 265, the *Federal Register*, and applicable state and local regulations for more complete sources of materials relating to this subject.**

INTERIM-STATUS TSDFS

As explained in Chapter 5, interim-status facilities are facilities that have not obtained a permit to operate as a TSDF, but which were already in the business of handling hazardous waste prior to passage of RCRA. This provision allowed owners and operators of facilities in existence on November 19, 1980, who met certain conditions, to continue operating as if they had permits until their permit applications were issued or denied.

When it passed the requirement that all TSDFs obtain an operating permit, Congress realized that it would take many years for the EPA to issue all permits, and this was justification to establish interim status under Section 3005(e) of RCRA. Therefore, at the time that RCRA was originally passed in 1976, the people involved with the development of the regulations were mindful of the needs of those existing facilities that had on-going operations and were businesses interested in maintaining a healthy fiscal status. The EPA required these facilities to file Part A permit applications, which established their interim status, until their Part B applications for permitted status could be filed and approved.

The interim-status requirements differ from those for permitted facilities because, in the absence of permits, the interim-status requirements need to be self-implementing. They have neither site-specific provisions nor provisions that require negotiations between the EPA or a state and the owner or operator.

Land-disposal facilities that did not apply by November 1985 lost interim status at that time. Incinerators that did not submit a final permit application by November 1986 lost interim status in November 1989; all other facilities lost interim status in November 1992, unless they submitted a Part B application by November 1988.

Today, several circumstances may require a company to file a Part A permit application and thus operate within the interim status as a TSDF:

▶ Currently, the EPA is developing guidelines for the operation and management of facilities involved with reactive materials. Therefore, companies involved in the treatment, storage, and disposal of reactive materials by open burning or open detonation still remain in interim status.

▶ The EPA is continuously studying and revamping its codes for identification of hazardous waste. With new codes, businesses not currently subject to RCRA permit requirements may become so. Thus, they would need to submit Part A and Part B permit applications and become interim status so they could continue to operate while awaiting their permit.

▶ Companies may alter their process and begin to handle new varieties of hazardous waste. Such companies would be required to submit Part A and Part B permit applications and would receive interim status for these particular processes while their applications were being considered by the EPA and any relevant state agencies.

Standards for interim TSDFs are primarily good-housekeeping practices that owners and operators must follow to properly manage hazardous wastes during the interim-status period. The permit standards, on the other hand, are a mix of performance standards and design and operating criteria that the permit writers include in facility-specific operating permits.

The purpose of the interim-status technical requirements is to minimize the potential for environmental and public health threats resulting from treatment, storage, and disposal of hazardous waste at existing facilities waiting to receive an operating permit. There are two groups of interim-status requirements:

◗ General standards applying to several types of facilities.
◗ Specific standards applying to each waste management method.

An owner or operator of an interim-status facility can find the applicable technical requirements in Subparts F through R of 40 CFR 265. Let's examine these in more detail.

General Standards (Subparts F–H)

The general standards cover the three areas of groundwater monitoring, closure and postclosure of the facility, and financial responsibility.

Groundwater Monitoring (Subpart F)

Groundwater monitoring is required only for owners or operators of surface impoundments, landfills, and land treatment facilities used to manage hazardous waste. The purpose of these requirements is to assess the impact of a facility on the groundwater beneath it. All or part of the requirements for groundwater monitoring may be waived if the facility can demonstrate that there is a low potential for migration of contaminants to the uppermost aquifer. For example, a surface impoundment located in highly adsorbent soils may qualify for this exemption. If wastes remain at the site, monitoring must continue for thirty years (or more) after the facility has closed. The interim-status groundwater monitoring program consists of the following:

1. **Development and installation of a monitoring system.** The groundwater monitoring program outlined in the regulations requires that a monitoring system of at least four wells be installed, one upgradient from the waste management unit and three downgradient. (It is important to note that these are the *minimum* required.) The downgradient wells must be placed to intercept any waste migrating from the unit, should such a release occur. The upgradient wells must provide data on groundwater that is not influenced by waste coming from the waste management unit (called *background data*). If the wells are properly located, comparison of data from upgradient and downgradient wells should indicate whether contamination is occurring.
2. **Background monitoring.** Once the wells have been installed, the owner or operator collects quarterly data for one year to establish background concentrations for selected chemicals. These data form the basis for all future comparisons. If

the TSDF suspects that contaminants are already migrating to the groundwater, this step may be skipped and the facility must comply immediately with the assessment monitoring requirements. There are three sets of indicator parameters for which background concentrations are established: drinking water parameters, groundwater quality parameters, and groundwater contamination parameters.

3. **Routine monitoring and evaluation.** Following the establishment of background levels, routine monitoring begins. Routine monitoring examines groundwater for elevated levels of indicator constituents that suggest contamination may be occurring. Indicator parameters, used to assess groundwater quality and potential contamination, are monitored annually. Drinking water parameters are not monitored routinely. The results of routine monitoring are compared to the background values and tested statistically to determine whether significant increases (or decreases in the case of pH) have occurred in the indicator parameters. If comparisons show a difference, then the regional administrator must be notified within seven days and an assessment program instituted.

4. **Assessment program.** If a statistically significant increase (or decrease in the case of pH) over background is detected for any of the indicator parameters, the owner or operator must implement a groundwater assessment program to determine whether hazardous waste is actually entering groundwater. The assessment program, based on a previously developed plan, requires the owner or operator to determine what is contaminating the groundwater, the extent of contamination, and the rate of the contamination migration. Within fifteen days of conducting this assessment, a report on groundwater quality must be submitted to the regional administrator. If the results of the groundwater assessment show no contamination by hazardous wastes, then the owner or operator resumes routine monitoring for the indicator parameters. However, if the assessment shows hazardous waste contamination, the owner or operator must continue assessing the extent of groundwater contamination quarterly until the facility is closed or further monitoring is required as a result of the permitting process. Corrective action may be required to remedy the release.

5. **Reporting requirements.** Several groundwater monitoring reports are required. During the first year, when initial background concentrations are being established, a report on each quarterly well analysis must be submitted within fifteen days of the analysis. From the second year on, an annual report must be submitted providing the results of monitoring for indicators of groundwater contamination, groundwater elevation, changes in background levels, and groundwater contamination assessments. An owner or operator may also use an alternate groundwater monitoring system if, given the facility's unique hydrogeological situation, the one prescribed in the regulations is not capable of yielding unbiased samples.

Closure/Postclosure (Subpart G)

Closure is the period when a facility can no longer accept wastes and during which time owners or operators of TSDFs complete treatment, storage, and disposal operations; apply final covers to or cap landfills; and dispose of or decontaminate equipment, structures, and soil. Postclosure, which applies only to land-disposal facilities, is normally a thirty-year period after closure during which owners or operators of disposal facilities conduct monitoring and maintenance activities to preserve the integrity of the disposal system. The EPA may either extend or shorten the time required for postclosure monitoring. The period may be shortened if the EPA finds that the reduced period will still protect human health and the environment. Conversely, postclosure may be extended if necessary to protect human health and the environment.

The purpose of the closure and postclosure requirements is to ensure that all facilities are closed in a manner that (1) minimizes the need for care after closure and (2) controls, minimizes, or eliminates the escape of waste, leachate, contaminated rainfall, or waste decomposition products to soils, ground or surface waters, and the atmosphere.

An owner or operator must develop a plan for closing the facility and keep it on file at the facility until closure is completed and certified. This plan must include the following:

1. A description of how the facility will be closed.
2. An estimate of the maximum amount of waste the facility will handle.
3. A description of the steps needed to decontaminate equipment and remove soils and debris during closure.
4. An estimate of the year of closure.
5. A schedule for closure.

The plan may be amended at any time during the active life of the facility. Furthermore, the plan must be amended whenever design and operation changes that affect the closure plan occur. Prior to closure, the plan is submitted to the EPA regional administrator for approval. The regional administrator, in turn, must provide both the owner or operator and the public an opportunity to comment on the plan. Following the comment period, the administrator must decide whether to approve, modify, or reject the plan. Closure activities occur according to a timetable outlined in the regulations. This timetable is subject to change by the regional administrator.

During closure, the owner or operator must treat, remove from the site, or dispose of on-site all hazardous wastes in accordance with the approved closure plan. Once closure is completed, the owner or operator certifies that the facility has been properly closed. As part of the closure activities, a survey plat indicating the location and dimensions of landfill cells or other disposal areas is submitted to the local land authority and the regional administrator. This plat preserves a record of the TSDF that can be referred to in future years. A notation on the deed to the facility property also must be made to notify potential purchasers of the property that the land was used to manage hazardous waste.

Postclosure is required for land-disposal facilities that do not "clean close" (discussed later). When a land-disposal facility is closed, it must be monitored for thirty years to ensure the integrity of any waste containment systems and to detect contamination. Postclosure care consists of at least the following:

1. Groundwater monitoring and reporting.
2. Maintenance and monitoring of waste containment systems.
3. Security.

Like the closure requirements, a postclosure plan outlining activities is developed and kept at the facility until postclosure care begins. This plan may be amended at any time, and an amendment is required if there is any change that affects the plan. Postclosure plans are submitted and reviewed in the same manner as closure plans. The postclosure care period may be lengthened or shortened by the agency if warranted.

Owners or operators of surface impoundments and waste piles that remove all contaminants from the unit may "clean close" the unit. This means that all wastes have been removed from the unit. If this is successfully demonstrated, postclosure care is not required.

At a minimum, owners and operators of surface impoundments and waste piles who wish to clean close their facilities must conduct soil analyses and groundwater monitoring to confirm that all wastes have been removed from the unit. Other requirements (e.g., contaminant concentrations) for each facility are set by the EPA and the affected state on a case-by-case basis.

Financial Requirements (Subpart H)

Financial requirements were established to ensure that funds are available to pay for closing a facility, rendering postclosure care at disposal facilities, and compensating third parties for bodily injury and property damage caused by sudden and non-sudden accidents related to the facility's operation. One important purpose of financial assurance is to prevent RCRA sites from requiring cleanup under SuperFund. Ensuring that funds are available for closure, postclosure, and liability helps minimize the need for future SuperFund action.

Under the Hazardous and Solid Waste Amendments of 1984 (HSWA), Congress mandated financial responsibility for other RCRA activities in addition to those outlined in Subpart H. This demonstrates the importance that Congress places on requiring financial assurance that proper hazardous waste management will be provided. For example, HSWA calls for financial responsibility for completing corrective action at TSDFs.

Subpart H contains two kinds of financial requirements:

1. **Financial assurance for closure/postclosure.** The first step owners and operators must take in meeting the financial assurance requirements is to prepare written cost estimates for closing their facilities. If postclosure care is required, a cost estimate for providing this care must be prepared as well. These cost estimates must reflect the actual cost of conducting all of the activities outlined in the closure and postclosure plans, and they must be adjusted annually for inflation. The cost estimate for closure is based on the point in the facility's operating life when closure would be the most expensive. Cost estimates for postclosure monitoring and maintenance are based on projected costs for the entire postclosure period.

   Following the preparation of the cost estimate, the owner or operator must demonstrate to the EPA the ability to pay the estimated amounts. This is known as financial assurance. There are six mechanisms for complying with closure and postclosure financial responsibility. All are adjusted annually for inflation, or more frequently, if cost estimates change. The six mechanisms are (1) trust fund, (2) surety bond, (3) letter of credit, (4) closure/postclosure insurance, (5) corporate guarantee, and (6) financial test. Any one of the mechanisms can be used in conjunction with the others to meet the financial assurance requirements. An owner or operator may also use one of the six mechanisms to meet the financial requirements of multiple facilities.

> NOTE: States and the federal government are exempted from these requirements.

2. **Liability coverage.** An owner or operator is financially responsible for bodily injury and property damage to third parties caused by a sudden accidental occurrence or a nonsudden accidental occurrence due to operations at a facility.

   a. **Sudden accidental occurrences.** An owner or operator of a TSDF must have liability coverage in an amount stipulated in Subpart H of 40 CFR 265, exclusive of legal defense costs. This liability coverage may be demonstrated using any of the six mechanisms allowed for financial assurance of closure and postclosure as just discussed. Sudden occurrences are usually due to an accident, such as an explosion or fire.

   b. **Nonsudden accidental occurrences.** Nonsudden occurrences take place over a long period of time, such as groundwater contamination. An owner or operator of a surface impoundment, landfill, land-treatment facility, or group of such facilities must maintain liability coverage for nonsudden accidental occurrences in an amount stipulated in the regulations. Liability coverage for nonsudden occurrences may be demonstrated in the same ways as for sudden occurrences, and the same mechanisms may be used to supply financial assurance for both sudden and nonsudden types of accidental coverage.

Specific Standards (Subparts I–R)

Subparts I–R of 40 CFR 265 consist of requirements tailored to the ten specific waste management methods discussed here.

Containers (Subpart I)

Drums and containers are frequently used to accumulate and store wastes. In the past, persons using waste drums often put them somewhere out of sight, without any further concern about what might happen to residues in them. The drums eventually weathered and corroded, releasing their contents and posing threats to human health and the environment. Recognizing that elementary and straightforward precautions may eliminate these problems, the EPA requires basic good management practices. The container regulations, therefore, require the following:

- Using containers in good condition. Wastes in leaking or damaged containers must be re-containerized.
- Ensuring the compatibility of the waste with the container (i.e., corrosive wastes should not be stored in metal containers).
- Handling containers properly to prevent ruptures and leaks.
- Preventing the mixture of incompatible wastes.
- Conducting inspections to assess container condition.

When closing a container storage area, the owner/operator must ensure that all hazardous waste residues (including contaminated soils) are removed.

Tanks (Subpart J)

Tanks are stationary devices, constructed primarily of nonearthen materials, that are designed to contain an accumulation of hazardous waste. Subpart J addresses tanks storing wastes that are hazardous under Subtitle C of RCRA. (Additional requirements have been developed for underground tanks storing petroleum or hazardous substances under Subtitle I of RCRA. See the section on Requirements for Recyclable Materials and Recycling Procedures found later in this chapter.) Regulations governing hazardous waste tanks were substantially expanded under HSWA. General operating requirements fall into five basic areas:

1. **Tank assessment.** An assessment must be completed to evaluate the tank system's structural integrity and compatibility with the wastes that it will hold. The assessment covers design standards, corrosion protection, tank tests, waste characteristics, and the age of the tank. (Most of the original interim-status facilities—those in operation before 1980—should have assessed their tanks by January 1988.)

2. **Secondary containment and release detection.** Unless the tank does not contain free liquids and is located in a building with impermeable floors, secondary containment and release detection is required. Secondary containment systems must be designed, installed, and operated to prevent the migration of liquid out of the tank system, and to detect and collect any releases that do occur. Commonly used types of containments include liners, vaults, and double-walled tanks. An example of a secondary containment structure is illustrated in Figure 6–1.

   Owners and operators of interim-status tank systems can demonstrate that an alternate design, location, and operating practice will prevent the migration of hazardous wastes or constituents while the tank system is in use. Alternatively, the tank system can be exempted if any release that might occur would not harm human health and the environment.

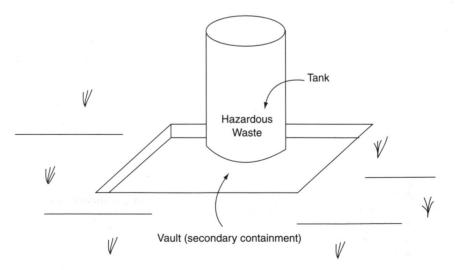

▶ **FIGURE 6–1**
Secondary containment structure.

3. **Operating and maintenance requirements.** Persons using tanks, either to store or treat wastes, must manage the tanks to avoid leaks, ruptures, spills, and corrosion. This includes using freeboard, or filling a containment structure only to a certain maximum level of any structure (tank, drum, dike, trench, etc.), to prevent and contain escaping wastes, and having a shut-off or bypass system installed to stop liquid from flowing into a leaking tank.

4. **Response to releases.** Tanks with leaks or spills must be emptied immediately. The area surrounding the tank must be visually inspected for leaks and spills. Based on the inspection, further migration of the waste must be stopped, and visibly contaminated soils and surface water must be properly disposed. All major leaks must be reported to the regional administrator, followed by a report describing the fate of the released materials.

5. **Closure and postclosure.** All contaminated soils and other hazardous waste residues must be removed from the tank storage area at the time of closure. If decontamination is impossible, the tank storage area must be closed following the requirements for landfills.

Surface Impoundments (Subpart K)

A surface impoundment is a depression or diked area (e.g., a pond, pit, or lagoon) used for storage, treatment, or disposal, with the following characteristics: (1) It is open on the surface, and (2) it is designed to hold an accumulation of waste in liquid or semisolid form.

The use of surface impoundments for managing hazardous wastes has given rise to great concern because wastes deposited in them tend to escape. The pressure of the liquids forces the contents to flow downward into surrounding areas, resulting in contamination, especially of subsurface waters. The initial requirements established for surface impoundments in interim status were not adequate to prevent contamination. They concentrated on general operating requirements to prevent overtopping (2 feet of freeboard was required) and containment of liquids (dikes were required to have protective covers, such as grass or rock, to preserve their structural integrity). Though groundwater monitoring was required, liners to prevent leakage were not, because it was deemed impractical and infeasible by the EPA for all surface impoundments to be retrofitted for the duration of the interim period before permitting. Congress, however, established minimum technological requirements for interim-status surface impoundments in HSWA.

HSWA increased the level of leak protection required at surface impoundments. Existing surface impoundments in interim status had to retrofit and install double liners and a leachate collection system by November 8, 1988, or stop receiving, treating, or storing hazardous waste. Currently, surface impoundments that come under Subtitle C regulation due to additional wastes being listed as hazardous have four years from the date the new wastes are added to meet the new requirements. Surface impoundments must be inspected once a week to determine if they have any leaks. If a leak is found, the surface impoundment must be taken out of service until it is repaired.

HSWA includes provisions for variances from minimum technological requirements, which had to be granted by November 1987. If any of the exempted impoundments are likely to leak or begin to leak, they must be retrofitted to meet the minimum requirements.

Waste Piles (Subpart L)

An owner or operator of a waste pile used for treatment or storage of a non-containerized accumulation of solid, nonflowing hazardous waste is given a choice regarding management requirements. The owner or operator may comply with either the waste pile or the landfill requirements. Waste piles used for disposal, however, must comply with the requirement for landfills. The requirements for managing storage and treatment waste piles include protecting the pile from wind dispersion. The pile must be placed on an impermeable base that is compatible with the waste being stored. If hazardous leachate or runoff is generated, control systems must be constructed, operated, and maintained.

Land Treatment (Subpart M)

Land treatment is the process of using soils and microorganisms as a medium to biologically treat hazardous waste. Land treatment has been successfully used for many years in the petroleum refining industry. However, it is highly regulated because it presents potential risks in the absence of operational controls (e.g., disposal of nondegradable waste types). These risks arise because land treatment involves the direct application of hazardous waste to soils.

An owner or operator may not place hazardous waste in or on a land-treatment facility unless the waste can be made less hazardous or nonhazardous. Ongoing monitoring and analysis of the soil beneath the treatment area is required. This allows for a comparison of soil conditions to detect the migration of hazardous constituents from any of the wastes. In addition, waste analyses must be conducted prior to placing wastes in or on the land to determine (1) if any substance in the waste is EP toxic (based on the extraction procedure discussed in Chapter 2), (2) what the concentration of hazardous waste constituents is, and (3) what the concentrations of arsenic, cadmium, lead, and mercury are, if food-chain crops are grown on the land. If the waste contains any of these compounds in concentrations that will prevent its degradation, immobilization, or transformation, then it cannot be treated in a land-treatment unit.

The requirements prohibit growing food-chain crops in a treated area containing arsenic, cadmium, lead, mercury, or other hazardous constituents. This prohibition may be waived if it is demonstrated that such elements or constituents would not be transferred to the food portion of the crop or, if transferred, would not occur in concentrations greater than would be expected in an identical crop grown on untreated soil in the same region. If food-chain crops are grown during postclosure, they must be raised in accordance with the requirements established in the regulations.

The owner or operator must continue to monitor soil, maintain runon and runoff management systems, and control wind dispersal after closure. In addition, access to the treatment unit must be restricted.

Landfills (Subpart N)

The landfill has historically been the cheapest, and thus the preferred, means of disposing of hazardous waste. Until recently landfill practices often focused only on burying the waste to get it out of sight and controlling surface problems such as wind dispersal of litter. As a consequence, landfills constitute a large number of sites on the National Priority List and are now being cleaned up under the Comprehensive

Environmental Response Compensation Liability Act (CERCLA) and the SuperFund amendments.

The EPA agrees with some who argue that, because wastes remain hazardous for very long periods, they should not be placed in landfills at all. Though in principle it is better to destroy or recycle hazardous wastes than to place them in a landfill, the fact remains that, for the foreseeable future, land disposal is necessary because it is technically infeasible at present to recycle, treat, or destroy all hazardous waste.

In recent years a number of techniques have become available for reducing potential adverse health and environmental effects arising from landfills. Many would argue that a wise investment would be to purchase landfills, which may, in the next decade or two, become a gold mine of resources as methods are further developed for extracting natural resources from the landfills themselves.

Problems that landfills accepting hazardous wastes have presented can be divided into two broad classes:

1. **Physical hazards.** These include fires, explosions, production of toxic fumes, and similar problems resulting from the improper management of ignitable, reactive, and incompatible wastes. Owners or operators are required to deal with the physical hazards by analyzing their wastes to provide enough information for their proper management, controlling the mixing of incompatible wastes in landfill cells, and accepting ignitable and reactive wastes only when the wastes are rendered nonignitable or nonreactive.

2. **Water contamination.** The second class of problems presented by landfills concerns the contamination of surface and groundwaters. Interim status regulations require the following:
   a. The diversion of runon (water flowing over the ground onto active portions of the facility) away from the active face of the landfill.
   b. Treatment of any liquid wastes or semisolid wastes so that they do not contain free liquids.
   c. Proper closure (including a cover) and postclosure care to control erosion and the infiltration of rainfall.
   d. Crushing or shredding most landfilled containers so that they cannot later collapse, thus leading to subsidence and opening of the cover.
   e. Groundwater monitoring to detect contamination.
   f. Collecting rainwater and other runoff from the active face of the landfill to control surface water pollution.
   g. Segregating waste, such as acids, that would mobilize, solubilize, or dissolve other wastes or waste constituents.
   h. Installing double liners and leachate collection systems (required for expanded or replaced interim-status landfills).

The official position of Congress and the EPA is to discourage land disposal. This stance is reflected in the proliferation of land disposal restrictions now identified as the *land ban.*

NOTE: The land ban is discussed throughout this text, specifically in Chapter 5, and the reader is further directed to research the regulations, particularly those found in 40 CFR Part 268.

Incinerators (Subpart O)

Incineration, a method of destroying primarily organic hazardous waste using flame combustion, can reduce large volumes of waste materials to ash and less toxic gaseous emissions. The interim-status incinerator general operating requirements include (1) achieving normal steady-state combustion conditions before wastes are introduced and (2) monitoring combustion and emissions.

The owner or operator must analyze the waste that is to be incinerated. Special requirements pertain to incinerators in addition to those required for all facilities under Subpart B. The waste analysis must determine the heating value of the waste, the total halogen and sulfur content, and the concentrations of lead and mercury, unless the facility can demonstrate that these elements are not present in the waste stream to be incinerated.

Interim-status facilities burning dioxin-containing wastes must meet additional requirements. Essentially, these requirements amount to meeting the permitted (40 CFR Part 264) incinerator standards. This includes destroying 99.9999 percent of the dioxins in the waste stream.

When closing an incinerator, the owner or operator must remove all hazardous waste and waste residues (including but not limited to ash, scrubber waters, and scrubber sludges).

Thermal Treatment (Subpart P)

Incineration is only one type of management process that can be used to thermally treat hazardous waste. Less conventional methods, such as molten salt combustion, calcination, wet air oxidation, and fluidized bed combustion, are regulated under this subpart. Owners or operators who thermally treat hazardous wastes (other than incinerators) must operate the unit following many of the requirements applied to an incinerator. The difference is that the thermal treatment standards prohibit open burning of hazardous waste except for the detonation of waste explosives. Standards for thermal treatment at permitted facilities under 40 CFR Part 264 have been incorporated under Miscellaneous Units, Subpart X.

Chemical, Physical, and Biological Treatment (Subpart Q)

Treatment—although most frequently conducted in tanks, surface impoundments, incinerators, and land-treatment facilities—can also be conducted in other ways through processes such as distillation, centrifugation, reverse osmosis, ion exchange, and filtration. Because there are many different types of treatment processes, and because the processes are frequently waste-specific, the EPA has not developed detailed regulations for any particular type of process or equipment. Instead, general requirements have been established to assure safe containment of hazardous wastes. In most respects, these other treatment methods are very similar to using tanks for treatment; therefore, they are essentially regulated the same way. The requirements that must be met concern avoiding equipment or process failure that could pose a hazard—e.g., reagents or wastes that could cause equipment or a process to fail must not be used in treatment. In addition, safety systems to shut down waste inflow in case of a malfunction must be installed in continuous flow operations. Standards for chemical, physical, and biological treatment methods for permitted facilities under Part 264 have been incorporated under Miscellaneous Units, Subpart X.

Underground Injection (Subpart R)

Underground injection is the disposal of fluids underground through a well. This technology is most appropriate for liquid wastes that are difficult or hazardous to treat (e.g., certain chlorinated pesticides). The wells are permitted at the state level under the Safe Drinking Water Act and granted a permit by rule under RCRA. Owners and operators of these facilities must meet the general standards outlined in Subparts A–E of 40 CFR Part 265, but are subject to the closure/postclosure or financial requirements (Subparts G and H of 40 CFR Part 265), since they are already regulated under the Safe Drinking Water Act. HSWA prohibits the disposal of hazardous waste by underground injection into or above a formation within a quarter mile of an underground source of drinking water.

Standards developed under the Safe Drinking Water Act consist of the following:

1. **Construction requirements.** New wells must be located so that they inject permitted hazardous substances and wastes into a separate formation from underground sources of drinking water, free of faults or fractures. Drilling logs and similar tests must be used to ensure that this requirement is met. Both new and existing wells must be cased and cemented to protect sources of drinking water so that the injection well does not create a significant risk to human health.
2. **Operating, monitoring, and reporting requirements.** The injection pressure of the well must not be so great as to fracture the disposal formation. The owner or operator must monitor the injection well to ensure the integrity of the well bore and must also periodically monitor the pressure, flow rate, and cumulative volume of the injected material. Monitoring information is submitted annually to the EPA.

RCRA Section 3004(0) requires the EPA to determine whether underground injection of hazardous wastes will endanger human health and the environment. In response to this requirement, the EPA has banned the underground injection of wastes that do not meet the applicable treatment standards of the land disposal restrictions. More specific information on the wastes banned from injection can be found in 40 CFR Part 148.

PERMITTED TSDFS

Permitted standards are more extensive than the general management practices detailed in the interim-status standards because they compel the owners and operators of the different waste management facilities to design their management units to prevent the release of hazardous waste. The permit standards (found in 40 CFR Part 264) also differ from the interim-status standards in that they are only a blueprint for the requirements applied to TSDFs. The specific requirements with which an owner or operator must comply are developed for each facility by permit writers, based on their "best engineering judgment" and the requirements of 40 CFR Part 264. Such requirements are then incorporated into the facility's operating permit.

For example, the groundwater monitoring requirements are found in 40 CFR Part 264, but the actual parameters that must be monitored are specified in each permit. Thus, although the technical requirements for permits are discussed here, each facility's permit must be consulted for the individual requirements that an owner or operator must follow.

The technical requirements for permitted facilities are structured similarly to the interim-status technical requirements. They too are divided into two groups: general standards and specific standards.

Because many of the interim-status technical requirements are the same or similar to permit requirements, the remainder of this chapter describes only these major provisions of Subparts I–O and X of 40 CFR Part 264 that are not found in, or differ from, Subparts I–R of 40 CFR Part 265.

General Standards (Subparts F–H)

The general standards for permitted TSDFs also cover three areas:

1. Groundwater monitoring requirements (Subpart F)
2. Closure/postclosure requirements (Subpart G)
3. Financial requirements (Subpart H)

> **NOTE: Closure/postclosure and financial requirements for permitted facilities are similar to the corresponding requirements under interim status. Thus, they are not repeated here.**

Groundwater Protection (Subpart F)

The groundwater protection requirements for permitted facilities are more specific than those found under interim status. They apply to surface impoundments, waste piles, land treatment units, and landfills. There are three parts to the groundwater requirements: a detection monitoring program, a compliance monitoring program, and a corrective action program.

1. **Detection monitoring.** Detection monitoring is conducted to determine whether hazardous wastes are leaking from a TSDF at levels great enough to warrant compliance monitoring. Detection activities are similar to those outlined under interim status, including background monitoring and semiannual monitoring for indicator parameters. Monitoring is conducted at a compliance point specified in the permit. This point is located at the edge of the waste management area, best envisioned as an imaginary plane on the outer limit of one or a group of disposal units. The indicator parameters and constituents that must be monitored are specified in the permit. If leakage is detected, then the owner or operator institutes a compliance monitoring program.

2. **Compliance monitoring.** The objective of the compliance monitoring program is to evaluate the concentration of certain hazardous constituents in groundwater to determine whether groundwater contamination is occurring at a level requiring corrective action. Each permit specifies the constituents and concentration limits that owners or operators must monitor for, according to the groundwater protection standard. The constituents are selected from

Appendix IX of Part 264 as those that could possibly originate from the TSDF. The groundwater protection standard can include any of the following:

a. Background levels.

b. The values in Table 1 of 40 CFR Part 264.94 (Maximum Concentration of Constituents for Groundwater Protection).

c. A site-specific alternate concentration limit approved by the regional administrator.

If compliance monitoring indicates a statistically significant increase in the concentration limits for those hazardous constituents specified in the permit, then corrective action must be instituted to bring the facility back into compliance with the groundwater protection standard.

3. **Corrective action.** Corrective action may be required under the authority of either Subpart F or HSWA. Subpart F corrective action applies only to cleaning up groundwater at a regulated unit at a TSDF. This type of corrective action program is incorporated into a facility permit. HSWA corrective action applies to releases to any medium from any unit at a TSDF. These two corrective action authorities are often used in combination at a TSDF.

Specific Standards (Subparts I–X)

The facility-specific standards for permitted TSDFs cover the following eight waste management methods.

Containers (Subpart I)

Permit requirements for containers are similar to the interim-status requirements, with the following exceptions:

1. Containers must be placed in a containment system that is capable of containing leaks and spills. This system must have sufficient capacity to contain 10 percent of the volume of all containers or the volume of the largest container, whichever is greater (this applies only to those holding liquids; containers holding solids are not factored into this volume determination).

2. When closing a container, all hazardous waste and hazardous waste residues must be removed, unless the container is to be disposed of as hazardous waste.

3. After closure, all contaminated equipment or soil must be decontaminated or removed.

Tanks (Subpart J)

Standards for tanks, detailed in 40 CFR Part 264, closely resemble the requirements for interim-status tanks described previously. Tank assessments, secondary containment and leak detection, operations and maintenance, response to releases, and closure/postclosure requirements are all the same. The difference is that new tanks (regulated under 40 CFR Part 264) must comply with these requirements *before* being put into use. Requirements for existing (interim-status) tanks are phased in. Additionally, owners and operators of permitted tank systems cannot obtain an exemption from the secondary containment and release detection requirements.

Surface Impoundments (Subpart K)

Prior to HSWA, the permit standards for surface impoundments required that a liner be designed, constructed, and installed to prevent migration of wastes out of the impoundment. In addition, double-lined surface impoundments meeting certain requirements were not subject to groundwater protection requirements. However, HSWA established minimum technology standards for land disposal facilities, including surface impoundments, that are more stringent. Existing requirements were considered inadequate to prevent hazardous waste from entering the environment.

All surface impoundments are required to have at least one liner and to be located on an impermeable base. Additionally, new surface impoundments, replacements, or lateral expansions of surface impoundments applying for a permit after November 8, 1984, must meet the minimum technological requirements added to Section 3004(0) of the Act by HSWA. These requirements are as follows:

1. The installation of two or more liners.
2. A leachate collection system between the liners.
3. Groundwater monitoring.

Variances for these requirements may be given by the administrator if the owner or operator demonstrates that alternative design and operation, together with location characteristics, will prevent migration of hazardous constituents into groundwater. Monofill surface impoundments containing foundry wastes and meeting certain conditions also may be issued a waiver. HSWA deleted the variances from groundwater monitoring standards for double-lined impoundments.

In addition to the HSWA minimum technology requirements, the original RCRA requirements calling for proper design, construction, and operation of surface impoundments still apply. These requirements include preventing liquids from escaping from the top (overfilling, runon) or sides (dikes) of surface impoundments. Liners must be properly constructed of appropriate materials and thicknesses. During construction and installation, liners and cover systems must be inspected for uniformity, damage, and imperfection. After installation all units must be examined weekly to ensure that the integrity of the unit is maintained and that no potentially hazardous situations have developed. If the liquid in a surface impoundment suddenly drops for no apparent reason, or if a dike leaks, the surface impoundment must be removed from service and, if the leak cannot be stopped, the impoundment must be emptied.

The closure and postclosure requirements for surface impoundments include removing or decontaminating all waste residues, and properly covering and maintaining the impoundments prevent leaks from occurring.

Waste Piles (Subpart L)

Unlike waste piles regulated under interim status, permitted waste piles must have an impermeable base with a liner designed and constructed to prevent any migration of wastes out of the pile into adjacent soil or waters. A leachate collection system also must be installed immediately above the liner. Owners or operators can be exempted from this requirement if alternate design and operation practices, together with location characteristics, will prevent the migration of hazardous wastes.

Owners or operators of waste piles can obtain a waiver from groundwater protection regulations if the waste pile is "an engineered structure" that the regional administrator finds does not receive or contain liquid waste, excludes liquids, and has a multiple-leak detection system that prevents waste migration.

Runon and runoff systems must be constructed to prevent water from flowing onto the active portion of the waste pile. Construction of liners and cover systems must be monitored to ensure that they are properly installed. During operation, the owner or operator must inspect the waste pile once a week to ensure that there is no deterioration and that the leachate collection system is functioning properly.

Land Treatment (Subpart M)

These standards require that an owner or operator establish a land treatment program to ensure that hazardous constituents placed in or on the treatment zone are degraded, transformed, or immobilized within the treatment zone. The elements of this program specified in the permit, include the following:

1. The wastes that can be treated.
2. Design and maintenance of the land treatment unit to maximize treatment.
3. Soil monitoring.
4. The hazardous constituents that must be degraded, transformed, or immobilized by treatment.
5. The size of the treatment zone.

The permit specifies the design and operating requirements that the owner and operator must use in the construction and maintenance of the land treatment unit. Before treatment can occur, a demonstration must be conducted to verify that the hazardous constituents are adequately treated by the unit. The regional administrator may allow the growth of food-chain crops in or on the treatment zone only if the owner or operator meets certain conditions outlined in 40 CFR 264.276.

The permitting standards for land treatment units include extensive unsaturated-zone monitoring requirements to determine whether hazardous constituents are migrating out of the treatment zone. Based on a sampling program outlined in the permit, if migration is detected, a permit modification must be submitted outlining changes in operating practices to maximize the success of treatment.

Landfills (Subpart N)

Landfills are regulated closely because of the potential impacts they may have on human health and the environment, and the regulations include numerous provisions that owners or operators must meet. Landfills (including expansions or replacements) permitted after November 1985 must install two or more liners and two leachate collection systems (one above and one between liners), and must conduct groundwater monitoring. Owners and operators are exempted if they can show that alternative design and operating practices, together with location characteristics, will prevent the migration of hazardous waste. Facilities permitted prior to November 1985 were required to install only one liner and a leachate collection system.

An important amendment affecting landfills is the "liquids in landfills restriction" in which bulk or non-containerized liquids (both hazardous and nonhazardous) are prohibited from placement in a landfill. In addition, the land disposal of containerized liquid hazardous waste or free liquids in containerized hazardous waste must be minimized. As with surface impoundments, HSWA requires that final permit applications for landfills be accompanied by information on the potential for public exposure to hazardous wastes or constituents from facility releases.

Incinerators (Subpart O)

By either conducting a trial burn or using alternate data, an owner or operator must determine the operating methods for his or her incinerator that will result in its meeting the following performance standards:

1. 99.99 percent of each principal organic hazardous constituent specified in the permit must be destroyed or removed by the incinerator (dioxins must meet 99.9999 percent).
2. Hydrogen chloride emissions must be minimized.
3. Particulate emissions must be limited.

The permit will specify the composition of waste feed that may be incinerated. Different waste feeds may be incinerated only if a new permit or permit modification is obtained. To prove that an incinerator can meet the required performance standards, a trial burn is usually conducted. Trial burns are temporary periods in which the owner or operator demonstrates the efficiency of the incinerator in destroying hazardous wastes. While incinerating hazardous waste, the combustion process and equipment must be monitored and inspected to avoid potential accidents or incomplete combustion. Incinerators may receive waste only after the destruction removal efficiency has been achieved and the unit is complying with its operating requirements. The regional administrator may ask for a sampling of the waste and exhaust emissions to verify that the operating requirements in the permit are being met.

Miscellaneous Units (Subpart X)

Until recently, although the EPA had issued regulations for all major waste management technologies, some gaps remained. Some technologies were difficult to fit into the framework of the prior regulations. To address these regulatory gaps, the agency issued regulations governing miscellaneous units.

Miscellaneous units are defined as any unit used to treat, store, or dispose of hazardous waste that is not a research, development, and demonstration unit, or not already regulated under 40 CFR Part 264 (i.e., a landfill, surface impoundment, incinerator, or tank). Miscellaneous units regulated under Subpart X include, but are not limited to the following:

1. Open burning/open detonation areas.
2. Thermal treatment units.
3. Deactivated missile silos.
4. Geologic repositories.

Requirements for miscellaneous units are based on technical performance standards. These units must be designed, constructed, operated, and maintained in a manner that ensures protection of human health and the environment. Requirements for each miscellaneous unit are determined on a case-by-case basis, including any of the technical requirements under 40 CFR Part 264 that may be appropriate.

REQUIREMENTS FOR RECYCLABLE MATERIALS AND RECYCLING PROCEDURES

As discussed in Chapter 2, certain materials that are hazardous in nature or exhibit a hazardous waste characteristic are exempt from the scope of hazardous waste regulation when recycled (40 CFR 261.6). These include the following:

▶ Reclaimed industrial ethyl alcohol.
▶ Used batteries returned to the manufacturer for regeneration.
▶ Used oil recycled but not burned.
▶ Scrap metal.
▶ Fuels from oil-bearing hazardous waste.
▶ Oil reclaimed from hazardous waste.
▶ Coke and coal tar from the iron and steel industry.

Consequently, handlers of these materials are not subject to generator, transporter, or TSDF regulations.

Similarly, regulations contained in 40 CFR Part 266 exempt recycling procedures of certain special materials from portions of the hazardous waste regulations. These procedures include:

▶ Hazardous waste fuels burned for energy recovery in boilers and industrial furnaces.
▶ Used oil burned for energy recovery in boilers and industrial furnaces.
▶ Hazardous wastes and waste-derived products used in a manner constituting disposal
▶ Spent lead-acid batteries that are reclaimed.
▶ Hazardous wastes from which precious metals are reclaimed.

In addition, units used to recycle hazardous wastes (e.g., stills that distill spent solvents) do not require a hazardous waste permit. However, owners or operators of recycling facilities must obtain permits for container or tank storage areas used to store the wastes prior to recycling, with the exception of facilities reclaiming lead-acid batteries and those engaged in precious metal recovery.

RCRA: UNDERGROUND STORAGE TANKS

Congress passed RCRA in 1976 to provide environmental protection from improper handling, storage, treatment, transportation, and disposal of hazardous wastes. As the '70s turned into the '80s, Congress came to realize that leaking underground storage tanks (USTs) were a national problem due to the threat they posed of contaminating soil, drinking water supplies, and air, as well as the additional danger of accumulation in confined spaces such as sewer lines, septic tanks, and the basements of nearby homes.

In 1984 the Senate established Subtitle I as a last minute (coattail) addition to RCRA to regulate all hazardous substances covered by SuperFund, including all petroleum products.

Provisions

The EPA is required by Subtitle I to develop regulations to protect human health and the environment from the dangers associated with leaking USTs. The goals of the UST regulations developed by the agency are to ensure the following:

▶ Prevent leaks and spills.
▶ Find problem USTs and UST systems.
▶ Correct the problems created by leaks and spills.
▶ Establish liability for correcting any problems.
▶ Provide oversight of state regulatory programs.

A UST is defined by the regulations as any tank or piping connection with at least 10 percent of its volume underground. UST regulations apply only to storage of petroleum products and those chemicals listed as hazardous under §101(14) of CERCLA. They do not apply to hazardous waste already covered by Subtitle C of RCRA.

Tanks installed after 1988 must meet standards for installation, spill and overfill protection, protection from corrosion, and leak detection. Tanks installed before December 1988 must meet three major requirements. In addition to immediately implementing proper tank filling procedures, operators and owners of old tanks must ensure that their tanks have the following:

▶ Corrosion protection for steel tanks and piping (to be installed before December 1998).
▶ Devices to prevent spills and overfills (to be in place by December 1998).
▶ Leak detection programs and systems (with a sliding scale of deadlines, depending upon tank installation dates). Tanks must be checked at least monthly to see if they are leaking.

NOTE: Appendix G, Leak Lookout (EPA/530/UST-88/006) provides an overview of external leak detectors for USTs.

All existing pressurized pipe in old tanks had to meet leak detection requirements of new pipe by December 1990. Existing suction pipe also had to meet requirements for new suction piping, on the same schedule as for leak detection in existing tanks.

When evidence of a leak is suspected, the owner or operator must take the following corrective actions:

▶ Immediately report the leak to the regulatory authority.
▶ Conduct tightness tests.
▶ Investigate the UST site for environmental damage.
▶ Clean up the site and inspect it afterward.

Tanks may be closed permanently or temporarily, and the statute provides guidelines to follow as appropriate.

NOTE: Appendix H, Straight Talk on Tanks (EPA/530/UST-90/012), provides a more detailed look at leak detection requirements as well as information on how to choose the most appropriate leak detection method for your UST.

The reporting requirements of the regulations mandate that owners or operators of USTs notify their state of any existing USTs they operate or any new USTs (upon installation), report suspected and confirmed leaks, and meet recordkeeping requirements. These recordkeeping requirements are very stringent, especially in the following areas:

▶ Leak detection performance and upkeep (see Appendix G).
▶ Certification of corrosion protection/inspection.
▶ Repairs and upgrades.
▶ All tank closures for three years.

General Requirements

Owners of USTs must demonstrate that they have the financial resources to clean up a site if a release occurs, correct environmental damage, and compensate third parties for injury to their properties or to themselves. The amount of coverage depends upon the size, type of tank, and contents of the UST. An owner can meet this requirement by:

▶ Obtaining environmental impairment liability insurance commercially.
▶ Placing the required amount in a third-party-administered trust fund.
▶ Demonstrating self-insurance by demonstrating financial worth.
▶ Relying upon coverage provided by a state financial assurance fund.

In addition to requirements that an owner must prevent and detect releases, they are also required to respond to a release by:

▶ Reporting a release.
▶ Removing the source of the release.
▶ Mitigating fire and safety hazards.
▶ Investigating the extent of the contamination.
▶ Cleaning up soil and groundwater as needed to protect human health and the environment.

SUMMARY

▶ The technical requirements for both the interim-status TSDF and the permitted TSDF are similar. Nevertheless, they have some very important differences.
▶ Technical requirements for each type of TSDF are broken down further into two main groups:
  1. **General standards,** which cover three specific areas:
     a. Groundwater monitoring.
     b. Closure and postclosure.
     c. Financial responsibility.
  2. **Specific standards,** which are requirements tailored to specific areas and include:
     a. Containers
     b. Tanks
     c. Surface impoundments

    d. Waste piles

    e. Land treatment

    f. Landfills

    g. Incinerators

    h. Thermal treatment

    i. Underground injection

    j. Chemical, physiological, and biological treatment

    The specific standards are extensive provisions, and a separate study of each is necessary to determine the full extent of the regulations.

▶ The regulations also include standards for the recycling of materials. These standards are identified specifically by material and specialized procedures.

▶ Tanks, and any underground piping connected to a tank that has at least 10 percent of its combined volume underground, must comply with Subtitle I of RCRA if the contents are either petroleum or certain hazardous materials. RCRA establishes standards for new tanks and pipes installed after 1988 and requires leak detection, spill protection, overfill protection, and corrosion protection. Rules governing USTs, their management, proper closure, and reporting and record-keeping requirements are developed by RCRA and implemented in most states by state UST program offices.

QUESTIONS FOR REVIEW

1. What is meant by "technical" requirements as opposed to "administrative" requirements for treatment, storage, and disposal facilities (TSDFs)?

2. Why are requirements for interim-status TSDFs different from those for facilities seeking new permits to operate a proposed TSDF?

3. What are the general standards required of an interim-status TSDF?

4. What are the specific standards required of an interim-status TSDF?

5. What financial assurances are required of an interim-status TSDF?

6. What are the general standards required of a permitted TSDF?

7. What are the specific standards required of a permitted TSDF?

8. What financial assurances are required of a permitted TSDF?

9. What materials and/or processes are exempt under RCRA recycling programs?

10. What is the definition of a UST? What are the general provisions of the programs designed for their management under RCRA?

ACTIVITIES

1. Contact your local agency (usually a state EPA division of solid and hazardous waste) and learn of the companies in your area that may have interim status as a TSDF.

2. Tour an interim-status TSDF and refine your definition of "interim" from information obtained and from company history.

3. Visit a recycling facility in your area.

4. As a group project, develop a list of every variety of business that has a storage tank and may have (or may have had) USTs.

7

The Permitting Process

Neal K. Ostler.

Upon completion of this chapter, you will be able to meet the following objectives:

▶ Describe the EPA and state programs that "permit" the generation, storage, treatment, transportation, and disposal of solid and hazardous waste.

▶ Define which facilities are subject to permitting requirements.

▶ Outline the six categories of permits issued under the Resource Conservation and Recovery Act (RCRA).

▶ Describe the actual process involved in application and approval of the permit itself.

▶ Outline the administrative and corrective processes involved in the compliance and enforcement of the permit.

OVERVIEW

Permit-by-rule refers to a provision of RCRA regulations which states that a facility or activity is deemed to have a plan approval if it meets the essential requirements and is regulated under another program. A *permittee* is any person who has received (1) an approval of a hazardous waste operation plan or (2) a federal RCRA permit to operate a treatment, storage, or disposal facility (TSDF). Operating with a *permit* then means that the owner or operator of a facility has submitted an application to conduct hazardous-waste treatment, storage, and disposal activities and has received approval of the plan.

A permit must identify the administrative and technical standards to which facilities must adhere and can be issued by the EPA or an authorized state. A state permitting program must meet national standards and, indeed, it must be no less stringent than a federal program. This chapter describes permitting as a federal program, and the procedures outlined will apply equally to permitting programs run by authorized states. It should be noted, however, that states may impose requirements that are more stringent or broader in scope than the federal program.

COVERAGE: WHO NEEDS A PERMIT?

Owners or operators of facilities whose operations meet the definitions of treatment, storage, or disposal of hazardous waste, as discussed in Chapter 6, must obtain an operating permit under Subtitle C of RCRA. An interim-status TSDF is an operation that was in existence on November 19, 1980, and if the owner or operator submitted a waste management plan, the facility may operate under such status until a final permit decision is made. "New" TSDFs (that is, any facilities or sites that did not exist prior to November 1980) are not eligible for interim status and must receive an RCRA permit before construction can commence.

Exceptions

In a limited number of circumstances, a person can treat, store, or dispose of hazardous waste without a permit. These are as follows:

▶ Generators storing waste on site for less than ninety days.
▶ Small-quantity generators that store waste on site less than 180 days.
▶ Farmers disposing of their own (hazardous) pesticides on site.
▶ Owners or operators of totally enclosed treatment facilities, waste water treatment units (tanks), and elementary neutralization units.
▶ Transporters storing manifested wastes at a transfer facility for less than ten days.
▶ Persons engaged in containment activities during an immediate response to an emergency.
▶ Owners or operators of solid waste disposal facilities handling waste only from conditionally exempt small-quantity generators.
▶ Persons engaged in SuperFund on-site cleanups and RCRA Section 7003 cleanups.

Any of the above individuals lose their exemption and become subject to mandatory permitting requirements if they treat, store, or dispose of hazardous waste in a manner not covered by their exclusion.

The permit application elaborates on how the applicant plans to accommodate all of the general and technical requirements for a TSDF (discussed in previous chapters), as well as requirements for corrective actions that require all TSDFs to clean up releases caused by facility operations.

TYPES OF HAZARDOUS WASTE PERMITS

Several categories of permits are issued under the RCRA Subtitle C program, and each defines operating requirements and various provisions specific to the permitting need.

Treatment, Storage, and Disposal Permits

Most commonly, RCRA permits are issued for treatment, storage, and disposal units. These units are containers, tank systems, surface impoundments, waste piles, land treatment units, landfills, incinerators, and miscellaneous units. These units are the most common way to treat, store, and dispose of hazardous waste. Minimum national standards have been promulgated for each of these methods at 40 CFR Part 264. As part of the permitting process, the Hazardous and Solid Waste Amendments of 1984 (HSWA) require facilities to correct releases to all media. Interim-status facilities or facilities permitted prior to HSWA are required to include provisions in their Part B permit application or to revise their permit, respectively, to comply with this requirement. The EPA must develop a schedule of compliance to address releases from all solid waste management units, as described later in this chapter.

Research, Development, and Demonstration Permits

The EPA encourages the use of alternate treatment technologies by issuing research, development, and demonstration (RD&D) permits for promising, innovative, and experimental treatment technologies.

For a treatment to qualify for an RD&D permit, a national standard must not exist for the treatment technology proposed. (For example, an incinerator could not be issued an RD&D permit because standards for incinerators have already been promulgated at CFR Part 264 Subpart O.) RD&D permits are for one year but may be renewed up to three times. RD&D facilities can receive only those wastes that are necessary to determine the efficacy of the treatment technology.

Issuance of RD&D permits follows a more streamlined process than a standard RCRA permit. The EPA may modify or waive the usual permit application and issuance requirements, with the exception of financial responsibility and public participation, as long as the agency maintains consistency with its mandate to protect human health and the environment.

Postclosure Permits

Facilities that plan to leave wastes on the land when they close must obtain a postclosure permit specifying the requirements for proper postclosure care.

Emergency Permits

In potentially dangerous situations when there is an "imminent and substantial endangerment to human health and the environment," the EPA can forego the normal permitting process and issue a temporary (ninety days or less) emergency permit to (1) a nonpermitted facility for the treatment, storage, or disposal of hazardous waste or (2) a permitted facility for the treatment, storage, or disposal of a hazardous waste not covered by its existing permit.

Permits-by-Rule

The EPA issues permits under a number of different laws. In some instances, the requirements of one statute's permitting regulations are quite similar to those of RCRA. To avoid duplication, the EPA has tried to abbreviate the application process for facilities that need to be permitted under two or more statutes. This is done through a *permit-by-rule.* A permit-by-rule eliminates the need for facilities to submit a full Subtitle C permit application when they are permitted under the following acts:

1. Safe Drinking Water Act (Underground Injection Control permit)
2. Clean Water Act (National Pollutant Discharge Elimination System permit)
3. Marine Protection, Research, and Sanctuaries Act (Ocean Dumping permit)

Facilities that are seeking a RCRA permit and that already have one of the three permits listed above need only meet a subset of the Subtitle C regulatory requirements. For example, an owner or operator of a barge or vessel that has an ocean dumping permit and complies with the appropriate conditions under Subtitle C (e.g., obtaining an EPA ID number, using the manifest system, and biennial reporting) will be considered to have a permit-by-rule under RCRA.

Trial Burn and Land Treatment Demonstration Permits

The construction of a new TSDF is a complex and costly undertaking, and before beginning such an undertaking, the owner or operator of the proposed TSDF must submit an application through the EPA for approval (permission). Such facilities cannot be constructed until a permit is issued.

For land treatment facilities and incinerators, there is an exception to this rule. This process requires that they go through a trial period during which their ability to perform properly under operating conditions is tested. This is called a "trial burn" for incinerators and a "land treatment demonstration" for land treatment facilities. The process involves obtaining temporary permits that are enforced while the facility is being constructed and tested. Once the facility adequately completes its test, the EPA can make decisions regarding the final permit and set the final operating conditions based on the data generated from these demonstrations.

The multimillion-dollar undertaking of design and construction of a full-service TSDF is very risky because special interest environmental groups are apt to scrutinize every step in the permitting process.

THE PERMIT PROCESS

When deciding whether to approve a permit application, the EPA must consider a number of federal laws, including these related environmental acts:

▶ Wild and Scenic River Act
▶ National Historic Preservation Act of 1966
▶ Endangered Species Act
▶ Coastal Zone Management Act
▶ Fish and Wildlife Coordination Act

Each of these acts has procedures that must be followed and often forbid approval of the permit application. Discovering that the proposed site is the habitat of an endangered species is one reason an application could be rejected. Or in the case of the Coastal Zone Management Act, the EPA may be prohibited from issuing a permit for an activity affecting land or water use in the coastal zone unless the proposed activity is agreed to by the state and complies with its program. (More information on these laws and their potential impact on Subtitle C's permitting process can be found in 40 CFR 270.3.)

TSDFs that are required to get an RCRA permit must go through the same basic permitting process. Exceptions are facilities that are issued a permit-by-rule or an emergency permit as discussed previously.

Submitting a Permit Application

The permit is a comprehensive application covering all aspects of the site location, design, operation, and maintenance of the facility. This application provides the EPA, and the state where the siting of the facility is proposed, with information needed to determine if the facility is in compliance with Subtitle C regulations and to develop a facility-specific permit. The permit application is divided into two parts:

1. **Part A.** The first part of the application is a short, standard form that collects general information about a facility such as the name of the applicant and a description of the activities to be conducted at the facility.
2. **Part B.** The second part of the permit application is much more extensive than Part A and requires the owner or operator to supply detailed and highly technical information regarding the proposal. There is no standard form for Part B; thus, the owner or operator must rely on the regulations (40 CFR Parts 264 and 270) to determine what to include in this part of the application.

In addition to the general Part B information stipulated in the regulations for all TSDF applications, unique information requirements are tied to the type of facility seeking a permit. Depending on the situation, Parts A and B may be submitted at different times. Existing facilities (i.e., those that received hazardous waste on or after November 19, 1980), submitted their Part A applications when applying for interim status. Their Part B applications can then either be voluntarily submitted or called in by the EPA.

Part B applications for interim-status TSDFs had to be submitted according to the guidelines outlined in Table 7–1. Facilities that failed to submit within these deadlines lost their interim status and were required to close. These deadlines did not apply to new facilities or facilities that gained interim status after November 8, 1984.

Any TSDF that comes under the jurisdiction of Subtitle C due to statutory or regulatory changes must submit its Part A application six months after the date of publication of the regulations in the *Federal Register,* or thirty days after the date they first become subject to the promulgated standards that are distinct and separate for land disposal facilities and incinerators.

Permit applications for a new facility must submit Parts A and B simultaneously at least 180 days prior to the date on which physical construction of the facility is expected to start. By requiring a permit to be issued prior to construction, the owner or

▶ **TABLE 7–1**
Part B application deadlines for TSDFs granted interim status before November 8, 1994.

Type of Facility	Lost Interim Status on	Unless Part B Submitted by
Land disposal	November 8, 1985	November 8, 1985
Incinerator	November 8, 1989	November 8, 1986
All others	November 8, 1992	November 8, 1988

operator does not risk losing a substantial financial investment by building a facility that fails to meet the more stringent permit (40 CFR Part 264) requirements.

Public Health Concerns

Permit applications for surface impoundments and landfills must be accompanied by information on the potential for public exposure to hazardous wastes or constituents from facility releases because such facilities pose a greater health risk than other types of disposal facilities. This information is made available to the Agency for Toxic Substances and Disease Registry (ATSDR), and if the EPA believes that the release poses a substantial risk to human health, it can request that the ATSDR perform a health assessment. The exposure information must at least address the following:

1. Reasonably foreseeable potential releases from both normal operations and accidents at the facility, including transportation to or from the facility.
2. The potential pathways of human exposure to hazardous wastes or constituents resulting from such releases.
3. The potential magnitude and nature of the human exposure resulting from such releases.

Reviewing the Permit Application

The EPA's first step upon receiving the application is to determine if all the required information has been submitted and, if it has not, send a notice of deficiency (NOD) letter to the owner or operator describing the additional information that is required for a complete application. The application is considered complete once the owner or operator has submitted all of the required information. Failure to provide this information can result in denial of the permit and/or enforcement action.

If the owners or operators considers certain information they are required to disclose in Part B of the application confidential, they can follow procedures described in 40 CFR 270.12 ("confidentiality of information"). Such claims of confidentiality are reviewed by the EPA's legal counsel to determine if the information can legitimately be claimed as confidential. If substantiated, the information is treated as confidential and not released. On the other hand, if a claim is denied, the information is made public.

Once the application is complete and the owner or operator is informed of such by letter, an in-depth evaluation of the permit application begins. The purpose of the evaluation is to determine if the application satisfies the technical requirements of RCRA and to give the EPA the ability to make a tentative decision to either issue or deny the permit and notify the owner/operator in writing.

The EPA must have either approved or denied the applications for facilities that received interim status on or before November 8, 1984, in accordance with the schedule set by HSWA.

> **NOTE: For facilities that submitted their applications after November 8, 1984, and for any new facilities, HSWA places no time limits on how long the EPA can take to evaluate the application. Generally, the process takes from one to three years.**

Drafting the Permit

If the EPA approves the application, it must prepare and draft a permit that incorporates applicable technical requirements and other conditions pertaining to the facility's operation. Some of these provisions are general in nature and some are applied on a case-by-case basis.

General Permit Conditions

The general provisions of the permit require that the facility do the following:

- Comply with all conditions listed in the permit.
- Notify the EPA of any planned alterations or additions to the facility.
- Provide the EPA with any relevant information requested and allow EPA representatives to inspect the facility premises under certain conditions.
- Certify annually that a program is in place to reduce the volume and toxicity of waste, and that the proposed method of treatment, storage, and disposal minimizes threats to human health and the environment.
- Submit required reports such as an unmanifested waste report, biennial report, and manifest discrepancy report.

Case-by-Case Permit Conditions

Permits are issued case by case, based upon different recipes of the following conditions for approval:

- **Compliance schedules.** These schedules are allowable only to bring a facility into compliance with corrective action requirements; all other permit conditions must be met prior to issuance.
- **Duration of permit.** The permit may be valid for up to ten years. Land disposal permits must be reviewed every five years; i.e., they are approved for five years only.
- **Technical requirements**. The facility must comply with the set of separate requirements specific to whichever type of unit (tank, waste pile, landfill, etc.) that the facility is operating. (See 40 CFR Part 264.)

Taking Public Comment

The EPA is required to give public notice and allow forty-five days for written comment; often, a public hearing about the permit application may also be held. The EPA must issue either a fact sheet or a statement of basis to inform concerned parties about the permitting process that is taking place. These documents are prepared for every major facility and any facility subject to widespread public interest, as determined by the EPA. These fact sheets include detailed information pertaining to (1) the nature of the facility, (2) the contents of the draft permit (or notice of intent to deny), and (3) the procedures to be used in reaching the final administrative decision on the permit application. Sometimes a statement of basis or a summarized version of the fact sheet is adequate, and all supporting documents are sent to the applicant and to any other interested person who makes a written request.

Should a controversy arise that cannot be resolved within the time allotted, the EPA may extend the period if the applicant can show good cause to do so.

Finalizing the Permit

After the comment period closes, final permit decision is reached, and a response to all significant public comments is prepared, the applicant or another petitioner may appeal the decision to the administrator. When administrative appeals are exhausted, the petitioner may seek judicial review of the final permit decision.

PERMIT ADMINISTRATION

As noted previously, RCRA permits are valid for up to ten years and land disposal permits have an additional requirement of being reviewed after five years. During the term of a permit, however, certain situations may arise that may cause the permit to be modified, revoked and reissued, or terminated.

Permit Modification

Permits may need modification for a number of reasons, including the following:

1. Substantial alterations or additions are made to the facility.
2. New information about the facility becomes available.
3. New statutory or regulatory requirements affect existing permitted activities.

In September 1988, the EPA published regulations (under 40 CFR 270.41 and 270.42) that revised permit modification procedures for changes that facility owners and operators may want to make. The EPA categorized selected permit modifications into three classes and established administrative procedures for approving modifications in each class.

**NOTE: It is important to remember that the previous
"major/minor" modification requirements are still implemented
by states that chose not to adopt these provisions.**

The revised permit modification regulations provide owners and operators more flexibility to change permit conditions, expand public notification and participation opportunities, and allow for expedited approval if no public concern exists for a proposed modification.

The classes are defined as follows:

Class 1: Routine changes and correction of errors.
Class 2: Common or frequently occurring changes needed to maintain a facility's capability to manage wastes safely or conform to new requirements.
Class 3: Major changes that substantially alter the facility or its operations.

The regulation also gives the EPA the authority to grant temporary authorization for facilities to respond promptly to changing conditions.

Revocation and Reissuance of the Permit

The EPA may revoke and reissue a permit in two circumstances:

1. When cause exists for terminating the permit, but the EPA decides that revocation and reissuance is a more appropriate step.
2. When the permit holder plans to transfer the permit.

Permit Termination

Occasions may arise when operators may not comply with the requirements stipulated in the permit, even after the EPA has taken enforcement action, and it may be necessary to terminate a hazardous waste permit. The EPA may terminate a permit or deny its renewal for three reasons:

1. Noncompliance by the permittee with any condition of the permit.
2. Failure on the part of the permittee to disclose any relevant information during the permit process or misrepresentation of facts at any time.
3. The permitted activity endangers human health and the environment and can only be regulated to acceptable levels by terminating the permit.

A facility whose permit is terminated must implement its closure plan as required under 40 CFR Part 264 Subpart G. If wastes remain on site, postclosure monitoring must also be done.

THE CORRECTIVE ACTION PROCESS

In HSWA, Congress expanded the EPA's authority to address releases of hazardous waste through corrective action beyond those contained in 40 CFR Part 264 Subpart F. The TSDF owner or operator is responsible for complying with any other requirements made under an EPA enforcement order. Permits issued to RCRA facilities must, at a minimum, contain schedules of compliance to address releases and include provisions for financial assurance to cover the cost of implementing the corrective measures.

The EPA has estimated that corrective actions at RCRA facilities could encompass about 2,000 TSDFs, take until the year 2025 to complete, and be as large and as costly as the current SuperFund program.

Scope of Corrective Actions

The following are key terms involved with corrective actions:

1. **Solid waste management units (SWMUs).** These are waste management units from which hazardous wastes or constituents may migrate, even if the unit was not intended for the management of hazardous waste. Additionally, any areas that become contaminated as a result of routine and systematic releases of wastes are SWMUs (e.g., spill areas).
2. **Regulated units.** These are a subset of all SWMUs. A regulated unit is any surface impoundment, waste pile, land treatment unit, or landfill that received waste after July 26, 1982.
3. **Hazardous constituents.** These are any of the substances listed in 40 CFR Part 261, Appendix VIII.

Corrective action for releases to groundwater from regulated units are addressed under 40 CFR Part 264, Subpart F. Corrective action for releases of hazardous wastes or constituents from any SWMU at a TSDF that is seeking or subject to an RCRA permit are found in 40 CFR 264.101, which also authorizes the EPA to impose corrective action requirements for releases that have migrated beyond the facility boundary.

This includes taking corrective action or other necessary measures through an enforcement order, whenever there is or has been a release of hazardous wastes or constituents from an interim-status TSDF.

The EPA can require permitted facilities with releases from regulated units to take corrective action on those releases to the uppermost aquifer (under 40 CFR Part 264 Subpart F), or to clean up any other contaminated media (under Sections 3004[u] and 3004[v]). The EPA decides which degree of corrective action to enforce on a case-by-case basis, taking into account the nature and magnitude of the release. For example, the EPA may require a TSDF to clean up other contaminated media if the media is a water resource.

Corrective Action Components

The corrective action process has four main components. Each component comprises a number of steps. The actual number of steps a TSDF must take and the complexity of the corrective actions or other enforcement actions it must adhere to depend on the extent and severity of the hazardous-waste releases. The decision on which steps to include is made on a facility-by-facility basis. The EPA also can require that facilities take interim corrective measures whenever necessary to protect human health and the environment.

RCRA Facility Assessment

Release determinations for all environmental media from SWMUs (i.e., soil, groundwater, subsurface gas, air, or surface water), are made by the regulatory agency primarily through the RCRA Facility Assessment (RFA) process. The major objectives of the RFA are to identify SWMUs and collect existing information on contaminant releases, and to identify releases or suspected releases needing further investigation.

The RFA process involves a review of pertinent existing information on the facility and a visual site inspection, if necessary, to verify information obtained in the preliminary review and to gather information needed to develop a sampling plan. A sampling visit is subsequently performed, if necessary, to obtain appropriate samples for making release determinations. The findings of the RFA will result in one or more of the following actions:

1. No further action since no evidence of a release(s) or of a suspected release(s) was identified.
2. An RCRA facility investigation by the facility owner or operator when the information collected indicates a release(s) or suspected release(s) that warrants further investigation.
3. Interim corrective measures by the owner or operator when the regulatory agency believes that expedited action should be taken to protect human health or the environment.

In cases where problems associated with permitted releases are found, the regulatory agency will refer such releases to the appropriate permitting authorities.

RCRA Facility Investigation

The conditions of the RCRA facility investigation (RFI) will generally be based on results of the RFA and will identify specific units or releases needing further investigation. The

RFI can range widely from a small, specific activity to a complex, multimedia study. Through these conditions of corrective actions or enforcement orders, the regulatory agency will direct the owner or operator to investigate releases of concern.

This evaluation is crucial to the corrective action process. The regulatory agency will ensure that data and information collected during the RFI adequately describe the release and can be used to make decisions regarding the need for a corrective measure study with a high degree of confidence.

Corrective Measure Study

If corrective measures are potentially identified during the RFI process, the owner or operator is then responsible for performing a corrective measures study (CMS). During this step of the corrective action process, the owner or operator will identify and recommend specific corrective measures that will correct the hazardous waste release. It is recommended that during the RFI and CMS, owners or operators collect data that may be needed to select and implement corrective measures.

Corrective Measures Implementation

Corrective measures implementation (CMI) includes designing, constructing, operating, maintaining, and monitoring selected corrective measures. If the remedy is not properly implemented, the EPA will direct the facility to take additional action on a site-specific basis.

Before CMI became part of the corrective action process, the regulations required that facilities demonstrate financial assurance to ensure that they had the necessary funds to clean up the site. Recently, the EPA has also finalized regulations that require financial assurance for corrective action.

THE ENVIRONMENTAL PRIORITIES INITIATIVE

Similar to the National Priorities List under SuperFund, the EPA has developed the Environmental Priorities Initiative (EPI) to ensure that those sites posing the greatest threat to human health and the environment are addressed first. Over 2,000 RCRA facilities are likely to require corrective action due to releases of hazardous materials and wastes; a similar number of sites must be addressed under the Super-Fund program.

Under the EPI, all RCRA facilities and SuperFund sites receive a ranking of their environmental priority based on the threat each site poses to human health and the environment. The highest priority facilities will then receive a site inspection, which, combined with the previous ranking, will provide the EPA with a basis for prioritizing the sites. The ranking of the EPI sites has already been determined as of this writing. The criteria used to rank sites for placement on the National Priorities List under SuperFund are outlined in detail in the SuperFund Amendment Reauthorization Act of 1986. A few examples of such criteria include the following:

- Proximity to population.
- Nature and severity of chemical hazards.
- Proximity to natural resources.
- Effects on human health.

SUMMARY

▸ Permits are the key to implementing Subtitle C regulations and detail the administrative and technical performance standards that TSDFs must adhere to. Owners and operators of existing or new facilities must (with a few exceptions) obtain an operating permit.

▸ Each TSDF permit must include provisions for corrective action to address releases from solid waste management units if a release has been detected.

▸ Special requirements pertain to permit-by-rule facilities and facilities demonstrating the efficacy of their treatment technology (trial burns and land treatment demonstrations).

▸ The permitting process has five steps:
1. Submitting a permit application.
2. Reviewing the permit application.
3. Preparing the draft permit.
4. Taking public comment.
5. Finalizing the permit.

▸ The final decision on the permit may be reviewed by the EPA administrator and appealed to the local U.S. District Court.

▸ Permit administration procedures include permit modification, revocation and reissuance, and termination.

▸ HSWA has greatly expanded cleanup requirements at RCRA facilities. Through a process called corrective action, facilities must remedy releases threatening human health and the environment.

Corrective action has four main parts:
1. RCRA facility assessment (RFA)
2. RCRA facility investigation (RFI)
3. Corrective measure study (CMS)
4. Corrective measure implementation (CMI)

Although usually done through the permitting process, corrective action may also be completed through an enforcement order.

QUESTIONS FOR REVIEW

1. What is meant by the term "permit-by-rule"?
2. What is meant when a facility is given an RCRA "permit" with regard to hazardous waste?
3. What types of facilities are subject to the "permit" requirements of the EPA and state-operated RCRA programs? Who needs a permit?
4. Prepare a list of types of hazardous waste permits, using single-sentence descriptions.
5. What are the steps involved in the permit process?
6. What types of public health concerns are addressed in the permit process?
7. How are concerns for confidentiality addressed to protect the interest of the permit applicant?
8. What terms or requirements are often made a condition of the issuance of a permit with regard to hazardous waste?
9. Briefly describe the system for the administration of permits and oversight of the management of operations pursuant to issuance of the permit.
10. When a permitted company or facility is not in compliance, what corrective process is implemented by federal and state RCRA agencies?

ACTIVITIES

1. Obtain the contact names and address of the office in your area that issues solid and hazardous waste permits under RCRA.
2. Contact the administrative secretary or other solid and hazardous waste board member in your area and obtain information, free to the public, with regard to individuals or companies currently in the RCRA permit application process.
3. As a group project, create an imaginary company with operations that will generate hazardous waste. Develop a checklist of information needed in the preparation of an application for a RCRA permit for each stage of the TSDF as it relates to your company.
4. Search the Internet or World Wide Web for environmental consulting firms that offer services in the preparation of permit applications. Make a list of those in your geographical area and compare them to those listed in the telephone directory. Contact them by e-mail or telephone and invite someone to meet with your group and evaluate the project your group developed in Activity 3.

8

Compliance and Enforcement

Neal K. Ostler

Upon completion of this chapter, you will be able to meet the following objectives:

▶ Summarize the goal and purposes of the Resource Conservation and Recovery Act (RCRA) compliance monitoring programs.

▶ Describe the purpose and different types of inspections under the compliance monitoring programs.

▶ Describe the purpose and different types of enforcement alternatives under the compliance monitoring programs, including administrative actions taken by the EPA or state agencies, civil actions, and criminal actions.

▶ Outline the compliance and enforcement mechanism that applies to federal facilities.

OVERVIEW

The effectiveness of almost any program depends upon the willingness of the participants to implement the provisions and stipulations of membership in the program. This is especially true of a program in which membership is mandated and which requires obedience to extensive rules and regulations.

The goals of the RCRA compliance and enforcement programs are to ensure that regulatory and statutory provisions of RCRA are met and to compel necessary corrective action. Accomplishing these goals requires close monitoring of all phases of hazardous waste handling activity, from the generator to the transporter and, finally, to the treatment, storage, and disposal facility (TSDF). It also necessitates prompt legal action when noncompliance is detected.

Inspections of facilities by federal and state officials are the primary tool for monitoring and compliance. When noncompliance is detected, quick and expeditious sanctions may follow. These actions may include the use of administrative orders, civil lawsuits, and/or criminal prosecution, depending on the nature and severity of the problem. The intent of such enforcement is to reduce the number of handlers not operating in compliance with RCRA's requirements and to deter potential violations by imposing penalties for noncompliance.

In this chapter, the two essential aspects of the enforcement program—compliance monitoring and enforcement actions—will be described. All of the provisions detailed are statutory, not regulatory.

> **NOTE: State requirements may be more stringent than those mandated by the federal government, and state enforcement authorities and procedures may differ from those of the EPA.**

COMPLIANCE MONITORING

The first phase of the enforcement program is monitoring facilities, which allows the EPA to evaluate the effectiveness of state programs and to monitor nationwide compliance with RCRA.

This monitoring serves several purposes:

1. It allows the EPA and states to find out which facilities are not in compliance.
2. It allows the EPA and the states to assess the effectiveness of specific legal actions that may have been taken against a handler.
3. It acts as a deterrent to noncompliant facility owner/operators who are subject to enforcement actions.

Inspections

The primary method of collecting data for compliance monitoring is through an inspection, which may include formally visiting a handler, reviewing records, taking samples, and observing operations. This data can also assist the EPA in developing RCRA regulations and help them to track program progress and accomplishments.

These inspections are conducted by state or EPA officials. In instances when criminal activity is suspected, the EPA's National Enforcement Investigations Center may become involved. The Department of Transportation (DOT) may also participate when waste transporters are involved. RCRA authorizes all of these agencies to use outside contractors for the actual inspection if they desire.

The Hazardous and Solid Waste Amendments of 1984 (HSWA) require that all federal or state-operated facilities must be inspected annually and all TSDFs must be inspected at least once every two years. These inspections may occur at any time if the EPA or the state has reason to suspect that a violation has occurred or whenever specific information is needed by the EPA to support the development of RCRA regulations.

Types of Inspections

A number of different types of inspections are conducted under the authority of the RCRA program.

Compliance Evaluation Inspection

A compliance evaluation inspection (CEI) is a routine inspection of hazardous waste generators, transporters, and TSDFs to evaluate compliance with the requirements of RCRA. A CEI comprises a file review prior to the site visit; an on-site examination of generation, treatment, storage or disposal areas; a review of records; and an evaluation of the facility's compliance with the requirements of RCRA.

Case Development Inspection

A case development inspection (CDI) is conducted when significant RCRA violations are known, suspected, or revealed. A CDI is performed to gather data in support of a specific enforcement action. Most of the activities conducted during a CDI are specific to the type of information required to document the violation (e.g., incinerator investigations and closure/postclosure investigations).

Comprehensive Groundwater Monitoring Evaluation

A comprehensive groundwater monitoring evaluation (CME) is conducted to ensure that groundwater monitoring systems are designed and functioning properly at RCRA land disposal facilities. In addition to the CEI activities, a CME includes sampling and analysis of the facility's groundwater monitoring system and hydrogeological conditions.

Compliance Sampling Inspection

A compliance sampling inspection (CSI) is an inspection in which samples are collected for laboratory analysis. A sampling inspection may be conducted in conjunction with a CEI or any other type of inspection except a CDI.

Operations and Maintenance Inspection

Many land disposal facilities close with waste in place. The purpose of an operations and maintenance inspection (O&M) is to ensure that groundwater monitoring and other systems continue to function properly after a land disposal facility has closed. O&M inspections are usually conducted at facilities that have already received a thorough evaluation of the groundwater monitoring system under a CME inspection.

Laboratory Audits

Laboratory audits are inspections of laboratories performing groundwater monitoring analyses. Audits ensure that these laboratories are using proper sample handling and analysis protocols.

Conducting the Inspection

Several steps are generally followed in RCRA inspections to ensure consistency and thoroughness; these steps are summarized here. (For more details on the inspection process, consult the *RCRA Inspection Manual* [OSWER Directive 9938.2A].)

1. **Preparing for the inspection.** The designated inspector must review handler records, develop an inspection plan, prepare a checklist, and pack appropriate safety equipment.
2. **Entering the handler's property.** The inspector identifies himself or herself and describes the nature of the inspection. In some circumstances, a warrant may be needed to gain entry to the facility.

3. **Holding an opening conference.** After the inspector has entered the property, he or she generally holds an opening conference with the owner or operator to discuss the nature of the inspection and to describe the information and samples to be gathered.

4. **Conducting the inspection.** The actual inspection involves checking hazardous waste generation, storage, treatment or disposal areas to ensure that hazardous waste is stored properly (e.g., no spills, leaks, or improper disposal), and reviewing records. The inspection may involve taking samples, copying documents, interviewing employees, and photography.

5. **Holding a closing conference.** The inspector holds a closing conference with the owner or operator to allow him or her to respond to questions about the inspection and provide additional information. The inspector usually summarizes his or her findings and explains any further action required by the handler.

6. **Examining facility reports.** The handler's permit mandates that the owner/operator keep a variety of reports such as biennial reports, corrective action schedules, and others that may contain information about the wastes being handled, the method of handling, and the ultimate disposition of wastes. These facility reports can assist the inspector in preparing the inspection report.

7. **Preparing the inspection report.** After the visit is completed, the inspector prepares a report that summarizes the records reviewed, any sampling results, and the handler's compliance status with respect to RCRA. Summary conclusions regarding inspections are tracked in some detail in the RCRA Information System (RCRIS) maintained by the EPA.

8. **Determining compliance.** The most important result of any inspection is the determination of whether the handler is in compliance with the regulations. If the handler is not complying with all of the appropriate state or federal requirements, enforcement action may be taken.

ENFORCEMENT ACTIONS

Enforcement is the second phase of the compliance monitoring and enforcement program. It involves actions necessary to bring handlers into compliance with applicable regulations. The goal of enforcement actions is to compel owners and operators to do the following:

 ▶ Properly handle hazardous waste
 ▶ Comply with RCRA's recordkeeping and reporting requirements
 ▶ Monitor and take corrective action in response to releases of hazardous and non-hazardous waste, and hazardous constituents

The EPA (or an authorized state agency) has a broad range of enforcement alternatives, and a decision to pursue one of these options is based on the nature and severity of the problem. Violations of RCRA requirements are grouped into classes and are discussed in the Enforcement Response Policy, OSWER Directive 9900.0-1A. The EPA may pursue one of three enforcement options: administrative actions, civil actions, or criminal actions.

Administrative Actions

An administrative action is nonjudicial enforcement action taken by the EPA or a state under its own authority. These actions can take several forms, ranging from informal notices of noncompliance to issuance of an administrative order accompanied by a formal public hearing. These actions are less complicated than a lawsuit and can often be quite effective in forcing a handler to comply with regulations or to remedy a potential threat to human health and the environment. The EPA can elect to take either *informal actions* or formal *administrative orders*. Each provides for enforcement response outside the court system.

Informal Actions

Informal administrative action is when an agency notifies the handler of a problem by any of a variety of communication media. This type of action is particularly appropriate when the violation is minor, such as a record maintenance requirement. Should the owner or operator not take steps to comply within a certain time period, he or she will receive a warning letter that details specific actions to be taken to move the handler into compliance. This letter also sets out the enforcement actions that will follow if the handler fails to take the required steps. An informal letter to the handler may be called a *notice of violation* or *notice of deficiency*. During the permitting process, a notice of deficiency is commonly issued to identify missing or deficient items in the facility's application.

Administrative Orders

Whenever the owner or operator does not respond to an informal notice, or when a more severe violation is detected, the agency can issue an administrative order. This is an order issued directly under the authority of RCRA that imposes enforceable legal duties on the handler and forces a facility to comply with the specific regulations. The four types of administrative orders issued under RCRA are as follows:

1. **Compliance order.** This is when the EPA issues an order requiring any person who is not complying with a requirement of RCRA to take steps to come into compliance. The compliance order may be immediate or it may set a timetable for moving toward compliance. The order can contain a penalty of up to $25,000 per day for each day of noncompliance and can suspend or revoke the facility's permit or interim status. The person to whom the compliance order is issued can request a hearing on any factual provisions of the order but, if no hearing is requested, the order will become final thirty days after it is issued.
2. **Corrective action order.** This is the issuance of an order requiring corrective action at an interim-status facility when there is evidence of a release of a hazardous waste or constituent into the environment. These corrective actions can range from investigations to repairing liners or pumping to treat a plume of contamination. These actions can be required regardless of when waste was placed at the facility. This mechanism is often used to clean up past problems at RCRA facilities. These orders can suspend interim status of a TSDF and can further impose penalties of up to $25,000 for each day of noncompliance with the order.
3. **Substantial hazard order.** If the EPA finds that a "substantial hazard" to human health and the environment exists, it can issue an administrative order. This order is used to evaluate the nature and extent of the problem through monitoring, analysis, and testing.
4. **Imminent and substantial endangerment order.** If the EPA finds a situation where "imminent and substantial endangerment to health or the environment" is caused by the handling of nonhazardous or hazardous wastes, the person

contributing to the problem can be ordered to take steps to clean it up. This order can be used against any contributing party, including past or present generators, transporters, or owners or operators of the site. Violations can result in penalties of up to $5,000 per day.

Civil Actions

A civil action is a formal lawsuit filed in court against a person who has failed to comply with some statutory or regulatory requirement or administrative order, or who has contributed to a release of hazardous wastes or constituents. Civil actions are generally employed in situations that present repeated or significant violations or serious environmental concerns. Attorneys from the Department of Justice handle RCRA civil cases for the EPA, while state attorneys general assume these roles for the states. Civil actions are useful in any combination of the following situations.

Compliance Action

The federal government can file suit to force a person to comply with any applicable RCRA regulations, and the court can impose a penalty of up to $25,000 per day per violation for noncompliance.

Corrective Action

The federal government can sue to have the court order the facility to correct any situation involving a release of hazardous waste from the facility. The court can also suspend or revoke a facility's interim status as a part of its order.

Monitoring and Analysis

The federal government can sue to get a court to force a person or facility to comply with an order in which the EPA required monitoring and analysis and with which the person or facility failed to comply. The court can assess a penalty of up to $5,000 for each day of noncompliance with the order.

Imminent Hazard

The federal government can sue to require action by a contributing person to remove a hazard or remedy any problem considered to be an imminent hazard to human health and the environment. If the agency had first issued an administrative order, the court can also impose a penalty of up to $5,000 for each day of noncompliance with the order.

It is common to employ several civil actions in the same lawsuit. This is particularly likely to happen when a handler has been issued an administrative order for violating a regulatory requirement, has ignored that order, and is in continued noncompliance. A lawsuit can be filed that seeks penalties for violating the original requirement, penalties for violating the order, a judge's order requiring future compliance with both the requirement and the administrative order, and substantial financial penalties.

Criminal Actions

A criminal action can result in the imposition of fines or imprisonment for certain acts considered by the legislature and upheld by the courts as "criminal" by intent.

RCRA identifies seven acts that are subject to criminal action and thus carry criminal sanctions and penalties. The financial penalties range from a fine of $50,000 per day up to a total fine of $1 million. A jail or prison sentence of up to five years can also be added to the financial penalty.

Criminal actions are usually reserved for only the most serious violations. The seven acts include the following:

1. Knowingly transporting waste to a nonpermitted facility.
2. Knowingly treating, storing, or disposing of waste without a permit or in violation of a material condition of a permit or interim-status standard.
3. Knowingly omitting important information from, or making a false statement in, a label, manifest, report, permit, or compliance document.
4. Knowingly generating, storing, treating, or disposing of waste without complying with RCRA's recordkeeping and reporting requirements.
5. Knowingly transporting waste without a manifest.
6. Knowingly exporting a waste without the consent of the receiving country.
7. Knowingly transporting, treating, storing, disposing of, or exporting any hazardous waste in such a way that another person is placed in imminent danger of death or serious bodily injury. This act carries a possible penalty of up to $250,000 or 15 years in prison for an individual or a $1 million fine for a corporation.

THE FEDERAL FACILITIES COMPLIANCE ACT OF 1992

The Federal Facilities Compliance Act of 1992 (FFCA) expanded the waiver of sovereign immunity that allows states to have full access to federal facilities. This means that federal facilities that treat, store, or dispose of hazardous waste need to obtain RCRA permits through state-authorized programs and, in most instances, must also comply with environmental statutes to the same extent as nonfederal facilities. However, enforcing compliance under RCRA is different at federal facilities. The EPA may only issue Section 3008(h) corrective action orders at federal facilities; no other orders may be used. States, however, may utilize the full range of their enforcement authorities at federal facilities.

Office of Federal Facilities Enforcement

This office was created to carry out the provisions of FFCA, with a charge to be even-handed in its approach. Otherwise, it closely follows the procedures employed in cases involving nonfederal parties. The FFCA requires the EPA to inspect all federal facilities at least once annually and gives any state with a hazardous waste program the option of also inspecting these facilities. The policy of the EPA is to coordinate closely with the Department of Defense, the Department of Energy, and the Department of Interior.

The criteria for determining the amount of penalty a noncompliant facility faces are based on the EPA's civil penalty policy. This policy takes into account (1) the gravity of the offense, (2) the economic benefit derived from noncompliance, and (3) any special circumstances, such as good faith (or lack of it), degree of negligence or willfulness, and history of noncompliance. Federal facilities, like their counterparts in the private sector, can negotiate settlement with the EPA.

Waste management activities at federal facilities are often managed by a private contractor. In this case, the EPA has full authority to take enforcement activities against the contractor for violations of RCRA.

NOTE: More information can be obtained by contacting the Federal Facilities Hazardous Waste Compliance Office, Office of Waste Program Enforcement, U.S. EPA, Washington, D.C. 20460.

RCRA AGENCY FUNCTIONS

RCRA enforcement responsibility is divided among different EPA headquarters' offices, the EPA regional offices, and state agencies. EPA headquarters is responsible for setting nationwide policy, monitoring regional and state activities, and providing technical support. EPA regional offices take the primary responsibility for performing inspections, issuing administrative orders, preparing civil actions, monitoring compliance with administrative and judicial orders, and providing support to the Department of Justice for ongoing lawsuits. Responsibility for enforcement is largely decentralized, with the states taking primary responsibility for enforcement. The EPA, however, retains its authority to take enforcement actions in authorized states if the state fails to do so, does not obtain acceptable results, or requests EPA assistance.

SUMMARY

▶ There are two essential elements to RCRA's compliance and enforcement program: compliance monitoring and enforcement actions.
▶ Compliance monitoring is used to determine a handler's level of compliance with RCRA's regulatory requirements through inspections and examination of reports that each handler is required to submit.
▶ Inspections by state or EPA officials are required annually at all federal or state-operated facilities and at least once every two years at each TSDF.
▶ Six types of inspections are conducted under the RCRA program, and either the EPA or an authorized state may lead the inspection:
  1. Compliance Evaluation Inspection (CEI)
  2. Case Development Inspection (CDI)
  3. Comprehensive Groundwater Monitoring Evaluation (CME)
  4. Compliance Sampling Inspection (CSI)
  5. Operations and Maintenance Inspections (O&M)
  6. Laboratory Audits
▶ The primary goal of enforcement actions is to bring facilities into compliance and keep them there. Enforcement actions may be taken when a facility is found to be (1) not complying with applicable Subtitle C regulations or (2) releasing nonhazardous or hazardous solid waste or hazardous constituents.
▶ Enforcement of RCRA is different at federal facilities. The EPA negotiates compliance agreements with federal facilities. Authorized states, however, may issue administrative orders or take other enforcement actions.
▶ The enforcement options available under RCRA are administrative actions, civil actions, and criminal actions.

▶ The responsibility for the various enforcement actions is divided among different EPA headquarters' offices, EPA regional offices, and authorized state agencies.

▶ Federal facilities are subject to the regulations promulgated in RCRA through state enforcement. The EPA can also step in if states request assistance or fail in their own enforcement efforts.

QUESTIONS FOR REVIEW

1. What is the goal of RCRA compliance monitoring programs?
2. What are the primary methods used by the EPA and the states in acquiring data for compliance monitoring?
3. Name and describe the six different types of compliance inspections and the purpose of each.
4. What are the general steps that are followed in RCRA inspections?
5. The EPA and state agencies have the following three types of enforcement alternatives available to them to ensure compliance with RCRA.

Briefly describe each:
   a. Administrative actions
   b. Civil actions
   c. Criminal actions
6. Outline the basic provisions of the Federal Facilities Compliance Act of 1992.
7. How is work and enforcement responsibility divided up amongst the various EPA headquarters, regions, and state agencies?
8. Under compliance actions, what is the limit the court can impose in financial penalties for noncompliance?

ACTIVITIES

1. Contact the office in your area that manages your state's RCRA compliance monitoring program. Obtain copies of all information that is free to the public.
2. Make contact with a RCRA compliance officer and invite that person to be a guest lecture at your school or other institution.
3. Telephone the RCRA Hotline and obtain a menu of documents available free to the public.
4. Access the EPA through a World Wide Web connection and search for information pertinent to the RCRA compliance monitoring programs.

9

Medical Waste Management

Neal K. Ostler

Upon completion of this chapter, you will be able to meet the following objectives:

▶ Provide an overview of the regulated medical waste community.

▶ Summarize the existing federal and state regulations.

▶ Recognize and identify the variety of regulated medical waste.

▶ Specify the basic components of a generic medical waste management plan.

▶ Identify the basic provisions of the Occupational Safety and Health Administration's (OSHA) Bloodborne Pathogens Standard.

▶ Identify the common methods of treatment, storage, and disposal of low-level radioactive and mixed waste.

OVERVIEW

The management of medical waste begins when and where an item ceases to be useful for its intended purpose and enters the waste stream. During the summers of 1987 and 1988, the concern about medical-related material and waste increased dramatically as the result of several incidents. The incident that brought the most attention from the media was the report that hypodermic needles and syringes had washed up on the beaches of five East Coast states. In another disturbing incident, children were found playing with vials of blood they had discovered in a dumpster.

119

The awareness of improper disposal of medical waste brought into question the adequacy of good medical waste management practices. These issues became a popular topic of debate and eventually brought forth a public outcry for more stringent controls, especially after the wastes have left the site of generation.

In response to this public outrage and fear, Congress passed the Medical Waste Tracking Act (MWTA), which was signed into law on November 1, 1988, as an amendment to the Resource Conservation and Recovery Act (RCRA). MWTA required the EPA to investigate and create a two-year demonstration on the tracking of medical waste from its point of generation to its disposal site. MWTA also required the EPA to establish uniform standards for the segregation, packaging, labeling, and reporting of infectious wastes.

In addition, MWTA required the Agency for Toxic Substances and Disease Registry (ATSDR) within the U.S. Department of Health and Human Services to release a report of the findings of these tracking efforts. In 1990, the ATSDR released "The Public Health Implications of Medical Waste: A Report to Congress," which contained an exhaustive discussion on the topic. Based on data developed in this report, the ATSDR reached a number of conclusions and made a number of recommendations, which are summarized as follows:

Conclusions: Medical waste adversely affects the environment and contributes to the overall environmental problem of solid waste disposal in the United States. Such waste can be effectively treated by chemical, physical, or biological means and disposed of in a sanitary landfill or by incineration. Medical waste is not to be considered "hazardous" by the definition of RCRA and, therefore, is not subject to these stringent controls.

Recommendations: The relevant federal, state, and local agencies should work to adopt regulatory definitions and sound public health principles consistent throughout the United States; to develop work practices for occupational subgroups frequently contacting medical waste and therefore at higher risk; to develop guidelines for in-home health care management of medical wastes; and to develop new technologies for waste management and treatment technologies, including waste reduction, recycling, reuse, and reclamation.

The outcome of MWTA has been the development of state and local regulatory mechanisms for handling and disposal of medical waste and the development of the Bloodborne Pathogens Standard by OSHA to protect workers at high risk of exposure.

Compared to the extensive regulations governing the handling of hazardous waste under RCRA, federal rules regarding medical waste are still in their infancy, and waste generators favor a more narrow definition of regulated medical waste.

REGULATED SOURCES OF MEDICAL WASTE

Before the reporting requirements of MWTA were mandated, it was believed that 75 percent to 80 percent of all medical waste was generated by the nation's 7,000+ hospitals, which are the most easily identifiable sources. As a result, hospitals are the primary target for regulation, but they are not the only generators of medical wastes. The range of potential generators includes the following:

1. **Hospitals.** This category includes facilities providing general medical and surgical care, psychiatric care, treatment of tuberculosis, and services in specialties such as obstetrics and gynecology, eye/ear/nose/throat, and rehabilitation.

2. **Intermediate care facilities.** This category includes nursing homes and facilities for in-patient care for the developmentally disabled.
3. **Clinics and physician offices.**
4. **Dental offices.**
5. **Laboratories.** This category includes the full spectrum of research, medical, industrial, diagnostic, manufacturing, and pharmaceutical preparation laboratories.
6. **Other.** For the purpose of this chapter, the remaining generators of medical waste are grouped together and include the following:

Funeral homes	Veterinarians
Agricultural facilities	Blood banks
Animal care facilities	Hospices
Emergency medical services	Illicit drug users
Home health care	Medical schools
Public health units	Nursing schools

The American Hospital Association recommends basing the regulations that govern generators of hazardous waste on the properties of the waste rather than the amount of waste generated or the size of the entity that generates it. In actual practice, most states specify that only entities generating beyond a certain amount are subject to regulations. Generally this minimum amount is greater than 100 kilograms (220 pounds) per month.

NONREGULATED SOURCES

Changes in the health care delivery system have resulted in dramatic increases in medical waste from nonregulated or residential sources. It has been reported that the number of injuries to refuse workers from sharp objects (often referred to simply as sharps) in residential solid waste is increasing and appears to coincide with the increasing trend to in-home health care.

There are actually several trends that have contributed to this increase in medical waste from nonregulated sources:

‣ Hospital patients are released on an outpatient basis more frequently.
‣ Many one-time use items that do not come within the jurisdiction of the regulations are available over-the-counter for small livestock operations or nonprofit entities.
‣ The use of disposable items in the home has increased.
‣ The number of users of illicit intravenous drugs has increased.

Because of these trends, efforts must be made to educate the public regarding safe disposal methods for home health care medical waste. Educating people about the ethics of good medical waste management is also imperative, as is making them understand that proper medical waste management is essential to protecting public health and the environment.

WASTE CHARACTERIZATION

For wastes to fall within the guidelines pertinent to medical waste, they must possess the potential to transmit infection or present a risk to public health or the environment for reasons other than their infectious potential. Before further discussion of these two waste characterizations, an understanding of the following terms is necessary:

1. **Medical waste.** The EPA defines medical waste as any solid waste that is generated in the diagnosis, treatment, or immunization of human beings or animals, in research pertaining thereto, or in the production or testing of biologicals (vaccines).
2. **Infectious agent.** The EPA defines an infectious agent as any organism (such as a virus or bacterium) that is capable of being communicated by invasion or multiplication in body tissues and capable of causing disease or adverse health impacts in humans.
3. **Infectious waste.** This is waste that contains pathogens with sufficient strength or quantity such that exposure to the waste could result in infectious diseases. At present, there is no definitive method to determine if a waste is infectious; therefore, a waste is defined as infectious if it is has the potential for disease transmission or infection.

Medical waste can be divided into ten categories, seven of which are characterized by the potential to transmit infection, and three of which are characterized by the risk they pose to human health and the environment. Let's look at all ten, by characterization and category.

The Potential of the Waste to Transmit Infection

These seven categories of wastes, by virtue of their characteristics, are universally handled as medical wastes regardless of their source. Because their infectious potential cannot be determined by their appearance and because the source of the item may not be identifiable, it is impractical and infeasible to test each item for its pathogen content as to quantity and type. These seven categories are as follows:

1. **Sharps.** This category includes hypodermic needles, syringes, scalpel blades, blood specimen tubes, pasteur pipettes, and broken glass (that has been exposed to infectious agents) regardless of whether the items have been used in animal treatment, patient care, research, or industrial laboratories. OSHA also includes orthodontic wires and sharps in this category for purposes of disposal.
2. **Cultures and stocks of infectious agents.** This category includes specimen cultures from medical and pathological laboratories; infectious agents from research and industrial laboratories; culture dishes; devices used to transfer, inoculate, or mix cultures; and discarded vaccines.
3. **Bulk human blood and blood products.** In addition to blood, this category includes items that are saturated and have the potential for dripping blood as well as blood products such as serum and plasma.
4. **Pathological wastes.** This category includes any human tissues, body parts and/or fluids, or any organs removed during surgery or postmortem procedures. Corpses and parts intended for interment or cremation are not included.
5. **Isolation wastes.** This category includes wastes contaminated with blood, excretions, exudates, or secretions from sources infected with highly communicable diseases.
6. **Animal waste.** This category includes body parts, contaminated carcasses, and fluids of animals that have been afflicted or purposely infected with agents during research, production of biologicals, or testing of pharmaceuticals.

7. **Unused sharps.** Even though these are possibly uncontaminated, unused sharps are included here because of the risk of the item having been used without the handler's knowledge. Other reasons are because of the potential for illicit use if the sharps are not disposed of as solid waste and because improper handling may cause physical injury.

Risk Other Than Infectious Potential

The following three additional categories of waste constitute a risk to human health and the environment if the wastes are not disposed of properly. These items are regulated as medical waste because they originate from sources identified as generators of medical wastes, but are not medical waste by definition.

8. **Low-level radioactive waste.** This category includes certain low-level radioactive wastes, such as those generated from nuclear medical procedures and administration of radiopharmaceuticals. These are wastes that are not regulated under the Nuclear Regulatory Commission, but that do constitute a risk to health and the environment.

9. **Cytotoxic and neoplastic drugs.** This category includes contaminated materials and human excreta that are not handled as RCRA hazardous waste, but that constitute a risk to health and the environment. Cytotoxic drugs are poisonous to living cells and are used in research and development of drugs to support and vitalize the human immune system in attacking viruses and bacteria. Neoplastic drugs are used in research and development of a cure for the growth or control of tumors.

10. **Small volumes of chemicals.** This category includes hazardous chemicals that are exempt from regulation under RCRA but evolve as waste products from a process or operation involving hazardous materials in a medical generator setting. These wastes demonstrate one or more of the four characteristics of a RCRA hazardous waste (i.e., ignitability, corrosivity, reactivity, and toxicity, as detailed in Chapter 2) or are listed by the EPA.

MINIMIZATION OF MEDICAL WASTE

In recent years, the shift to disposable products for health care and the implementation of universal precautions have significantly increased the quantity of medical wastes. The quantity of wastes requiring special handling could be greatly reduced by educating the public so that they can identify which wastes can be handled as solid wastes and which wastes must be segregated and handled as medical wastes. The minimization of medical waste can also be accomplished through reuse, recycling, and source reduction.

Reuse

Arguments have been made for the use of disposable health care products. These arguments suggest that disposables reduce liability issues, control infection, save labor costs for reprocessing, and minimize exposure to chemicals used in the sterilization process. However, the identification of items that can be segregated for return, reuse (by cleaning, disinfecting, or sanitizing), or reprocessing can significantly reduce the total volume of waste generated.

Recycling

Waste minimization can occur through recycling of the many components involved with medical waste. These components include the plastics and metals in syringes, the glass in tubes and vials, and fabrics from contaminated laundry, such as gowns, gloves, bedding, and personal protective equipment. Due to the risk of exposure to infection, however, recycling of regulated medical wastes involves substantial reprocessing to sanitize or disinfect the original components.

Source Reduction

Source reduction efforts must focus on reducing either the toxicity or the quantity of discarded products. Manufacturers can consider new designs of medical products for the consumer, and consumers can make their purchases based on ease of use as well as a regard for waste reduction. PVC plastics can be replaced with nonchlorinated plastics, materials that are safe to burn can be used, and materials management can be improved to eliminate overpurchasing of items with a limited shelf life or to eliminate other losses through improper storage or handling methods. Purchasing recyclable, less toxic, or reusable items will expedite waste minimization efforts later in the waste handling process.

THE MEDICAL WASTE MANAGEMENT PLAN

Any medical waste management program must have a plan. While small generators may not require a written plan, they should still train their personnel in the proper handling, packaging, and transport of medical waste generated in the course of their duties.

Large generators should prepare a written management and operations plan outlining policies and procedures consistent with OSHA's Bloodborne Pathogens Standard (to be discussed later in this chapter). The plan should be reviewed and updated as necessary.

The EPA recommends that states require each regulated generator of medical waste to establish an infectious waste management plan with the following basic components.

Identification

The waste management plan should provide procedures for the identification of items in the waste stream that will be designated as medical or infectious waste. (A brief summary of types and categories of medical or infectious waste was discussed previously.)

Segregation

The plan should include operational procedures for separating infectious wastes from noninfectious wastes at the point of generation. Segregation is important, especially for generators who plan to transport their waste off site for disposal. The cost of handling regulated medical waste is much more involved and expensive than handling nonregulated medical waste or other solid waste streams. Cost savings should encourage generators to segregate their wastes even if they are not required by state regulations to do so.

Proper Collection

The plan should include procedures for properly collecting and depositing infectious wastes into special primary containers and then placing them into more durable secondary packages for transportation. Secondary packaging should be mandatory for off-site transportation.

Packaging

The plan should include purchase of primary and secondary packaging that is consistent with OSHA's Bloodborne Pathogens Standard. This means that (1) sharps should be placed in impervious, puncture resistant, rigid containers to eliminate the hazard of physical injury, (2) liquids should be placed in capped or tightly stoppered bottles, flasks, or containment tanks, and (3) solid and semisolid wastes should be placed in plastic bags or lined corrugated boxes. Packaging should be strong and durable enough to ensure the exclusion of vermin or rodents and maintain its integrity during storage and transportation. Secondary packaging may consist of either corrugated boxes or reusable containers.

Labeling

Primary and/or secondary containers must be properly labeled in accordance with the following guidelines:

1. Information must be permanently printed in indelible ink on all sides and must be clearly and easily readable.
2. The international biological symbol must be displayed in red or orange with a contrasting background and must be of a size suitable to the package or container.
3. The words Biohazardous Waste, or a similar phrase, in red or orange with a contrasting background must be permanently imprinted on the packages.
4. Packages containing infectious substances and regulated medical waste must contain the following statement: In case of damage or leakage, immediately notify public health authorities, in the U.S.A., notify the director of the Centers for Disease Control, Atlanta, Georgia, 404-633-5313.

Storage

During storage, the packaging method should ensure that no vermin or rodent can get access to the waste. If the waste cannot be treated immediately after generation, the time in storage should be minimized and the waste should be kept at a low enough temperature (refrigerated) to slow the waste's decaying and putrefaction processes. Storage areas should be designated, enclosed, and ventilated. Depending upon the state, the regulations also may describe how the storage areas must be constructed and secured, and may provide limits on the length of storage time. Many states require that storage permits be obtained. These permits can be issued on a conditional basis until the storage facility complies with these regulations.

Spill Response

Procedures should be in place in the event that a spill of regulated medical waste occurs, and employees should be trained to implement these procedures. One method is to cordon off the area and douse the spill with a high-powered bleach or disinfectant spray.

Transportation

Once the primary waste containers have been double bagged and/or placed in secondary containers, they are generally moved (within the generator facility) using carts identified with the universal biological hazard symbol. Before being loaded onto a vehicle, waste to be transported from generators to off-site treatment facilities should be placed in containers that are rigid, leakproof, and resistant to bursting, punctures, moisture, and tears. Figure 9–1 shows how medical waste should be packaged and readied for transport. The Department of Transportation (DOT) regulations require that packages conform to Packaging Group II standards. If the generator does not transport the waste itself, it should make arrangements for pickup with responsible third-party transporters or the treatment, storage and disposal facility (TSDF) that will be receiving it.

▶ **FIGURE 9–1**
Packages for shipment of medical waste.

Transport vehicles may be required to be refrigerated, specially placarded, provided with security, and in compliance with other specific requirements. In many states, the transporter may be required to be permitted through a process that requires bonding, insurance, safety standards, and waste management plans in the event of an accident.

Treatment, Destruction, and Disposal of Medical Waste

The EPA recommends that all infectious waste be treated or disinfected prior to disposal in a sanitary landfill. Once disinfected properly, the waste no longer poses a danger of dispersing hazardous materials into the environment and can be handled and disposed of as general waste.

Treatment Methods

Treatment is any method, technique, or process designed to change the biological character or composition of waste to reduce or eliminate pathogens so that the waste no longer poses a hazard to persons who may be exposed to it.

The primary methods of treatment are incineration, autoclaving/steam sterilization, irradiation, and thermal inactivation. The choice of which treatment method to use depends on the waste's physical composition and the different types of packaging selections. Each of the ten categories of medical waste described previously may require a different method of treatment, destruction, and disposal suitable to its own peculiarities and in compliance with any applicable regulations.

Incineration   An incinerator is a multichamber facility with a secondary chamber that operates at temperatures of 1,800–2,200°F. At these high temperatures, the pathogens are removed and the high volume of plastics is reduced. In addition to the removal of infectious agents from medical waste at high temperatures, an incinerator also minimizes acid emissions and results in an 85 to 92 percent volume reduction of the waste. Incineration, therefore, not only largely eliminates the hazards of pathogens, but also significantly reduces the amount of waste going into the nation's landfills.

> **NOTE: The federal government at this time has preferred to permit each state to develop its own medical waste management program, and despite objections and condemnation by environmental activists, medical waste may be landfilled without treatment unless prohibited by law.**

Although incineration is a suitable treatment for most types of infectious waste and reduces landfill and transportation costs, it also creates air emissions problems. If not done properly and completely, incineration may allow infectious agents to be released in stack emissions and wastewater effluents, and the ash residue may become hazardous and subject to regulations under RCRA. Proper control of air emissions can be achieved by using air pollution control devices, such as scrubbers and electrostatic precipitators.

A state-of-the-art incinerator converts the heat produced from combustion into steam, which can then be used to heat structures or to generate electricity and thus lower the cost associated with incineration.

Steam Sterilization/Autoclaving   This method of treating waste involves the utilization of saturated steam within a pressure vessel at sufficient temperatures and duration to kill infectious agents that are present. This method is efficient with small quantities of low-density and low-water-content wastes, but is not effective for wastes such as large body parts and fluids. This method also should not be used to treat toxic chemicals, radioisotopes, and chemicals that may become volatile in steam treatment. While steam sterilization does not produce pollutants, it also does nothing to reduce mass of material that must be landfilled.

Irradiation   For certain materials that cannot be thermally treated, it may be possible to expose these wastes to ultraviolet or ionizing radiation in an enclosed and shielded chamber. While this method requires very little expenditure of electrical or steam energy, it requires a high cost for initial equipment purchases, operator training, ongoing acquisition of radiation sources, and ongoing disposal of the decayed radiation source. This method also provides very little penetration of the waste with ultraviolet exposure and those areas shadowed by other waste will not be effectively treated.

Thermal Inactivation   This method reduces or eliminates infectious agents in waste by transferring heat to continuously fed batches of the waste. This batch-type unit is a vessel heated by exchangers or a jacket filled with steam. Thermal inactivation of solid waste is accomplished by the application of dry heat in an electrically operated oven. Air circulation in the vessel, adequate temperatures, and adequate duration are important to proper operation of the unit. The wastes should be mixed before being treated with this method to obtain maximized homogeneity during both the loading and heat-application steps.

Alternative Technologies   The EPA and the Department of Health and Human Services encourages the development of alternative technologies for regulated waste treatment systems. Three such processes include the following:

1. **Chemical decontamination.** This is a method being developed that uses an electrocatalytic system that purports to destroy any living organism by a combination of temperature, acidity, and chemical activity.
2. **Microwave.** Traditional thermal treatment heats wastes externally, while microwave heating occurs inside the waste material. Generally the waste is shredded, injected with steam, and rotated on an auger while heated under a series of microwave units. Treated waste may be landfilled. This method is primarily suitable for treating blood-soaked items such as bandages, napkins, secretions, and single-use hypodermic needles.
3. **Macrowave.** With this technology, heat produced by low-frequency radio waves kills disease-causing pathogens by electrothermal deactivation.

Destruction Methods

Two principal methods used to reduce the volume of medical waste are grinding and shredding, and compaction.

Grinding and Shredding   The process of grinding and shredding treated and untreated medical wastes reduces the overall volume of the waste, but not the weight. This process allows the waste to be handled more easily to facilitate treatment, makes it more homogenous for uniform treatment, and renders it unrecognizable. This process also reduces storage, transportation, and handling costs. Grinding and shredding equipment should be operated under negative pressure, and a high efficiency particulate air filter (HEPA filter) is recommended to ensure that no materials escape from the device.

Compaction   Generally a hydraulic ram is used to compress the waste against a rigid surface area within an impervious container such as a storage drum. This method reduces waste volume and renders the waste unrecognizable, but does not reduce the hazards of disease transmission. Because the containers can be damaged, the potential for accidental releases of the contaminated waste increases. Therefore, this method is not recommended for untreated waste.

Disposal Methods

Regulated medical wastes or infectious wastes that have been treated by an effective method are no longer biologically hazardous. Once such treated wastes have been packaged so that their treatment is evident, they are no longer subject to management as medical waste and may be collected, mixed with other wastes, and disposed of as ordinary waste. Treated liquid wastes may be released into the sewer system, but these are subject to the local water treatment authority. Solid wastes that have been shredded and ground as well as treated may also be flushed into the sewer system, but generally sewers are suitable for the disposal of liquid wastes only. Incinerator ash and other solid wastes can be disposed of in a properly sited and constructed sanitary landfill.

OSHA'S BLOODBORNE PATHOGENS STANDARD (29 CFR PART 1910.1030)

The Bloodborne Pathogens Standard (BPS) is a program developed by OSHA to ensure that workers in medical and other related fields who have any possibility of contact with bloodborne pathogens are fully advised of how these agents constitute a hazard to their health and are fully trained to implement a proper program for the handling and disposal of such hazardous materials. Every employer with one or more employees who are exposed to potentially infectious body fluids must develop a written exposure control plan.

The Exposure Control Plan

BPS requires that every employer have a written plan designed to eliminate or minimize employee exposure to bloodborne diseases. OSHA requires that the plan include the following elements.

Potential Exposure Determination

This essentially includes two employer-prepared lists: (1) a list of job classifications in which all employees in the classification have risk of exposure and (2) a list of job classifications in which some of those employees in the classification have a risk of exposure. To prepare these lists, the employer must be aware of the activities and tasks that put employees at high risk. Such jobs include the following:

▸ Emptying trash and other cleaning duties.
▸ Changing linens.
▸ Refilling first-aid supplies and cabinets.

▶ Processing soiled laundry.
▶ Nursing activities in medical offices.

Emergency responders such as firefighters, EMTs, police officers, and medical teams are among those workers whose jobs present an obvious risk of exposure to bloodborne pathogens due to the very nature of the duties they perform.

Schedule of Risk Management Techniques

This is essentially the employer's schedule for implementing OSHA requirements for each of the following:

1. **Barrier techniques.** This refers to the use of personal protective equipment (PPE) to protect employees. Because the principal routes of entry for bloodborne pathogens are by skin, mouth, nose and eyes, it follows that such PPE should include masks to cover the nose and mouth, gloves when hand contact is anticipated, gowns to protect clothing, and splash-protective eyewear .

   The employer must make PPE available, train employees in proper use of the equipment, and assure that the employees use it. The only exceptions to regular usage of PPE are rare and extraordinary circumstances when such usage would have further jeopardized the employee or the victim and the intended emergency response.

2. **Universal precautions.** These includes such basic safety and hygiene measures as keeping hands away from the face, and removing PPE and washing the hands before eating, drinking, smoking, applying cosmetics and lip balms, or handling contact lenses.

3. **Work practice controls.** This refers to identifying and implementing the safest means to perform certain tasks. This can be accomplished through task analysis activities involving the workers.

4. **Engineering controls.** This refers to the use of designated tools and equipment to handle certain tasks. For example, a company may designate a particular broom and dust pan to be used to clean up spilled materials or broken containers.

5. **Housekeeping practices.** This includes written plans for the following tasks:

   ▶ General cleaning (specifying decontamination methods).
   ▶ Proper cleaning and disinfecting of any surface immediately after it has come into contact with blood or other materials.
   ▶ Replacement of contaminated containers, bags, tools, etc.
   ▶ Discarding of contaminated equipment, such as PPE or linens, unless the plan includes decontamination procedures.
   ▶ Proper disposal according to state and local health ordinances.

Vaccination Program

BPS requires every employer to provide a free Hepatitis B vaccination to every employee and to provide proof that an employee has refused the vaccination if he or she has not had one. The refusal form must be signed by the employee. If an employee is exposed to a pathogen, BPS requires the employer to provide that employee with a series of vaccinations and a follow-up medical consultation.

Procedures for Exposure

If an employee is exposed to medical waste, BPS requires that the employer have procedures in place for handling such exposure. In most cases, the first line of treatment is that the employee must immediately wash the exposed area with nonabrasive soap and water or flush the eyes with water.

If an employee has been exposed to another person's bodily fluids by accident, the employee is required to report the exposure to the supervisor, who then is required to advise the safety department and the company's nurse or medical program manager. The report should include routes of exposure, circumstances of exposure, and the identity of the source individual.

Laboratory tests should be made of the source person's blood and the exposed employee's blood, at the employee's discretion. The employee should be invited to have free vaccinations, and when evaluations have been completed, he or she should be informed of the medical condition and the need for further vaccinations. All of this information is to be kept confidential between the employee, the employer, and the medical professional.

Training

Employers must train their employees so that they know (1) the hazards of bloodborne pathogens; (2) what to do in the event of an exposure; (3) how to clean up and disinfect a contaminated area; (4) what to do with contaminated objects including PPE, sharps, bandages, and clean-up materials; and (5) their rights of accessibility to vaccinations, medical care, and follow-up treatment.

Regulated Waste Requirements

Employers must adhere to strict labeling and packaging requirements. BPS has made certain signs and warning labels mandatory when a bloodborne pathogen is suspected or known. Every container must be marked with the biohazard label, which is fluorescent orange/red in color and has a contrasting (white or black) biohazard symbol. The requirement for labeling with appropriate colors and symbols also applies to any equipment or material suspected of being contaminated (which, for these purposes, means an infectious substance is present on the material's surface).

In addition, the PBS labeling requirements apply to secondary packages necessary for the disposal of the waste. These secondary packages must be color coded, labeled, closable, leakproof, and puncture resistant. All contaminated materials and known bloodborne pathogens must be disposed of according to the rules, procedures, and policies of the state and local health departments.

Recordkeeping Requirements

BPS requires that employers keep medical records, training records, accident records, and copies of any exposure reports. It also requires that employers make all of this information, including the written Exposure Control Plan, accessible to their employees and that employers update the plan at the end of each year.

BPS covers full-time employees, temporary and part-time employees, "per diem" health care workers, employees trained in first aid and designated as being responsible to render medical assistance as part of their job description, and employees in any industry who may be exposed to blood or other potentially infectious materials.

OTHER FEDERAL REGULATIONS

Though programs for the management of medical waste are essentially run by the states within guidelines provided by the EPA and OSHA, a large number of other federal laws affect the overall medical waste management plan and must be considered by the generator. The following is a brief list and explanation of these federal regulations, broken down by the government agency or department that administers them.

EPA Regulations

RCRA (40 CFR Parts 24–246, 258–259)

As a result of the Medical Waste Tracking Act, an amendment to RCRA in 1988, the EPA issued the "EPA Guide for Infectious Waste Management," which outlines the EPA's perspective on environmentally safe techniques and policies in the management of infectious waste. While medical waste is specifically excluded under RCRA, certain medical wastes contain hazardous constituents that may be subject to RCRA; therefore, the waste itself is subject to RCRA.

The EPA has also issued guidelines for the land disposal of solid wastes. These guidelines include requirements and recommendations for facilities accepting infectious waste.

Under RCRA, the EPA can bring suit against any person who merely contributes to the improper handling, storage, treatment, transportation, or disposal of solid waste, including medical waste or hazardous waste, if there is evidence that the waste presents an imminent and substantial danger to public health or the environment.

Comprehensive Environmental Response Compensation Liability Act (CERCLA)

Under CERCLA, the EPA is authorized to remove or arrange for the removal of any pollutant or contaminant, including disease-causing agents, that presents an imminent and substantial danger to the public health or welfare. The EPA can take any action consistent with the National Contingency Plan (discussed at length in Volume 2 of this Environmental Technology Series) that is deemed necessary.

The Clean Air Act

Incineration is currently the most popular method for treating regulated medical waste and, consequently, the regulated industry is greatly affected by the Clean Air Act and numerous subsequent amendments. The 1990 Clean Air Act sets air emission limits for medical waste incinerators, and the states are required to include EPA performance standards in their regulations. The act is designed to encourage the development of large, regional incinerators with state-of-the-art air pollution control equipment and to eliminate many on-site incinerators (those located at the site of the hospital, research laboratory, or other medical-waste generator) that cannot meet EPA performance standards.

The Clean Water Act (40 CFR Parts 60.50–.65)

The Clean Water Act establishes programs to control discharges of wastes into navigable waters and outlines standards for acceptance of solid wastes by publicly owned treatment works (POTWs). The Clean Water Act issues permits that allows the direct disposal of certain types of waste, but it also forbids the discharge of many other types of wastes, including any medical waste.

U.S. Public Vessel Medical Waste Antidumping Act (33 CFR Part 151 Subpart A)

Passed in 1988, this act mandates that vessels not dump medical waste unless necessary. If such dumping is deemed necessary, the waste must be disposed of at least 50 nautical miles from shore and only after the waste has been properly sterilized, packaged, and weighted to prevent it from coming ashore.

Marine Protection, Research, and Sanctuaries Act (33 CFR Part 152 Subpart B)

This act prohibits transport of, among other things, medical waste for the purpose of ocean dumping. The Shore Protection Act of 1988 extends this prohibition to private vessels hauling wastes, including medical waste, unless permitted by the U.S. DOT, which enforces shipping regulations. The DOT can impose heavy civil penalties of up to $10,000 for each day that a vessel is in violation of regulations and can also subject the violator to a possible criminal sentence of up to three years imprisonment.

Department of Transportation (49 CFR Parts 100-177)

The DOT is responsible for regulation of hazardous materials transported by road, water, rail, or air. Current regulations classify etiologic agents as hazardous and define such agents as any viable microorganism, or its toxin, that causes or may cause disease in humans. These include agents listed by the Public Health Service as bacterial, fungal, viral, and rickettsial. The terms etiologic, medical waste, and infectious substances are now considered synonymous by DOT regulations. The DOT regulations provide separate requirements for proper packaging and labeling of both infectious waste and regulated medical waste.

Department of Health and Human Services (42 CFR Part 72)

The Public Health Service forbids any person from knowingly transporting any etiologic agent unless it is properly packaged. It has more restrictive requirements for certain specifically listed agents.

Food and Drug Administration (21 CFR Part 58)

The Food and Drug Administration (FDA) has set standards for laboratory research practices and for practices in handling and preparing drug products. It also sets strict facility standards for storage and disposal of animal waste and general refuse.

U.S. Department of Agriculture (9 CFR Parts 103–112)

Under regulations promulgated by the U.S. Department of Agriculture (USDA), contaminated animals must be cleaned and disinfected, or even destroyed and then disposed of by incineration. The USDA also regulates veterinary biological products. Animal carcasses are not exempt from regulations governing medical waste.

U.S. Postal Service (39 CFR Part 111)

U. S. Postal Service regulations place restrictions on mailing sharps and other medical devices. These items must be packaged according to strict guidelines and mailed either first class or priority mail. Such packages must be accompanied by a four-part manifest, placed in an envelope and attached to the outside of the container. Distributors and manufacturers must obtain approval from the Postal Service. Labels must be placed on all primary and shipping containers.

Nuclear Regulatory Commission (10 CFR Parts 20 and 61)

The Nuclear Regulatory Commission (NRC) specifically regulates certain medical waste resulting from the use or research of nuclear medicine. These regulations are stricter than those governing medical waste. The NRC permits the incineration of radioactive waste only if it is below permitted levels of radioactivity. The NRC enforces regulations imposing requirements for packaging and handling of radioactive waste. It also requires infectious or pathogenic waste mixed with radioactive isotopes to be treated to reduce as much as possible any potential hazards from the nonradiological materials. (The DOT still provides standards for movement of such wastes by any mode of transportation.)

Department of Commerce (15 CFR Part 799)

The Department of Commerce restricts the export of certain viruses, bacteria, protozoa, and fungi to prevent these substances from being used in chemical or biological warfare programs. The department's Bureau of Export Control manages and enforces this oversight.

Department of Defense (32 CFR Parts 626–627)

The U.S. Department of Defense (DOD) includes a program of research and development of biological warfare, and this program is governed by a number of regulations that affect etiologic agents and waste. The purpose of these regulations is to (1) minimize the potential exposure of personnel, (2) establish procedures for reporting mishaps, (3) provide hazard analysis to determine safety precautions, (4) train personnel, (5) establish medical surveillance programs, (6) develop emergency preparedness and contingency planning, (7) enforce labeling and posting of the biohazard symbol, and (8) develop policies for the packaging, treatment, storage, transportation, and disposal of hazardous waste. The DOD manages its own programs, but does so in accordance with DOT, EPA, and Health Department standards.

National Institute for Occupational Safety and Health

The National Institute for Occupational Safety and Health (NIOSH) is an organization within the U.S. Department of Health and Human Services. A primary function of NIOSH is to make recommendations to OSHA to help them set rules and standards. NIOSH has published guidelines for the proper disposal of infectious waste in order to protect workers and the environment. NIOSH was a principal contributor to the development of OSHA's Bloodborne Pathogens Standard.

RECENT INTEREST IN MEDICAL WASTE

In recent years Congress has shown increasing interest in passing legislation governing the generation, treatment, handling, transportation, storage, and disposal of medical waste. Congress has introduced several bills that have sought to restrict the interstate transportation of waste, including medical waste, for purposes of disposal. Other bills have sought to restrict generators from mailing medical waste to out-of-state disposal sites.

Perhaps one of the most important bills introduced in recent years concerning medical waste was Senate Bill 976 in the 102d Congress. This bill would have been an amended RCRA and would have developed a full blown "cradle-to-grave" system of regulations for medical waste similar to those now enforced in the management of hazardous waste. In concert with both OSHA and the Centers for Disease Control, the EPA would have also been responsible for setting educational standards for households and other nonhospital sources of medical waste.

Passage of these and similar bills that have been introduced but not acted upon in Congress will depend upon a pro-environmental Congress and thus will rest upon the shoulders of the U.S. electorate.

SUMMARY

▶ The Medical Waste Tracking Act (MWTA) was passed in 1988 as an amendment to RCRA. The purpose was to track medical waste in several states and compile data from which recommendations could be made regarding the management of medical waste.

▶ From MWTA data, the EPA concluded that medical waste did constitute a risk to public health and the environment. It further recommended that certain medical wastes be regulated, and the EPA released a document to provide guidelines for state programs.

▶ Before the reporting requirements of the MWTA, it was believed that between 75 and 80 percent of all medical waste is generated by the nation's 7,000+ hospitals, which are the most easily identifiable sources. As a result, they are the primary target for regulation, but they are not the only generators of medical wastes. The range of potential generators includes a wide variety of sources from medical home care to users of controlled substances (drug abusers).

▶ A number of definitions are important when considering the management of medical waste:

Medical waste
Regulated medical waste, including sharps, culture wastes, bulk blood wastes, pathological waste, isolation waste, animal waste, and unused sharps
Infectious agents

Infectious waste

Medical waste, other than infectious waste, including low-level radiological, cycto-
toxic and neoplastic drugs, and small volumes of chemical wastes

▶ OSHA's Bloodborne Pathogens Standard (BPS) seeks to ensure the protection of
workers at all levels where contact with bloodborne pathogens is possible by requiring employers of such to develop a written exposure control plan. The employer also has strict recordkeeping responsibilities relative to BPS.

▶ Large generators of medical waste should prepare a written management and operations plan outlining policies and procedures consistent with BPS. The written
management plan should address the following issues at a minimum:

Identification of the waste
Segregation of regulated and nonregulated wastes
Proper collection procedures, times, and containers
Packaging and labeling
Storage
Spill response procedures
Transportation
Treatment, destruction, and disposal

▶ Although small generators may not be required to prepare a written plan, they
must train their personnel to properly handle, package, and transport medical
waste generated in the course of their duties.

▶ Several methods for the treatment of medical waste have been approved. The most
common are incineration, steam sterilization, irradiation, and thermal inactivation.

▶ The EPA also encourages the development of new treatment technologies.
Grinding and shredding medical waste both during and in addition to other treatment methods provides a more homogenous, unrecognizable waste. Compaction
may reduce the total volume of waste but does not decrease the potential hazards
and is inappropriate for certain wastes.

▶ Properly treated liquid and some well-ground and shredded wastes can be disposed of directly into the sewer if the locally publicly owned treatment works permits such disposal.

▶ Once treated, solid and semisolid medical wastes can be disposed of in a landfill
that has been properly constructed and permitted to accept such wastes.

▶ Most states in the U.S. currently have either state or city-county regulatory systems
for medical waste because numerous federal laws and regulations mandate proper
management and control. The EPA, OSHA, the FDA, the U.S. Postal Service, the
NRC, NIOSH, and the Departments of Health and Human Services, Transportation, Commerce, Defense, and Agriculture, all have established regulations for
medical waste management and control.

QUESTIONS FOR REVIEW

1. What was the purpose and outcome of the
Medical Waste Tracking Act?
2. What are some examples of generators of regulated and nonregulated medical waste?

3. Write a paragraph that briefly identifies and
differentiates between the following terms:
a. medical waste
b. infectious waste

    c.   infectious agents

    d. other than infectious waste

    e. low-level radioactive and mixed waste

    f. small volume chemicals.

4. Develop a list of the ten different categories of medical waste (those that are potentially infectious and those without such potential).

5. Describe in a brief paragraph what can be done to minimize medical waste.

6. What are the basic components of a generic medical waste management program as recommended by the EPA?

7. Describe the management of medical waste in terms of treatment possibilities, storage, and disposal.

8. What are the basic provisions of OSHA's Bloodborne Pathogens Standard?

9. Describe what protective measures can be taken to prevent human contamination from medical wastes.

10. Describe the principal federal agencies that govern some form of the management of medical waste.

ACTIVITIES

1. Contact the health and safety officer within a hospital in your area and attempt to obtain a copy of their written medical waste management program.

2. Contact the health department or other office in your area that manages medical waste and obtain the names of contact persons to discover the current compliance mechanism. Explore the possibility of attending or sponsoring a hazard communications training session on OSHA's Bloodborne Pathogens Standard.

3. If your state has a TSDF that handles regulated medical waste, contact the public information officer or other health and safety officer at the facility. Make arrangements to visit the facility or to have that person make a presentation at your institution.

4. Research the current status of federal and state regulations regarding the management of medical waste.

10

Low-Level Radioactive and Mixed Wastes

Neal K. Ostler

Upon completion of this chapter, you will be able to meet the following objectives:

▶ Differentiate between the management of purely radioactive wastes and radioactive wastes that are mixed with other hazardous components.

▶ Outline the major federal acts that govern the management of radioactive wastes.

▶ Identify the basic provisions of the Atomic Energy Act.

▶ Describe the meaning of and provide examples of high-level radioactive wastes and low-level radioactive wastes (classes A, B, C, and greater-than-C).

▶ Outline the basic health effects of human exposure to radioactive wastes.

▶ Define two different principal means of minimizing low-level radioactive and mixed waste.

▶ Identify the common methods of treatment, storage, and disposal of low-level radioactive and mixed waste.

OVERVIEW

Exceptions to the Resource Conservation and Recovery Act's (RCRA's) definition of hazardous waste include wastes that are radioactive or have radioactive components as their primary hazard. Purely radioactive wastes, both high-level and low-level, are strictly managed under rules and regulations of the Department of Energy (DOE) and the Nuclear Regulatory Commission (NRC).

An issue of critical concern is the treatment and disposal of *mixed waste,* which comes about when low-level radioactive waste is mixed with other components that are "hazardous" according to RCRA definitions. Such waste may then come under management of both the EPA and the NRC. The range of controls and the responsibility for the management of nuclear waste is very complex, depending on whether the waste is high-level, low-level, or mixed. These controls are further complicated by whether the waste is generated from commercial industry or from Department of Defense (DOD) activities. Each major branch and subcategory of waste is managed by a different mechanism and system of controls.

This chapter will touch on all of these types of waste, but will focus primarily on the management of low-level radioactive waste and mixed waste.

FEDERAL LEGISLATION OVERVIEW

The Atomic Energy Act

The Atomic Energy Act of 1954 authorized the creation of the Atomic Energy Commission. This commission had full authority over all radioactive materials and other special nuclear materials until Congress separated the research and regulatory functions by passing the Energy Reorganization Act of 1974. This act abolished the Atomic Energy Commission and created two new ones: the Energy Research and Development Administration, later renamed Department of Energy, and the Nuclear Regulatory Commission.

When it initially passed the Atomic Energy Act, Congress declared that atomic energy is valuable for peaceful as well as national defense purposes. Therefore, its use and control should be directed towards making a maximum contribution to the general welfare of U.S. citizens (commercial uses), or towards promoting world peace, general welfare, and a competitive free market, and increasing the standard of living. Some basic provisions of the Atomic Energy Act included the following:

1. No person may transfer, receive, deliver, acquire, own, possess, import, or export any special nuclear materials.
2. Both military and nonmilitary applications of nuclear energy must be managed.
3. Atomic weapons must be managed in cooperation with the DOD.
4. Users of atomic energy for commercial, medical, and research purposes must be licensed.
5. Patents and inventions of a nuclear nature must be regulated.

As part of its enforcement mechanism, the Atomic Energy Act empowers the president to use the services of the FBI and the Department of Justice to protect against unlawful dissemination of information and to safeguard facilities, equipment, materials, and other property of the NRC and the DOE.

The Nuclear Waste Policy Act

The Nuclear Waste Policy Act of 1982 focuses on the management of high-level radioactive wastes, spent fuels, and transuranic waste. These wastes are regulated by either the NRC or the DOE, depending upon whether they are commercial or defense wastes.

High-level radioactive wastes include used nuclear fuel that is not to be re-processed and the waste generated by reprocessing of used nuclear fuel by means of chemical separation of uranium and plutonium from other elements. Most of this source is from reprocessing used fuel from weapons production reactors to obtain material for use in the fabrication of nuclear weapons. It is usually stored in liquid form, or it may be solidified or altered to a dry granular form before shipment.

High-level wastes pose a serious enough hazard that a barrier of concrete, water, or lead must be placed between the waste and any person for protection against external radiation. These wastes must be handled remotely or controlled by humans from behind a barrier or from a considerable distance.

Spent fuel in the United States consists mainly of fuel removed from commercial nuclear reactors. Such fuel is highly radioactive and generates much heat. It requires heavy shielding and remote handling.

Transuranic wastes come primarily from the reprocessing of spent fuels and the use of plutonium. These wastes emit less intense radiation (usually alpha emissions) and less heat than fission products, but they typically remain toxic for centuries and thus require isolation as high-level waste. Some transuranic wastes must be handled remotely and require protective shielding.

Byproduct materials are any materials that are contaminated or made radioactive during the production or use of special nuclear material.

Along with the substantial development of nuclear energy as a resource for domestic utilities comes the problem of what to do with wastes that these utilities generate. Congress adopted the Nuclear Waste Policy Act and numerous complementary acts to enable the NRC and the DOE to develop rules and regulations adequate to address this problem.

> NOTE: Section 1004(27) of RCRA specifically excludes from the definition of solid waste any byproduct, special nuclear, and "source" materials regulated under the Atomic Energy Act, which includes the Nuclear Waste Policy Act.

The Uranium Mill Tailings Radiation Control Act

The purpose of this act, passed in 1978, was to provide for the remediation of uranium mill tailings, which make up the largest volume of any category of radioactive waste, but for many years were the most neglected. Uranium mill tailings are what remains after the mining and extraction of uranium from ores. Mill tailings are earthen residues and are usually in the form of fine sand. These tailings are produced in large volumes and usually contain low concentrations of naturally occurring radioactive materials. Radon is the main radiological hazard from mill tailings.

Because such tailings do not contain a high enough concentration of radiation to be considered a "source" of radioactive waste, the Atomic Energy Commission insisted it had no jurisdiction over these wastes, so they were abandoned and unregulated, despite the fact that nearly all of the tailings were accumulated during the mining of uranium for sale to the government.

The regulations pertinent to the cleanup and disposal of uranium mill tailings fall under the Remedial Action Program, which authorizes the DOE to conduct cleanup activities in accordance with the EPA's SuperFund standards at designated uranium-mill-tailings sites. The purpose of this remediation is to stabilize the mill tailings and control the site so that it does not pose a hazard to human health or the environment.

It should also be mentioned that the NRC has jurisdiction at active sites where mining for peaceful uses creates mill tailings. The NRC is responsible for ensuring compliance with EPA standards for the protection of public health, safety, and the environment from radiological and nonradiological hazards.

The Low-Level Radioactive Waste Policy Act and Amendments

Today, NRC regulations guide the siting, operation, and closure of low-level-waste disposal facilities. But in the late 1970s, all low-level wastes were being shipped to just three states for disposal: Nevada, Washington, and South Carolina. To make all of the states accept the disposal burden, Congress passed the Low-Level Radioactive Waste Policy Act, making each state responsible for providing disposal capacity for the commercial low-level wastes generated within its borders.

This act defines *low-level radioactive waste* as essentially a catchall category that is actually best defined by saying what it is *not* rather than what it is. Low-level waste includes radioactive wastes other than the following:

▶ High-level waste
▶ Nuclear fuel
▶ Transuranic waste
▶ Uranium mill tailings

Most low-level wastes remain radioactive for only a short time and have low levels of radioactivity, although a few types present greater radiation hazards. These wastes are generated by a wide range of institutions and facilities using radioactive materials, including nuclear power plants, defense laboratories, industrial plants, and hospitals. The waste is found in a variety of forms, including medical treatment and research materials, contaminated wiping rags, used filters and sludge, hand tools and equipment, and personal protective clothing.

Some of the largest potential sources of low-level radioactive waste are parts and pieces of nuclear power plants that have either shut down or have been scheduled to shut down in the near future. These sources usually also contain large quantities of high-level radioactive wastes.

Low-level waste is classified by the NRC into four categories, according to the degree of hazard it poses and thus, the type of management and disposal it requires. The categories include class A waste, which is least hazardous; class B and C waste, which can be disposed of by shallow land burial; and greater-than-C waste, which is a fourth class for waste that is too hazardous to be disposed of in a near-surface facility. This waste must be disposed of in a geologic repository licensed for disposal of high-level waste, which requires more stringent disposal procedures than for class C waste. The federal government is responsible for disposal of all greater-than-C class wastes.

Mixed waste is usually treated as low-level and is waste that contains both hazardous chemical components as defined by RCRA and radioactive components subject to the Atomic Energy Act. Under the Atomic Energy Act, the NRC regulates the radioactive components of mixed waste from commercial sources, and the DOE regulates the portion of mixed low-level waste generated by weapons production facilities. The EPA is required by the Federal Facilities Act of 1992 to develop treatment requirements for mixed waste.

The Low-Level Radioactive Waste Policy Act was amended in 1985 to give states more time to comply with the law, but by 1992, no state or regional group of states had actually sited, licensed, and constructed a low-level waste disposal facility. Government's purpose with this act is to make each state responsible for the wastes it produces. However, it remains questionable whether each state will be able to do this without making a compact with other states for disposal of their wastes.

Although mixed low-level waste comprises less than 10 percent of all low-level waste, it has been identified by states as one of their major problems and concerns in managing low-level waste. In cooperation with the states, the EPA, DOE, and NRC are working together to streamline regulations for mixed waste.

Department of Transportation Regulations

The Department of Transportation (DOT) currently has responsibility for ensuring safety in the transportation of all hazardous materials, which includes radioactive materials and wastes. The NRC is responsible for regulating safety in receipt, possession, use, and transfer of radioactive materials. The NRC also reviews and approves/rejects packaging designs for containment of wastes with high concentrations of low-level radioactive materials.

Now that you have an understanding of the basic laws covering management of radioactive wastes, let's take a closer look at the primary concern of this chapter: low-level wastes.

THE BASICS OF LOW-LEVEL WASTES

As already discussed, the Low-Level Radioactive Waste Policy Act actually stipulated what low-level waste is *not* rather than what it is. To further help you understand low-level waste, this section discusses its characteristics and health effects, the variety of sources of low-level wastes, and how the NRC classifies commercial low-level waste along with the special category of mixed low-level waste.

Characteristics and Health Effects

The phenomenon of ionizing radiation has been present on this earth as long as there has been life. It is a form of energy generated by the activity of atoms, which are the basic building blocks of matter. All people are routinely exposed to a number of sources of ionizing radiation. The amount of exposure from all of these sources makes up a person's total average effective dose (TAED) of radiation. Such sources include the following:

1. **Natural sources.** These sources of radiation account for 82 percent of a person's exposure and include cosmic rays from the sun and stars; natural elements in the earth such as radium and potassium; radioactive elements in food, water, air, and building materials; and natural background levels in nature.
2. **Manmade sources.** These sources vary greatly from individual to individual and constitute about 15 percent of a person's total exposure. Examples include radiotherapy for treatment of disease and X-rays for medical and dental tests.
3. **Industrial uses.** These sources make up the remaining 3 percent of a person's exposure and include emissions from certain consumer products, previously conducted nuclear testing, nuclear power plant operations, and miscellaneous activities.

Low-level waste accounts for less than a third of the ionizing radiation coming from industrial sources, or less than a fraction of 1 percent of the TAED.

NOTE: Figure 10–1 shows sources of radiation, and Figure 10–2 shows their pathways through the *biota* (natural food chains and environment) to the human population.

In nature, some elements contain unstable atoms that spontaneously change into another form. Such element are said to be *radioactive* and the process of changing form is called *radioactive decay.* Atoms that spontaneously produce this radiation are known as *radionuclides* and within a certain period called a half-life (which can be from fractions of a second to billions of years), they will decay and give off half of their radiation. When an atom is in this process of change, it may lose or gain an electron and will release an excess of energy in the form of waves or very fast-moving particles. *Ionizing radiation* is the name given to the energy generated by the activity of these unstable atoms.

Three types of ionizing radiation result from the decay of radionuclides: alpha, beta, and gamma. Their radiation is measured in terms of *curies,* with 1 curie being the amount of radiation from 1 gram of radium in 1 second (37 billion/second).

Alpha particles have high energy and are positively charged, but lose their energy rapidly upon passing through matter. They are emitted from naturally occurring elements (radium and uranium) and from manmade elements such as plutonium. These particles are the largest and heaviest and, though they can be stopped by a sheet of paper or the human skin, if ingested or inhaled would damage internal organs and tissue with grave consequences (because of their half-life).

Beta particles are smaller and travel much faster than alpha particles. As a result, they can penetrate several layers of skin. Most low-level waste contains beta particles. These can be reduced or stopped through the use of a few millimeters of aluminum, glass, plastic, or other tightly woven fabric.

Gamma radiation is in wave form and consists of small packets of energy that can travel great distances and penetrate matter. These packets of energy are known as photons and can pass through the human body or be absorbed into bone and tissue. Three feet of concrete or 2 inches of lead will stop 90 percent of typical gamma radiation. Gamma-emitting radionuclides are found in most low-level waste.

▶ **FIGURE 10–1**
The percentage contribution of various radiation sources to the total average effective dose equivalent in the U.S. population. *Source: National Council on Radiation Protection and Measurements, Ionizing Radiation Exposures of the Population of the United States, NCRP Report 93, Bethesda, MD, 1987, p. 55.*

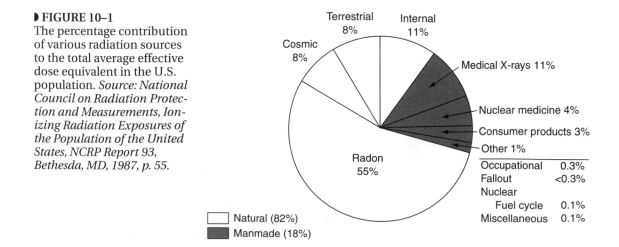

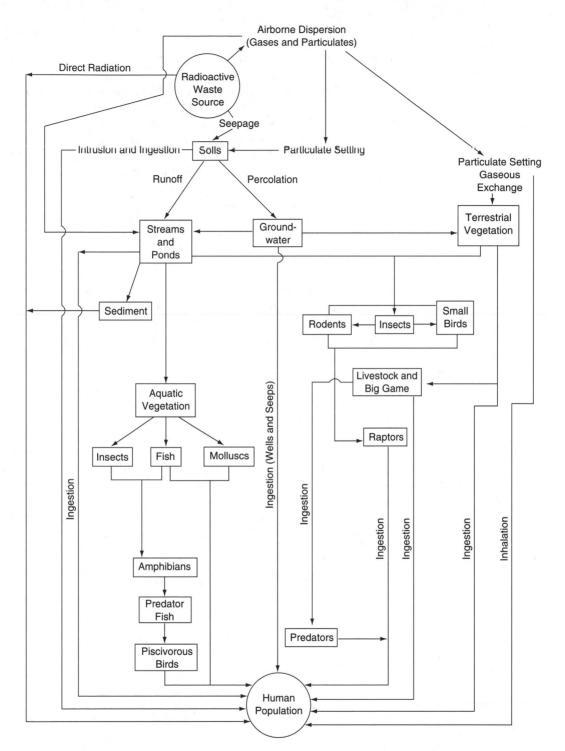

▶ FIGURE 10–2

Pathway analysis to biota and man: generation and disposal locations on common site. *Source: Robert E. Berlin and Catherine S. Stanton, Radioactive Waste Management (New York, NY: John Wiley & Sons, Inc., 1989), p.128. Reprinted by permission of John Wiley & Sons, Inc.*

145

Because different types of radiation produce different amounts of ionization, they are measured in terms of *exposure*, with the common unit for measuring exposure expressed as a *roentgen*. When radiation penetrates biological material, it gives up its energy in a series of collisions with the atoms of the material being irradiated and causes dislocation of atoms, the breaking of chemical bonds, or the loss of electrons. Whichever occurs, the result is alteration of the material (flesh, bone, etc.) and an impairment of its biological function. The amount of radiation actually deposited in the material is measured by the rad, which is the abbreviation for *radiation absorbed dose.*

NOTE: *Rad* describes something that happens to a particular
material when it absorbs (is irradiated by) the radiation;
the term *exposure* describes the condition of being
subject to the effects of radiation.

The amount of damage that is done to the biological organism, as the result of the particular absorbed dose (rad), is often referred to as the *dose* but more correctly is the *dose equivalent.* This dose depends on the kind, amount, and rate of the radiation; the nature of the organism exposed or irradiated; the organism's age, sex, state of health, and surroundings; and the particular biological effect being considered. A dose is measured in *rems* (roentgen equivalent man) and the international unit is the *sievert,* which is 100 rems.

The average annual whole body dose, from all sources, is about 0.36 rems (or 360 millirems) and an individual may accumulate between 5 and 10 rems over the course of an entire lifetime. (This total is broken down into percentages from different sources and was discussed previously in the section on TAED.)

A comprehensive study of the health effects of radiation is a book in itself. For the purpose of this chapter, you are reminded that an excessive dose of radiation can result in damage to the cells of the body that compose the tissues and organs *(somatic damage)* and damage that can become hereditary *(genetic damage).* If cell production is impaired by radiation, it can also result in cancer, with cancer of the blood (leukemia), lung, breast, and thyroid being the most common types. Table 10–1 provides additional information on the health effects from different levels of exposure to radiation.

NOTE: Because there is no scientific consensus to determine
a threshold limit of exposure, the NRC has developed a policy
that requires that exposure at all times be kept "as low as
reasonably achievable" (ALARA). The ALARA concept has
influenced the imposition of engineering controls,
improvement in validation instruments, and the evolution of
radiation protection programs throughout the DOE.

Sources of Low-Level Waste

Generators of commercial low-level waste can be grouped into three general categories:

1. **Nuclear power plants.** The wastes generated from these facilities include dry solids such as rags, paper, and clothing, as well as used equipment, sludges, and other materials that have become irradiated.
2. **Industries.** This group includes fuel fabrication facilities and research reactors, industrial plants using radioactive materials, and manufacturers of radioactive instruments and radiopharmaceuticals.

▶ TABLE 10–1

Acute health effects estimated from whole body irradiation.

Dose (rems)	Health effect
5–20	Possible late effect; possible chromosomal aberrations
25–100	Blood changes
>50	Temporary sterility in males (>100 rem–1 year duration)
100	Double the incidence of genetic defects that are normally expected
100–200	Malaise, vomiting, diarrhea, and tiredness in a few hours; reduction in infection resistance; possible bone growth retardation in children
200–300	Serious radiation sickness; bone marrow syndrome; hemorrhage; LD10–35/30*
>300	Permanent sterility in females
300–400	Resulting loss of blood defenses and vascular integrity; electrolyte imbalance; marrow/intestine destruction; LD50–70/30
400–1,000	Acute illness; early death; LD60–95/30
1,000–5,000	Acute illness; early death in days; intestinal syndrome; LD100/10
>5,000	Acute illness; early death in hours to days; central nervous system syndrome; LD100/2

*Lethal dose to percentage of the population in number of days (for example LD10–35/30 = lethal dose in 10 percent to 35 percent of the population in 30 days.

3. **Academic and medical institutions.** This group includes facilities such as hospitals, clinics, and other university and private sector research laboratories. Wastes generated from this group includes scintillation fluids, lab equipment, animal carcasses, research materials, and gaseous wastes.

Classes of Low-Level Waste

In light of the wide range of low-level radioactive materials, their half-lives, and the type of radiation they emit, the NRC has developed a four-tiered classification system for low-level waste.

Class A Waste

Class A wastes are the least radioactive, but must meet numerous minimum requirements for packaging to facilitate handling of the waste and to protect the health and safety of workers at disposal sites. Most of the volume of low-level waste is class A and is usually segregated from other low-level waste at a disposal site.

Class B Waste

Class B wastes have intermediate levels of radioactivity that are ten to forty times higher than class A wastes. Thus, the package requirements for class B wastes must also include an ability to be structurally stable for at least 300 years to prevent leakage. These high-stability packages are used for both B and C class wastes.

Class C Waste

The levels of radioactivity in class C wastes are generally ten to a hundred times higher than levels for class B wastes. Although packaging requirements are the same, some shielding and remote handling is required for C wastes, which must also be buried at minimum of 16 feet below the earth's surface or covered with a thick intrusion barrier such as a slab of concrete.

Greater-than-Class C Wastes

Although less radioactive than spent fuels, greater-than-class C waste is more radioactive than C waste. Under the Low-Level Radioactive Waste Policy Act, the DOE has responsibility for disposal of greater-than-class C waste. The DOE is currently investigating deep-geologic repository sites for such waste as well as for commercial spent fuels and defense-produced high-level radioactive wastes. At present, greater-than-class C waste is being stored on location at several hundred generation sites and is of serious concern because the total radioactivity of this waste is equivalent to all other commercial low-level waste generated and disposed of in NRC designed and specially selected sites.

Special Categories of Low-Level Wastes

Several categories of low-level wastes do not fall neatly into commercial low-level waste definitions, but can be considered special because of their composition, volume, or unique characteristics.

Mixed Low-Level Waste

Waste that contains both low-level radioactive and chemically hazardous constituents is regulated jointly by the NRC and the EPA. This waste is produced by a full range of generators, with the typical hazardous constituents including the following:

- Organic liquids such as scintillation fluids used in diagnostic testing, which typically contain toluene and xylene (other industries use chemicals such as chlorofluorocarbons, and acetone)
- Metallic lead, which becomes contaminated when used to store radioactive materials and may be in the form of foil, sheets, bricks, or containers for storage or shipping
- Cadmium wastes generated at nuclear power plants
- Chromate wastes generated by water circulation systems within nuclear power plants, used to inhibit corrosion
- Waste oils from pumps and other equipment used in radioactive areas (the EPA is studying such waste oil to consider if it should be an RCRA-listed waste)

Most commercially generated mixed-low-level waste that cannot be treated and disposed of as trash will have to remain in storage until mixed-low-level waste disposal facilities are developed by states or compacts.

NORM and NARM Wastes

Naturally occurring and accelerator-produced radioactive materials (NORMs and NARMs) are mainly composed of short-lived radionuclides and are often recycled in an accelerator for medical purposes. NARM is usually treated as low-level waste and most sources are from naturally occurring radiation materials (NORMs) such as radium-226, radium-228, and radon-222, and other materials that exist throughout the earth's crust.

Discrete NORM is a small volume of highly radioactive material such as that found in industrial gauges, radium needles, and resins used to remove radium from drinking water. *Diffuse NORM* has low concentrations of radioactive material, but a high volume of it, and was not regarded until just recently as posing much potential health threat. Diffuse NORM is found in metal tailings, coal ash, phosphate waste from fertilizer production, contaminated water from oil and gas drilling, and sludge from drinking water treatment. At present, the EPA is studying the NORM waste problem and considering regulating it under the Toxic Substance Control Act and RCRA, much like asbestos and lead are regulated. At the time of this writing, only one facility was accepting NORM wastes for disposal, the Envirocare facility located in Tooele, Utah.

Other Low-Level Waste

Certain special projects generate a special type of low-level waste. Two examples are the decontamination of Unit 2 at Three Mile Island and cleanup operations at no-longer operating commercial spent-fuel reprocessing plants.

MANAGEMENT OF LOW-LEVEL WASTE AND MIXED WASTE

The management of low-level radioactive waste, including mixed waste, consists of three main phases: minimization, treatment, and disposal.

Minimization Techniques

Waste minimization, for the purpose of this section, is defined as in-plant practices that reduce, avoid, or eliminate the generation of harmful waste to reduce risks to human health and the environment. In this context, then, waste minimization applies to pregeneration of waste.

The driving force behind efforts to reduce low-level wastes (efforts that have not been required of waste generators) has been the costs associated with new disposal regulations and disposal surcharges established in the Low-Level Radioactive Waste Policy Act. If unit disposal costs continue to decrease during the 1990s, as smaller yet more expensive disposal facilities come on line, the trend of minimizing low-level waste will likely continue. Meanwhile, the NRC also encourages waste minimization through a variety of technologies. The two principal means of waste minimization are material substitutions and good housekeeping practices.

Material Substitutions

The use of nonradioactive substitutes for many industrial research applications can be an effective method of reducing low-level waste. Scintillation vials are one of the largest contributors to the overall volume of mixed low-level waste generated in the United States. By substituting nonradioactive tracers, a generator produces only a hazardous waste and not a mixed low-level waste, nor even a low-level waste. In some cases, this may lead to equivalent or superior results.

It may be possible to substitute a nonhazardous liquid, called an "environmentally benign" or "biodegradable" scintillation liquid, and then the waste may not even be an RCRA-listed hazardous waste. It may also be possible to substitute both the nonhazardous liquid and the nonradioactive tracer. Thus, the result would be neither RCRA hazardous waste, nor mixed low-level radioactive waste, nor even a low-level radioactive waste.

Good Housekeeping Practices

Good quality process techniques can be utilized in the management of hazardous and radioactive materials. Examples of this include improving the scheduling of material use, reducing excess purchases, and coordinating purchases through a clearinghouse.

Segregation of radioactive and nonradioactive material can also lead to significant waste reductions. An example is covering lead shields with a plastic-like material called Herculon to ensure that the lead does not become contaminated. Yet another example of using good housekeeping practices would be in packaging methods, such as using plastic bags as inner liners of shipping containers so the container itself does not become a low-level waste in the event of a spill.

Treatment Techniques

Treatment, in contrast to waste minimization, is applied to post-generation of waste before it is disposed of. RCRA defines treatment as "any method, technique, or process, including neutralization, designed to change the physical, chemical, or biological character or composition of any hazardous waste so as to neutralize such waste or so as to render such waste nonhazardous, safer to transport, store, or dispose of; or amenable for recovery, amenable for storage, or reduced in volume." The NRC also expands this to include several other techniques, such as decontamination.

In some instances it is not possible to use material substitution or good housekeeping techniques to avoid generating a particular low-level waste or mixed waste. However, a few techniques can reduce the volume and/or toxicity of the waste after it has been generated.

Post-Generation Good Housekeeping

Substantial reductions in waste volume can be achieved by concentrating liquid wastes through evaporation or boiling off water to leave only suspended solids; ion-exchange removal of dissolved radionuclides by absorption; filtration of waste oils rendering the filter as low-level waste but releasing the oil to a recycler, and precipitation, centrifuging, and distillation to reduce the waste volume (but not eliminate it).

Neutralizing corrosives by raising or lowering their pH can allow them to be handled as purely radioactive wastes, rather than mixed wastes. Other steps in reducing waste volume include sorting nonradioactive wastes from radioactive ones as well as sorting them into different categories; market development for reducing, reusing, and recycling materials; and development of cleaning techniques to remove surface contaminants.

Storing for Decay

Short-lived radionuclides generated by medical and biomedical research institutions must be stored for time periods of ten to twenty half-lives to allow the radioactivity of the material to decline to below background levels. This waste can then be regulated without regard to its radioactivity. Attention must be paid to segregating radionuclides and to using shielded storage containers that are able to withstand degradation to assure quality control over long-term storage.

> NOTE: An RCRA storage permit must be granted for storage periods of ninety days or longer and, in most cases, because of the Land Ban, these materials can only be stored to accumulate sufficient quantities to facilitate proper recovery, treatment, or disposal of the waste. Remember, this only applies to mixed low-level waste because it is considered hazardous according to RCRA as well as radioactive.

Compaction and Shredding Techniques

Today's supercompactors are capable of producing 1,000 tons of force or greater and are thus capable of crushing containers in minutes (e.g., they can turn 55 gallon drums into hockey pucks). Stationary devices capable of 5,000 tons of force can even compact the so-called "noncompactable" wastes.

Shredders are other size-reduction devices that can reduce waste volumes by tearing, ripping, shattering, and/or crushing waste materials into smaller sizes and thus permit more materials to be placed in containers for storage or transportation. Shredders can also be used to prepare the waste before incineration to provide a more uniform feed into the incinerator.

Incineration

One of the most efficient ways of reducing waste volumes is by incineration. At the time of this writing, there was no commercial off-site incinerator for low-level radioactive waste and most permitted ones were used mainly to treat municipal solid waste. But the technique could be used to treat combustible liquid and solid, dry low-level radioactive waste.

Incineration can potentially reduce waste volumes by at least a ratio of 25:1, or about 96–97 percent, by converting mixed-low-level waste into carbon dioxide and radioactive ash.

The three major types of incinerators currently being used worldwide are

1. **Rotary kilns.** These type of incinerators operate by burning in a slowly rotating combustion chamber mounted at a slight incline to gradually move the ash toward a discharge point. The combustion chamber contains more oxygen than needed to fully oxidize the waste, but there is usually a secondary combustion chamber to increase the time that wastes are subjected to extremely high temperatures; this is called an afterburner. Often, liquid injection units are coupled with this design and the wastes may be mixed with highly flammable materials to assure their complete combustion.
2. **Controlled air.** Depending on the waste, this type of incinerator can operate under either a starved-air or excess-air condition. Waste is fed onto a platform at the bottom of a combustion chamber, often coupled with liquid injection attached to the side of the chamber. After burning in the primary chamber, the particles of incomplete combustion are fed into a secondary high-combustion oxygen-enriched chamber. Consequently, it burns at even higher temperatures.
3. **Fluidized-bed.** This system uses a layer of small particles suspended in an upward flowing stream of air (like a fluid) to help burn highly viscous liquid sludges that are not easily burned by other incinerators. The sludge hangs on to the sand or limestone particles, which are then lifted by the air stream and subjected to a flame. This system is also effective against acid gases.

All three systems also have disadvantages. They all produce a relatively large amount of ash and particulates, which must be trapped by an air-pollution control device (bag house, electrostatic precipitator, or HEPA filter, for example). Such air pollution technologies must be incorporated to control the emission of gases and radioactive particulates. The collected ash, trapped particulates, and air-pollution emissions sludges must all be classified as either low-level waste or mixed waste.

Waste Stabilization

This waste-reduction technique involves fixing the constituents of the waste in a solid form that is inert, has low solubility or release rates in water, and can be safely transported or stored. An example is stabilizing incinerator ash in a cement-based waste form that will fix the ash and retard the migration of the radioactive constituent. The Code of Federal Regulations (10 CFR Part 61. 28,30) lists the requirements for all classes of low-level waste. Essentially these requirements focus on packaging, the use of absorbents, the reactivity of the waste (it must not be pyrophoric, explosive, or capable of generating poisonous gases), its structural stability, and the proper disposal of the container.

Waste Packaging Materials

Different classes of radioactive materials have different levels of radioactivity and, therefore, require packages that can be made of different materials. These packaging materials must be designed to retain their physical and chemical integrity for at least 300 years. Containers can be made of a variety of materials, and often a combination is the most suitable. An example is an external skin made of stainless steel to resist corrosion and maintain integrity, combined with a liner of polyethylene, all of which is then packed with absorbent material to reduce or eliminate any void space and keep the waste material stable.

As with other waste forms, more research is needed on packaging and packing materials that may be appropriate for both low-level waste and mixed waste.

The future of waste minimization and treatment technologies is very optimistic as generators are pressured to maximize their use of these techniques. Their primary reasons for doing so will be economic, but once the EPA has fully enforced its RCRA regulations for all mixed-waste treatment and storage facilities, the owners and operators of these facilities will try to change their practices and will either not generate mixed waste or they will try to treat all of it so that it is either radioactive or entirely hazardous, but not both.

Disposal Technologies

Disposal, for the purpose of this section, is defined as the isolation of low-level waste and mixed waste during the time it poses an undue risk to humans and the environment. Since the longevity and toxicity varies with each waste (as the constituents vary), the required time and level of containment also varies.

Most low-level radioactive waste generated in the United States between 1950 and 1990 has been disposed of at the surface. Low-level waste generated from civilian and defense activities was buried in shallow, unlined trenches at sites owned and operated by the federal government during the 1940s through the 1960s. Some low-level wastes were packed into 55 gallon drums and dumped at sea, but this practice ended in 1970. Throughout this entire range of time, there was little recognition of the possible need for treatment and little concern for the presence of chemical contaminants mixed with the waste. Some low-level waste facilities even accepted these "mixed wastes."

Over the years, containers have collapsed or broken, and waste has contaminated surrounding soil. Some of these burial grounds were located in areas with high water tables, providing the means (groundwater) for radionuclides to leach and move toward surface waters or wells. In many of these instances, water infiltrated the shallow trenches and caused radioactive contaminants to migrate into the surrounding environment.

In the past, low-level waste disposal sites were selected using very little or inadequate earth-science criteria. Additionally, there were no uniform regulations for the design and operation of these facilities.

As discussed previously, in the 1960s, the government restricted its burial sites to federal use only, and private companies opened and operated disposal facilities. By the late 1970s, all low-level wastes were being shipped to just three states: Nevada, Washington, and South Carolina.

Pressured to potentially make all of the states accept the disposal burden, Congress passed the Low-Level Radioactive Waste Policy Act, making each state responsible for providing disposal capacity for the commercial low-level wastes generated within its borders. The act was amended in 1985 to give states more time to comply with the law, but by 1992, no state or regional group of states had actually sited, licensed, and constructed a low-level waste disposal facility.

By the end of 1992, the site at Beatty, Nevada, was closing and the site at Richland, Washington, was accepting wastes from only a limited number of states. At the beginning of 1993, Barnwell, South Carolina, was the only destination for most of the country's low-level radioactive wastes.

Containment Periods

The NRC has established requirements for the time of containment for each classification of low-level waste and mixed waste, as well as high-level waste. These containment periods generally involve burial of the waste beneath the earth's surface. The following is a list of these containment periods:

Class A	100 years
Class B	200 to 300 years
Class C	up to 500 years
Greater-than-class C	longer than 3,000 years
High-level radioactive waste	longer than 3,000 years

Standards for Siting Repositories

The current selection process of an appropriate site for a waste disposal facility is stipulated by NRC regulations found in 10 CFR Part 61.50, summarized in the following list:

▶ Primary emphasis is on the suitability of the site for total isolation of the waste.
▶ The site must be capable of being fully modeled, analyzed, characterized, and monitored.
▶ Projected population growth and urban sprawl must not affect the project.
▶ Areas of known natural resources must be avoided if their exploitation would damage site performance.
▶ The site must be free of ponding, totally drained, away from wetlands or coastal high-hazard areas, and be above the 100-year flood plain.
▶ Drainage areas upstream from the site must be minimized to decrease water runoff to the site.
▶ The groundwater level must be well below the level of the depth of storage packages.
▶ Groundwater cannot be discharged to the surface or accessed by the site.
▶ Areas vulnerable to seismic or tectonic processes must be avoided.
▶ Areas subject to active geologic processes such as slumping or erosion must be avoided.

◗ The location cannot be near any facility or activity that could potentially damage the performance of the site.
◗ The site must be stable so as to support engineered containment structures.
◗ Areas of highly vulnerable hydrology must be given special attention.

Facility Design for Radioactive Waste Disposal

Although there is no single disposal technology that can be unequivocally judged as best for all situations, four types of facilities, based on their location relative to the earth's surface, could probably provide acceptable levels of waste isolation if they were properly implemented. A number of studies done by the government have compared each of the these types:

1. **Above-ground disposal in concrete vaults.** This method involves building a concrete structure with reinforced walls up to 3 feet thick. Waste would be isolated as long as the building maintained its integrity (Figure 10–3).
2. **Intermediate-depth disposal in augured holes.** This technology involves boring 8- to 10-foot diameter holes as a depth of 20 to 50 feet. The hole is then lined with about 1 foot of cement grout, and a concrete filler and lid is poured over the waste with the remaining hole backfilled with soil (Figure 10–4).
3. **Deep disposal in geologic repositories.** These repositories are located a few hundred to a few thousand feet below the earth's surface, in formations providing natural barriers to migration of radioactive nuclides from groundwater.

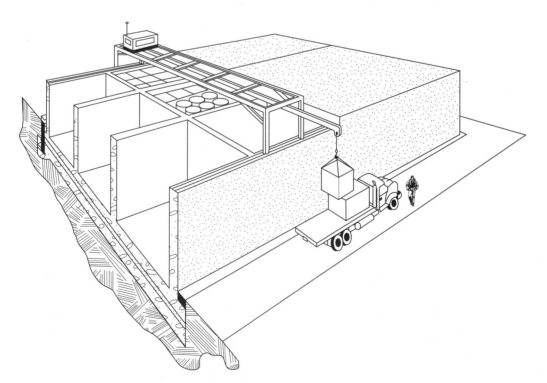

◗ **FIGURE 10–3**
Above-ground disposal in concrete vaults.

This is a technology preferred by the scientific community for disposal of high-level and transuranic wastes, but it is not practical for low-level waste. The DOE's massive project at Yucca Mountain in southern Nevada is an example of deep disposal in a geologic repository. At Yucca Mountain, high-level radioactive waste will be stored under many hundreds of feet of solid rock. Scientists believe this site may protect the waste from distrubance for up to 10,000 years.

4. **Near-surface underground disposal facilities.** This is the most commonly implemented technology in the United States to date. If sited away from surface water and in areas where the water table is substantially below the surface, it can provide good isolation measures. Figures 10–5 and 10–6 illustrate the construction and configuration details of near-surface underground storage vaults.

Other Generic Technologies

In addition to criteria for the selection of repository sites and optimal facility design, additional generic technologies can be used for disposal of radioactive wastes.

1. **Form of the waste.** Water infiltrating into trenches can be minimized by compacting the volume of all low-level waste and mixed waste to the maximum extent practical prior to disposal.

▶ **FIGURE 10–4**
Intermediate-depth disposal in augered holes.

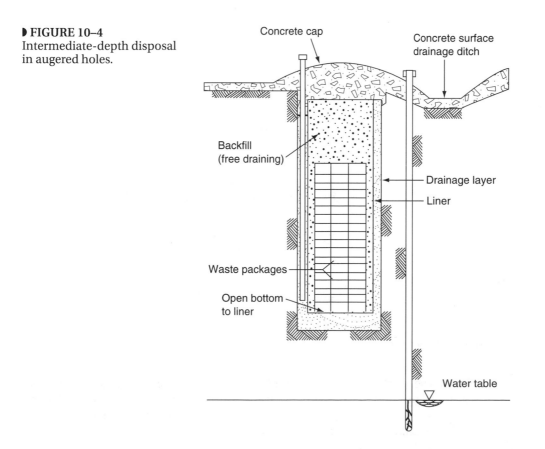

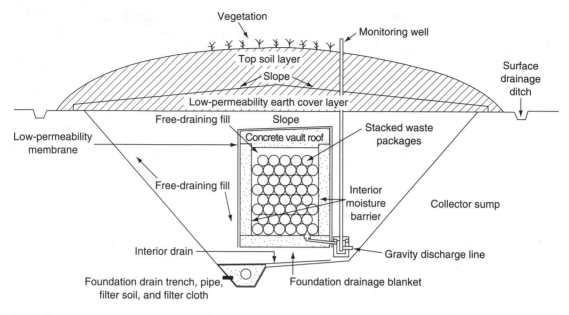

▶ **FIGURE 10–5**
Below-ground vault cross-section.

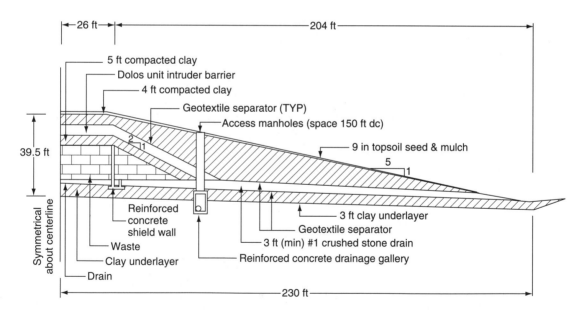

▶ **FIGURE 10–6**
Above-grade tumulus cross-section.

2. **Packaging.** High-integrity containers and concrete overpack containing waste canisters can provide added structural stability, water resistance, and shielding for the waste (see Figure 10–7). Packages must be properly marked and labeled for identification as illustrated in Figure 10–8.

▶ **FIGURE 10–7**
Concrete overpack for low-level radioactive waste disposal.

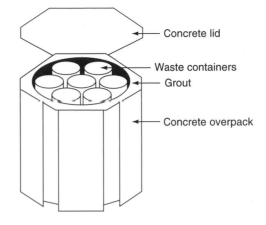

Concrete lid

Waste containers

Grout

Concrete overpack

3. **Optimal materials.** Concrete containment vaults add structural stability to a disposal unit and help prevent water from infiltrating. Space between packages can also be filled with sand, soil, gravel or even cement grouting. The outside and insides of vaults, caps, and packages can be lined with resins, membranes, and panels of materials that would increase their resistance to water.

4. **Cap/lid design.** The design and construction of the cap is critical in diverting precipitation away from the disposal units and also is the easiest and cheapest

▶ **FIGURE 10–8**
Radioactive waste label.

RADIOACTIVE WASTE

HANDLE WITH CARE!

Generator_____ Phone #_____

Date_____

Contents _____

_____

_____

Sample analysis #_____

Disposal requisition #_____

HWH Received Date_____

Contact Reading

_____ mR/hr.

☐ β γ   ☐ γ   ☐ NEUT.

part of the containment structure to repair. These caps consist of multiple soil layers and impermeable synthetic membranes (see Figure 10–9). For class A and B wastes, the total cap must be 3 to 6 feet in thickness; for class C wastes, it must be at least 16 feet thick.

5. **Monitoring systems.** Past problems highlight the need for lifetime monitoring of movement of precipitation at disposal facilities. These monitors in sumps or wells can trigger pumps to remove the water, but they should only be a backup to the basic structure design allowing passive drainage.

Because of differences in site characteristics, annual precipitation, and time-of-travel of groundwater, there is no one design for a disposal facility that is optimum for all regions of the country. Both below-grade and above-grade designs have advantages and disadvantages when used in parts of the country with either high or low annual precipitation.

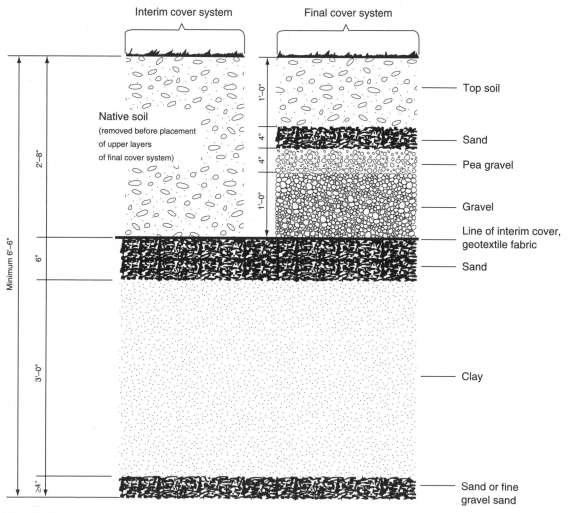

▶ **FIGURE 10–9**
Typical multilayered cap.

Facility Design for Mixed-Waste Disposal

Through 40 CFR Part 264, the EPA regulates the disposal of hazardous waste in landfills and the agency's goal is to totally contain hazardous wastes. To do this, the bottom and sides of facilities are lined with layers of clay and double layers of synthetic materials, thus forming a sort of double-lined bathtub. Collection systems for leachates are located between the double liners to prevent escape of leaking contaminants.

The EPA and the NRC have developed joint guidelines for siting and facility design of mixed-low-level waste disposal facilities. They propose an above-grade design with a multilayered cap to form an umbrella over the waste, as well as a bathtub beneath it. Infiltrating water is channeled via gravity to collection basins for monitoring, treatment, and off-site discharge. These facilities may or may not have concrete vaults, and monitoring systems and other institutional controls must be in place for 30 years and possibly even 100 years after closure.

One area of improvement in facility design, craftsmanship, and building materials that should have an immediate payoff is the engineering of the cap. Better combinations of soil layers and synthetic membranes, improved monitoring systems, and the integrity of drainage systems are a few areas holding the greatest potential for improving the performance of past and future near-surface disposal facilities.

SUMMARY

- Federal legislation regarding radioactive waste is made up largely of the Atomic Energy Act, which created the Department of Energy and the Nuclear Regulatory Commission; the Nuclear Waste Policy Act, for the management of high-level radioactive wastes; the Low-Level Radioactive Waste Management Act; the Uranium Mill Tailings Radiation Control Act; and Department of Transportation regulations, which govern the transport of hazardous materials. RCRA specifically excludes purely radioactive wastes from its definition of hazardous waste.
- High-level radioactive wastes include spent nuclear fuel, the waste generated by reprocessing of spent nuclear fuel and plutonium (called transuranic wastes), and byproduct materials, those contaminated or made radioactive during production of nuclear material.
- Uranium mill tailings are what remains after the mining and extraction of uranium from ores. They are regulated separately from high-level and low-level radioactive waste.
- Low-level radioactive waste is defined by what it is not: that is, it is any radioactive waste other than high-level radioactive waste, spent nuclear fuel, transuranic waste, and uranium mill tailings.
- Mixed waste is waste that contains both low-level radioactive materials that are subject to Atomic Energy Act regulations and hazardous components that fall under RCRA's regulatory umbrella.
- Radiological hazards comprise alpha particles, beta particles, and gamma waves. Radiation comes from natural sources such as uranium; from manmade sources such as X-rays; and from industrial uses. Different amounts of radiation are measured in terms of exposure, with the common unit expressed as a roentgen. The radiation absorbed dose (rad) is a measurement of the actual amount of radiation deposited in the material, and the amount of damage resulting from the rad is

called the dose equivalent. A dose is measured in rems (roentgen equivalent man) and also by the sievert, which is 100 rems.

▶ The main sources of low-level radioactive wastes are nuclear power plants, industries, and academic institutions.

▶ The NRC has classified low-level waste in four tiers: class A, class B, class C, and greater-than-class C wastes.

▶ Management of low-level waste and mixed waste is accomplished through the waste minimization techniques of materials distribution and good housekeeping practices; and the treatment techniques of good housekeeping, storing for decay, compaction and shredding, incineration, waste stabilization, and waste packaging materials.

▶ Disposal technologies for radioactive wastes include containment periods, standards for siting of repositories, facility design and operating guidelines, and other generic technologies such as waste forms, optimal package designs, optimal use of material ingredients, cap design, and monitoring systems.

▶ Optimal facility design for mixed-low-level waste disposal includes the use of both a double-lined bathtub positioned under the waste (EPA) and a well-designed cap (DOE).

▶ There is both a need and technology available for optimal management of low-level radioactive waste, and mixed waste.

QUESTIONS FOR REVIEW

1. What are the basic provisions of the Atomic Energy Act?

2. What agencies are the principal players in the management of low-level radioactive and mixed waste and by what legislation are they empowered or supported?

3. What are some examples of high-level radioactive wastes and what are the primary sources of their generation?

4. What are some examples of low-level radioactive waste and what are the primary sources of their generation? Include in your answer the descriptive differences between class A, class B, class C, and greater-than -class C waste.

5. What does the acronym TAED stand for and from what sources are humans routinely exposed to radiation each day?

6. Into what three general categories can the generators of commercial low-level waste be grouped?

7. What are the two principal means of waste minimization used by the generators of low-level radioactive and mixed waste?

8. What treatment techniques can used by generators in reducing the volume of their wastes?

9. What are the principal issues considered by the NRC in locating a disposal facility?

10. What are the four types of disposal facility technologies, described in the text, that can probably provide acceptable levels of radiological waste isolation?

11. Where in the Code of Federal Regulations does the EPA regulate the design and criteria for the site locating of disposal facilities for mixed waste?

ACTIVITIES

1. Using the Yellow Pages, compile an alphabetical list of companies in your area that are potential generators of low-level radioactive and/or mixed waste.

2. Contact the office in your area that manages low-level radioactive and mixed waste, obtain the names of contact persons, and find out the current compliance mechanism. This is likely

to be a state environmental or health agency with a division, bureau, or office of radiation control or safety. Explore the possibility of attending or sponsoring a Radiation Worker I training session.

3. If your state has a treatment, storage, and disposal facility for low-level radioactive and mixed waste, make contact with the public information officer or other health and safety officer. Make arrangements to visit the facility or have that person make a presentation at your institution.

11

Phase I and II Environmental Site Assessments

Michael A. Williams

Upon completion of this chapter, you will be able to meet the following objectives:

▶ Understand the legal conditions giving rise to the environmental site assessment process.

▶ Define the primary site assessment tools.

▶ Understand the importance and relevance of environmental site assessments.

▶ Describe the components of the site assessment tools.

▶ Gain a general knowledge of the results, implications, and recommendations from the site assessment process.

This chapter is not intended to provide guidance on how to conduct an environmental site assessment; rather, it describes the environmental site assessment's purpose and familiarizes you with the components, procedures, and ramifications of each level of investigation.

OVERVIEW

The potentially devastating effects of environmental contamination on the actual value of real estate, coupled with potential liabilities under state and federal environmental law, are major considerations for lending institutions making loans secured by real property. The *environmental site assessment* (ESA) has evolved as the primary tool used during the property transfer process to evaluate the potential for soil and groundwater to be contaminated with hazardous materials or petroleum products.

ENVIRONMENTAL SITE ASSESSMENT PROCESSES

The primary tools that are most commonly employed to evaluate potential environmental risk and compliance status of a property encompass the transaction screen; Phase I, II, and III environmental site assessments (ESAs); and the environmental audit. This chapter focuses on an explanation of the purpose, procedures, and components of Phase I and II ESAs. Only minor consideration regarding the Phase III ESA and environmental audit is given to familiarize you with the subsequent actions that may occur as a facility tries to maintain compliance with an assortment of environmental regulations.

The *transaction screen* process represents the least complex level of inquiry that may be applied to certain properties to satisfy the needs of lending institutions and prospective buyers and sellers. The transaction screen (essentially defined by the American Society for Testing of Materials [ASTM] as Standard E1528-93) is intended to rapidly evaluate the potential for hazardous materials or petroleum products to have impacted an undeveloped site, residential parcel, or low-risk commercial property. The transaction screen standard practice includes guidance for the minimum scope of work and a questionnaire that may help to insulate the purchaser or lender from the potential liability associated with the cleanup of a preexisting environmental problem.

The *Phase I ESA* represents a significantly more rigorous level of inquiry than the transaction screen. The Phase I ESA commonly is performed on sites that have served as commercial or industrial facilities. The ASTM has published a standard (E1527-94) defining the minimum scope of work to satisfy the needs of interested parties. Other organizations have prepared guidelines and recommended procedures for the Phase I ESA; however, they have generally been supplanted by the ASTM standard. Other Phase I ESA guidelines may provide additional guidance in evaluating specific situations because ASTM acknowledges that E1527-94 is intended to outline only the minimum scope of work.

The *Phase II ESA* is conducted when a Phase I ESA identifies certain environmental concerns. The purpose of the Phase II investigation is to confirm or deny the presence of hazardous materials, waste, or petroleum product contamination. The Phase II investigation is characterized as an intrusive study involving samples of materials stored on-site, as well as the collection of air, soil, groundwater and/or building materials which are then subjected to laboratory analysis for quantification of the presence of regulated substances. The scope of the Phase II investigation is generally targeted to evaluate the specific areas of concern identified in the Phase I evaluation. ASTM is currently preparing another standard practice tailored to provide a framework for conducting an intrusive evaluation.

A *Phase III ESA* is conducted by a responsible party (or parties) when contamination in excess of regulatory action levels is encountered at a site. The primary objectives of the Phase III ESA are to delineate the vertical and horizontal extent of the contamination, evaluate risk to human health and the environment, and propose remedial actions that may be appropriate to clean up the site or sufficiently reduce the level of risk to the satisfaction of the regulatory agency.

The *environmental audit (EA)* generally is performed by companies that want to proactively pursue compliance with an evolving set of environmental regulations. It focuses on evaluating the facilities' compliance with federal, state, or local regulations. Compliance with the federal Clean Air Act, Clean Water Act, Resource Conservation and Recovery Act (RCRA) and state environmental legislation are a few examples of the scope of an EA investigation. Its results are intended to serve as a self-regulating guide and may lead to significantly reduced penalties or the cost of compliance if the conditions revealed during the audit were not detected internally.

TYPES OF ENVIRONMENTAL PROFESSIONALS

Most states do not have a specific certification program that qualifies an individual to conduct ESAs; however, several states require that the environmental professional have a bachelor's degree in a field related to the natural or physical sciences or engineering. ASTM has defined an *environmental professional* as a person with the training and ability to satisfactorily complete the components of the investigation, make appropriate conclusions, and recognize potential issues of environmental concern (later referred to as *recognized environmental conditions*). As long as the individual possesses the necessary skills, the environmental professional may be an employee of the property owner or an outside contractor, such as an environmental engineering consulting firm.

Geologists, toxicologists, industrial hygienists, and engineers are commonly retained to complete ESAs for a particular property of interest. Those individuals with a professional registration such as a Professional Geologist, Certified Industrial Hygienist, or Professional Engineer are generally regarded as being highly qualified to supervise the completion of an ESA. To reduce costs, junior personnel are commonly charged with specific tasks and reporting requirements under the direct supervision of the registered professional.

THE PURPOSE OF ENVIRONMENTAL SITE ASSESSMENTS: A LEGAL OVERVIEW

The primary reason for preparing an ESA is to provide the purchaser with a defense for avoiding the tremendous liability and cost associated with the cleanup of sites contaminated with hazardous substances or petroleum products. The ESA was developed after passage of the Comprehensive Environmental Response Compensation Liability Act (CERCLA) of 1980, also known as SuperFund. CERCLA's main purpose is to mitigate and clean up abandoned sites on which there has been a release or threatened release of hazardous materials that may impact human health and the environment. Because the intent of CERCLA is to fund the cleanup of abandoned contaminated sites through cost recovery from a responsible party (or parties), rather than from the taxpayer, the act provides the EPA with guidance on assigning such financial responsibility.

If there is an actual or threatened release of hazardous materials to the environment from a facility, the EPA may evaluate a site through the *Hazard Ranking System* to gauge the degree of endangerment to human health and the environment. If the site scores high enough or there is an imminent and substantial risk to the environment, the EPA may proceed with further characterization and remediation (cleanup) of the site. It then may attempt to recover costs from potentially responsible parties

(PRPs) after the potential danger has been mitigated. The EPA can sue to recover the costs for the initial response, site characterization, source removal, and remedial actions as long as such actions are consistent with the National Contingency Plan. The EPA also will pursue recovery of attorneys' fees, interest, natural resource degradation fees, costs for risk assessment and studies of health effects, and other third-party costs or losses.

Definitions of CERCLA Liability

The following definitions provide a basic understanding of CERCLA liability and the implications for a potentially responsible party (PRP) (Brown, 1994):

Strict liability. An entity that uses hazardous materials is strictly liable (without any regard for fault) for the harm those compounds may pose to human health or the environment. For example, an innocent party may be held liable for the cleanup of a release site simply because his or her hazardous waste was encountered at the contaminated facility.

Joint liability. A single entity can be held liable for the entire cost of the cleanup. The EPA commonly does not attempt to name all PRPs because on identifying a single party, that entity will seek others to distribute the burden of the cleanup costs. The potential expenditure for the resulting legal battle can exceed a party's portion of financial liability for cleanup; therefore, subsequent PRPs are commonly advised by legal counsel to settle without going to court.

Several liability. The degree or extent of environmental liability is not proportional to the volume or toxicity of the material that the PRP delivered to the facility. When more than one PRP is named in a suit, the courts have ruled that the liability may be divisible by the number of parties or the relative contribution of each entity.

Retroactive liability. Parties can be held liable for contamination resulting from acts that were committed prior to the passage of CERCLA.

Prospective liability. The liability for a PRP extends to future damages from ongoing releases from the facility or contaminated media generated at the facility.

CERCLA Defenses

CERCLA provided the following statutory defenses (ASTM, 1993) that a defendant being named as a PRP could use. The release or threatened release and the resulting damages must be attributed solely to the following:

1. An act of God
2. An act of war
3. An act or omission of a third party
4. Any combination of the above

The third-party defense was the strategy most likely to be relied upon to deflect the potential liability for a contaminated site. Under CERCLA, the third-party defense was available only to entities who could establish the following (Prescott, 1990):

1. No direct or indirect contractual relationship existed between the defendant and the party that caused the release.
2. The defendant exercised due care to appropriately manage the hazardous materials involved in the release.
3. The defendant took appropriate precautions to protect himself or herself from foreseeable acts or omissions of the third party responsible for the release.

The difficulty in establishing the third-party defense under CERCLA is that the EPA (and the courts) generally interpret the deed of a property to represent a contractual relationship between the entity responsible for the release and subsequent holders of the title to a contaminated property. Many legal battles have resulted in which entities listed in a chain of title subsequent to the party responsible for a release were named as PRPs. For example, in the case New York v. Shore Realty Corporation (759 F.2d 1032 [2d Cir. 1985]), the court held Shore Realty liable for the cleanup of a property even though it was not directly responsible for the release of hazardous waste at the site. The third-party defense was not honored because Shore Realty owned the property at the time the release was discovered and failed to take precautions to properly handle hazardous wastes before or upon property acquisition after learning that the previous owner was involved with the improper disposal of hazardous waste on the property.

SARA's Protective Measures

Congress soon recognized that CERCLA needed a provision to protect innocent people or entities who purchased contaminated property but were not liable for the contamination that was the result of site activities by previous owners. An important component of the 1986 SuperFund Amendments and Reauthorization Act (SARA) further defined the conditions necessary to establish a third-party defense to CERCLA liability. SARA confirmed CERCLA's stance that a deed represents a contractual relationship that may invoke liability to subsequent owners of a release site. However, SARA carefully defined the conditions that could be used by a third party to avoid CERCLA liability. If the purchaser of a property exercised due diligence prior to acquiring a property and, therefore, did not know or have reason to know of the presence of hazardous chemicals on the property, the *innocent landowner defense* is available. SARA states that to establish the innocent landowner defense, the defendant must "have undertaken, at the time of acquisition, all appropriate inquiry into the previous ownership and uses of the property consistent with good commercial or customary practice in an effort to minimize liability" (42 U.S.C. Sec. 9601[1994]). The court also may consider any specialized knowledge or experience of the defendant, the purchase price relative to the value of uncontaminated property, the availability of pertinent site information, the observability of the evidence of contamination, and the ability to detect the contamination as a defendant tries to establish the innocent landowner defense.

RCRA Requirements

Passed in 1976, RCRA (42 U.S.C. Sec. 1601) imposes strict requirements that apply to the generation, transportation, use, storage, and disposal of hazardous materials and petroleum products. The purpose of RCRA is to impose specific liability and require pervasive management of hazardous waste from the point of generation through the ultimate disposal facility (cradle-to-grave documentation). Closure of an RCRA treatment, storage, or disposal facility may require corrective actions to remediate the site and the surrounding properties. A lending institution may be held liable for an RCRA closure and the subsequent cleanup if it forecloses on a property before the property owner secures formal site closure as required under RCRA statutes.

State legislatures have fashioned their respective environmental regulations to include the principles of liability formulated under CERCLA, SARA, and RCRA. Therefore, state courts have held that the federal statutes can be applied to establishing responsible parties for other environmental cleanup projects (for example, state SuperFund, RCRA, or petroleum underground storage tank [UST] sites). Since the SuperFund's implementation, lenders have generally applied environmental policies based on CERCLA liability. After SARA's enactment, the criteria necessary to successfully establish the innocent landowner defense have naturally formed the framework for various ESA models.

POTENTIALLY RESPONSIBLE PARTIES

This section details who may be held financially liable for an environmental cleanup under CERCLA or other regulations. As a consequence of their potential liability, these entities create the market demand for the various ESA products.

As specified in CERCLA, the list of PRPs from whom the EPA may recover costs for the cleanup of contaminated sites includes the following:

1. The owner or operator of a facility on which there is a threatened or has been an actual release of regulated substances (hazardous material or petroleum products) to the environment.
2. An entity who at the time of a release owned or operated the facility where regulated substances were improperly disposed.
3. Any person or third party who made arrangements for the improper disposal or treatment of hazardous substances or petroleum products.
4. Any transporter of regulated substances who selected the disposal or treatment site.

In U.S. v. Fleet Factors (19 E.L.R. 20529 [S.D. Ga. 1988]), the financial institution became the owner/operator of a contaminated property and the associated equipment through foreclosure. While preparing some of the on-site equipment for sale, Fleet Factors' contractor relocated barrels that may have contributed to the release of hazardous materials. Because Fleet Factors was regarded as an owner who contributed to the release, the courts held it liable for cleanup costs. In U.S. v. Mirabile (15 E.L.R. 20992 [E.D. Pa. 1985]), a lender was named as a PRP because of the bank's pervasive involvement in financial aspects of the bankrupt company's operations. In that case, the lender was regarded as actively participating in the management of the facility and was essentially acting as an owner of the property.

The following parties are excluded from liability for the cleanup of a contaminated facility (Prescott, 1990) under CERCLA:

1. Financial institutions that simply hold indicia of ownership (security interest) are excluded from liability and satisfy the lender liability rule.

2. An innocent landowner who, at the time of purchase, made all appropriate inquiry into the past and present use of the property and found no evidence of contamination or a threatened release. The innocent landowner must establish that he or she did not contribute to the contamination, handled hazardous materials appropriately, and took appropriate precautions to guard against foreseeable acts or omissions from third parties.
3. State and local government agencies that involuntarily acquire ownership of a contaminated property through abandonment, tax delinquency, or bankruptcy. The agency must demonstrate that it took sufficient care after acquisition to appropriately manage hazardous materials and guard against the foreseeable acts of others.
4. A person who acquires a contaminated property through inheritance or bequest, as long as the recipient takes sufficient care after acquisition to appropriately manage hazardous materials and guard against the foreseeable acts of others.

The formidable liability associated with being named a PRP in an environmental cleanup clearly has established the need for the purchaser, lender, and property owner to satisfy the due diligence clause of the innocent landowner defense. The Phase I ESA has evolved as the primary tool for conducting all appropriate inquiry prior to acquiring a property. The transaction screen also satisfies the requirements of due diligence for certain low-risk properties.

TRANSACTION SCREEN

As discussed previously, the ASTM developed the transaction screen to establish a basis to gauge good commercial and customary practice and constitute all appropriate inquiry to be used as a component of the innocent landowner defense. The transaction screen is intended for use on commercial properties as is the Phase I ESA; however, the transaction screen allows only minimal flexibility to discount recognized environmental concerns. Another substantial difference between the transaction screen and the Phase I ESA is that the transaction screen process does not need to be completed by an environmental professional. Employees, agents, owners, or independent contractors may prepare it. Of course, an environmental professional could be relied upon to complete the transaction screen if the user desires.

Components of the Transaction Screen

The purpose of this section is to provide the reader with a description of the scope of work involved with the completion of a transaction screen. The primary components of the transaction screen process can be divided into three general activities:

1. Interviews are conducted with persons familiar with the property to gain a historical perspective of site use and facilities.
2. Governmental databases are searched to determine whether the subject site or other facilities within a certain radius of it have registered USTs, landfills, or other regulated activities.
3. The site visit is conducted to document the presence or absence of various observable environmental concerns.

The interviews required for the transaction screen involve the owner, occupant, and occasionally a representative of the local fire department. A major component of the process is the completion of a brief but succinct questionnaire; a yes, no, or unknown response is required for each question. A summary of the environmental concerns specifically addressed in the questionnaire follow:

1. Past or present industrial activities of the subject site and the adjacent properties.
2. Past or present commercial use (for gasoline station, service garage, printer, dry cleaner, photo developer, etc.) on the site or the adjacent properties.
3. Past chemical storage and waste disposal practices on the subject site.
4. Presence of fill dirt from an unknown source on the site.
5. Current or previous existence of stained soils, treatment or disposal pits, ponds, or lagoons.
6. Current or previous existence of USTs, product piping systems, and electrical or hydraulic oils containing polychlorinated biphenyls (PCBs).
7. Current or past drains, walls, or floors that are stained or exhibit unusual odors.
8. Notification of contaminated well water if the site is served by a private well.
9. Knowledge of environmental liens or lawsuits historically placed on the property, hazardous substance or petroleum product storage, or any previous environmental site assessments of the subject site.
10. Current and past waste water discharge practices.
11. Historical disposal, burial, or burning of regulated wastes.

The ASTM transaction screen standard practice document provides additional explanation, photographs, and specific guidance to assist in completing the questionnaire. The standard's guidance and interpretations provide a minimum level of quality control so that a person who would not ordinarily be considered an environmental professional has sufficient guidance to accurately collect and record the necessary data.

Reasonably available government records must be reviewed during the transaction screen to identify potential environmental concerns surrounding the property being evaluated. At a minimum, the databases listed in Table 11–1 should be reviewed for sites within the specified radius (ASTM, 1993).

A written request to obtain the required information may be submitted to specific government agencies under the Freedom of Information Act. Generally, a minimum

▶ TABLE 11–1
Required transaction screen database and minimum search radius inventory.

Governmental Database	Radius (miles)
SuperFund (National Priorities List) sites	1.0
Sites investigated under CERCLA (CERCLIS)	0.5
RCRA treatment, storage, or disposal (TSD) sites	1.0
State SuperFund sites	1.0
Sites investigated under the state CERCLA	0.5
State leaking UST (LUST) sites	0.5
State solid waste disposal or landfill sites	0.5

of six to eight weeks is required for an agency to respond directly to such a request. Commercial services also are available to expedite the process. Various environmental search companies subscribe to an assortment of federal and state databases and can typically provide all available data within one to three working days. Most commercial services also provide a detailed map showing the location of the subject property, major cultural or natural features, the specified search radius, and an indexed location map of sites within the prescribed radius.

Another component of the environmental records search is the review of available fire insurance maps. Detailed maps showing various features of concern have periodically been prepared for insurance companies since the 1800s. Most major cities have extensive fire insurance map coverage that documents property use, flammable liquid storage areas, building construction, and utility locations through time. The maximum area of concern for the fire insurance map review is the subject site and adjacent properties within a radius of 660 feet. Public libraries and historical societies have commonly retained fire insurance map collections for their respective cities. Providers of commercial environmental database searches also have access to the vast collections of maps of fire insurance companies.

Other facilities are commonly identified within the specified search radius that may or may not represent a significant environmental concern to the integrity of the subject site. The environmental professional can determine whether adjacent properties or other facilities identified in the database search pose a significant threat to the site. He or she may rely on known hydrogeologic conditions and site-specific information on the subject property and suspect facility to determine whether additional investigation is warranted.

The last major component of the transaction screen process requires the preparer to physically and visually observe the property. The preparer of the transaction screen must physically observe the perimeter of the site, dead-end roads, paths, overgrown areas, and other locations on which improper waste disposal may have occurred. The preparer also is required to observe the interior spaces, custodial areas, mechanical rooms, occupied areas, and perimeter(s) of any structure(s) on the property. During the site visit, the preparer must also answer yes, no, or unknown to each question in the transaction screen. A yes to any of the questions may require additional investigation to conduct all appropriate inquiry that may be used in part to establish the innocent landowner defense.

Transaction Screen—Results and Recommendations

As noted, a single affirmative answer to any of the twenty site-specific questions requires further investigation or documentation to confirm or discount a recognized environmental condition. ASTM (1993) defines a *recognized environmental condition* as the following:

> *The presence or likely presence of any hazardous substances or petroleum products on a property under conditions that indicate an existing release, a past release, or a material threat of a release of any hazardous substances or petroleum products into the structures on the property or into the ground, groundwater, or surface water of the property. The term includes hazardous substances or petroleum products even under conditions in compliance with laws. . . .*

The findings and recommendations of the transaction screen process could include any of the following:

1. The determination that all appropriate inquiries have been conducted with no environmental conditions recognized. Such findings could be used to establish a portion of the innocent landowner defense.
2. The scope of the transaction screen was insufficient to establish due diligence; therefore, additional investigation may be needed in the form of a more comprehensive Phase I ESA.
3. Environmental conditions that could be adequately evaluated with additional records searches or interviews exist. The information collected during the additional review may be included in the document and could help establish all appropriate inquiry, depending on the nature of the recognized environmental condition. For example, during one site visit, a pole-mounted electrical transformer was observed on the site. Upon further investigation, the environmental professional acquired additional imformation from the local electrical utility to determine that the unit did not contain hazardous substances.
4. The discovery of a recognized environmental condition, facilitating preparations for an intrusive study. As an example, the subject site might not exhibit any recognized environmental condition; however, during an interview with the local fire marshal it could be learned (and later confirmed) that an agricultural fuel storage tank was once located adjacent to the site. The environmental professional would be relied upon to formulate an assessment plan for an intrusive study, if necessary.

Transaction Screen Process—Summary

The completion of the transaction screen can satisfactorily constitute an appropriate inquiry on clean commercial, residential, agricultural, or undeveloped properties. Economic factors may make the transaction screen an attractive alternative to a typical Phase I ESA because cost savings are likely to be realized, as an environmental professional does not have to be involved with the preparation of the document. However, the potential environmental liability may far outweigh the initial savings if an unskilled preparer does not recognize or adequately evaluate an environmental concern.

PHASE I ENVIRONMENTAL SITE ASSESSMENTS

As financial institutions became aware of their potential liabilities under CERCLA, RCRA, and SARA, the demand for comprehensive environmental site assessments escalated. Through the 1980s and early 1990s various guidance documents were published to assist the environmental professional with the nebulous pursuit of the due diligence standard. Lenders and purchasers required "all appropriate inquiry," yet the environmental professional did not have a standard to certify the satisfaction of such inquiry. Until 1993, a disparity evolved in the quality and philosophy of the preacquisition site assessment (PSA), the precursor to the Phase I environmental site assessment, because some investigations were clearly deficient while others were overly zealous. Lending institutions, federal loan programs, various states, and professional organizations eventually developed independent standards and requirements for the PSA that environmental professionals were asked to satisfy.

The Association of Soil and Foundation Engineers (ASFE) published the first widely adopted PSA standard in 1989. This served as a checklist to guide the environmental professional in completing the investigation. The ASFE guidance document identified three main components that should be included in the PSA—state and federal records review, an aerial photograph review, and a site inspection. A significant portion of the guidance document was devoted to discussing the potential liability the preparer is exposed to as a result of the PSA (McHugh, 1992).

A National Ground Water Association (NGWA) committee composed primarily of geologists developed a guidance document that focused on protecting groundwater resources. The 1992 NGWA standard was organized by property type (vacant land, agricultural property, and commercial and industrial facilities) and outlined their common areas of environmental concern, available databases, and site inspection protocol to assist the preparer in evaluating the property.

The ASTM released its transaction screen process and Phase I ESA guidance documents in 1993. The ASTM committee was composed of a diverse group of members; the multidisciplinary perspective of the committee resulted in guidance documents generally superior to the others available. The factors that made this guidance unique include site inspection and records review protocol, interviewing guidance, and clear and concise definitions in addition to a succinct discussion regarding the scope, purpose, and goals of the Phase I ESA and the transaction screen processes (as defined by ASTM).

Phase I ESA—Introduction

ASTM's transaction screen evaluation can be applied successfully to only a limited number of properties because most commercial and industrial facilities cannot adequately be evaluated through this process. The Phase I ESA (ASTM E1527-94) provides immensely more flexibility to address site-specific environmental concerns and, as a result, significantly greater opportunity to interpret the collected data. Consequently, ASTM requires that the Phase I ESA be prepared by an environmental professional who may be an employee of the purchaser or lender or an outside consultant retained by the user. Again, the primary objective of the Phase I ESA is to establish all appropriate inquiry for the innocent landowner defense by identifying, to the extent possible, recognized environmental conditions that may potentially impact the property's integrity. Recall that recognized environmental conditions are actual or threatened releases of hazardous substances or petroleum products into structures, soil, groundwater, or surface water of the subject site. Because contaminants can be transported to the subject site by wind, surface drainage, groundwater, or surface water, recognized environmental conditions on adjacent properties also may be of concern. From this perspective, a recognized environmental condition may be located on site or off site.

Components of the Phase I ESA

The primary components of the Phase I ESA are as follows:

1. **Records review.** Compared with the transaction screen, the list of federal, state, and local government databases to search has expanded significantly. A

description of the physical setting and historical use of the subject site and vicinity also is required.

2. **Site reconnaissance.** A thorough examination of the subject site and observations related to the adjacent properties must be conducted.
3. **Interviews.** The current owner and occupant, site manager, and local agency officials are interviewed routinely.
4. **Final report.** A recommended report format is presented to standardize report organization and content. Specific language must be used in certain sections of the report, as detailed in the standard.

Phase I ESA—Records Review

The records review process for the Phase I ESA is intended to identify environmental concerns that may impact the property through the examination of reasonably available federal, state, and local records. Four types of records can be relied upon to provide information on potential environmental concerns associated with a property (ASTM, 1994):

1. Standard federal and state environmental records.
2. Additional state and local environmental records.
3. Resources to describe the physical setting.
4. Records that track the property's historical use.

Table 11–2 indicates the minimum search radius for the standard federal and state environmental database resources (ASTM, 1994). Depending on site-specific conditions, the environmental professional may wish to expand or reduce the search radius; however, justification for such modifications must be presented in the final report.

ASTM Standard E1527-94 indicates that only those records that are reasonably ascertainable, practically reviewable, and publicly available are to be reviewed to satisfy the intent of this requirement. If there is ample time to complete the Phase I ESA, the required data may be obtained through the Freedom of Information Act. Most commonly, an environmental database search service is retained to compile and provide the information indicated in Table 11–2 within a few days.

▶ TABLE 11–2
Phase 1 ESA minimum search radius.

Governmental Database	Radius (miles)
SuperFund (National Priorities List) sites	1.0
Sites investigated under CERCLA (CERCLIS)	0.5
RCRA treatment, storage, or disposal (TSD) sites	1.0
RCRA generators list	(a)
Federal ERNS* list	1.0
State SuperFund	1.0
Sites investigated under the state CERCLA	0.5
State leaking UST (LUST) sites	0.5
State solid waste disposal or landfill sites	0.5
State UST list	(a)

(a) Subject site and adjoining properties
*Emergency Response Notification System

The data obtained during this phase of research can provide insight into the environmental concerns of the subject and adjacent properties. The environmental professional should tailor personnel interviews and site reconnaissance with these data in mind.

Additional state and local environmental records can provide valuable data regarding specific environmental concerns in the vicinity of the site. These sources enhance and supplement data obtained from the federal and state databases. Databases that may typically be available and applicable to the Phase I ESA include the following:

▶ Emergency release response reports filed pursuant to SARA.
▶ Analytical data indicating water quality from local wells.
▶ Fire department response records.
▶ Building permit and inspection records.
▶ Mineral claims, mine tailings, and underground mine sites.
▶ Local landfill and solid waste disposal sites.
▶ Public utility company records for PCB occurrence.
▶ Other records that are reasonably available and reviewable.

Mandatory and discretionary physical setting sources are incorporated into the Phase I ESA. The only required physical setting source is the current 7.5-minute topographic map showing the site and the surrounding area. The environmental professional can include other physical setting maps (groundwater, bedrock geology, soil, and surficial geology maps) if desired. The physical setting maps provide valuable information for evaluating how contaminants might be transported in the groundwater, the depth to groundwater, and bedrock characteristics. This information assists the environmental professional with the preparation of a drilling program for an intrusive investigation.

Various records are available for evaluating the historical use of the property and vicinity to help identify environmental concerns. The following list includes those resources commonly employed to develop a historical-use perspective for the subject site:

1. Aerial photographs that are of sufficient scale to determine property use of the subject site and surrounding vicinity.
2. Fire insurance maps that provide important information on property use and potential environmental concerns.
3. Property tax files that contain information on property use, maps, and building sketches.
4. Land title records that can be used to supplement other historical-use records.
5. Topographic maps of the site and vicinity produced by the U.S Geological Survey at a 7.5-minute scale.
6. Reverse directories or local street directories published periodically indicating the occupant and property use according to street name and address.
7. Building permit records that indicate the property owner, document construction, and demolition activity.
8. Property zoning or land use records available from municipal or county planning departments.
9. Maps, newspaper archives, personal records, or other credible sources of historical property use data.

The protocol for ASTM E1527-94 requires the environmental professional to present a history of the obvious property use of the subject site from 1940 or the time when the property was undeveloped, whichever is earlier, to its current use. To accomplish this the environmental professional should review as many of the suggested information sources as necessary to create a summary of developed site use at five-year intervals.

Phase I ESA—Site Reconnaissance

The purpose of the site reconnaissance phase is to physically and visually observe the subject property to document any recognized environmental conditions that may impact the site. The environmental professional also should make an effort to confirm or dispute the environmental conditions identified during the records review and interview activities. ASTM E1527-94 recommends several on-site reconnaissance procedures.

Description of General Property Setting

In reviewing the site, the environmental professional should note current or observable past uses of the subject site, adjacent parcels, and the surrounding area, taking particular notice of any chemical storage and waste disposal practices. A topographical description of the site and surrounding area as well as a general description of the structures on the property should be made.

Roads and parking areas on and adjacent to the subject property should be identified, as should the potable water supply and sewage disposal system that services all structures.

Observation of the Building Interior and Exterior

The current and past uses of the facility should be identified. Any activities that involve the use, treatment, storage, disposal, or generation of hazardous substances or petroleum products should be investigated in detail. Any above-ground storage tanks, USTs, fill pipes, vent pipes, or associated equipment-access covers should be investigated to determine the type and quantity of material stored, its approximate age, and its compliance status. Interviews of site personnel and a review of on-site records should assist the preparer in collecting this information.

The source of any strong or noxious odors detected during site reconnaissance should be noted. Areas of standing surface water, collection sumps, or pools that may contain hazardous substances or petroleum products also should be described in the Phase I ESA report. Storage drums or containers that have indications of containing hazardous substances, petroleum products, or unknown materials are environmental concerns and should be described in the report. The location, quantity, contents, and condition of the storage containers also should be noted. In addition, electrical or hydraulic equipment that may contain oils mixed with PCBs should be described in the report.

Interior Site Structure Observations

The heating and cooling systems and the fuel source should be described. Stains or corrosion of the interior floor, walls, or ceiling may indicate a hazardous substance or petroleum product spill; these should be noted during site reconnaissance. The purpose and discharge point of interior drains and sumps also should be described.

Exterior Site Structure Observations

Pits, ponds, and lagoons are commonly associated with waste disposal or treatment and may constitute an environmental condition. Areas of stained soil or pavement and stressed vegetation may indicate improper disposal practices. The environmental professional should note any such conditions on the subject site and adjacent properties, and then investigate further through site interviews to determine whether a recognized environmental condition exists.

Areas of earth that have been regraded or filled may suggest improper waste disposal practices. Such areas should be investigated and described in the report. Waste water and stormwater disposal systems may represent environmental concerns if not properly managed. In addition, any active or abandoned wells (monitoring, water supply, irrigation, injection, or dry wells) also may indicate potential environmental conditions.

Phase I ESA—Interviews

The interviews required for the Phase I ESA are intended to gather information about potential environmental conditions resulting from the site reconnaissance and records review. Generally, a key site manager, property occupant, and a local public official are interviewed to meet the minimum requirements of ASTM E1527-94.

During or prior to the interview with the key site manager or owner, the environmental professional may ask to review available environmental legal proceedings, previous environmental or geotechnical reports, violation notices, environmental permits and registrations, material safety data sheets, and RCRA waste manifests in addition to emergency response and spill prevention plans. Interviews with the site manager and occupants should focus on potential environmental conditions identified. The environmental professional may fashion these interviews to be similar to the transaction screen questionnaire.

ASTM also requires the environmental professional to interview at least one public official who may have knowledge of recognized environmental conditions associated with the property. Public officials may be from the local fire department, health agency, or the agency responsible for regulating hazardous substance or petroleum product release sites.

Throughout the entire interview process, the environmental professional should focus on obtaining details regarding the potential recognized environmental conditions, compliance issues, or legal concerns that solidify or discount the need for an intrusive investigation.

Phase I ESA—Final Report

ASTM includes an appendix to E1527-94 recommending a format for the Phase I ESA report. The intent of the format is to unify the report's organization to allow the user to efficiently review and present the findings.

The credentials of the environmental professional(s) performing the investigation must be provided to the user as the standard requires. Individual and corporate qualifications statements can be incorporated into the report or may be provided under separate cover.

ASTM requires that the Phase I ESA have a findings and conclusions section that includes a statement of compliance with E1527-94, along with any deviations from the scope of the standard. Nonscope issues also should be specified as appropriate. ASTM also requires a statement that specifies whether recognized environmental conditions exist and the environmental professional's opinion of the impact that those conditions may have on the environmental integrity of the property.

As with the transaction screen, the outcome of the Phase I ESA commonly includes one of the following:

1. An indication that all appropriate inquiry has been conducted and no recognized environmental conditions were encountered. The Phase I ESA could be used to establish a portion of the innocent landowner defense.
2. An indication of environmental concerns that, in the opinion of the environmental professional, do not pose a threat to the property's environmental integrity. In this case, the Phase I ESA also could be used to establish a portion of the innocent landowner defense with no further investigation.
3. A list of recognized environmental concerns that may require an intrusive investigation (Phase II ESA) to determine whether the property's environmental integrity has been impacted.

Phase I ESA—Process Summary

The protocol established for the transaction screen and the Phase I ESA represent two specific tools that, based on the needs of the user, may be utilized to constitute all appropriate inquiry (assuming that no recognized environmental conditions were identified). When the other provisions of the innocent land owner defense have been satisfied, CERCLA liability may be avoided.

PHASE II ENVIRONMENTAL SITE ASSESSMENTS

The results of the transaction screen or Phase I ESA evaluations may identify environmental conditions that could affect a property's environmental integrity. To assist the user (potential purchaser, lender, or property owner) in conducting "all appropriate inquiry" and making an informed business decision about a property, a Phase II ESA is often conducted. It has become established as the primary tool for evaluating whether recognized environmental conditions have actually contaminated the site. Phase II ESA involves these primary goals:

1. Confirming, or discounting, the site's potential to be impacted by hazardous substances or petroleum products.
2. If contamination is encountered, delineating the vertical and horizontal extent of the contaminated media.
3. Completing the due diligence necessary to qualify the purchaser for the innocent landowner defense if no contamination is detected in association with the recognized environmental condition.

At a minimum, the Phase II ESA scope should be sufficient to allow the environmental professional to offer one of four conclusions at the completion of the investigation: (1) the property has not been impacted by recognized environmental conditions; (2) contamination was encountered in excess of federal or state action levels (further investigation in the form of additional site characterization and corrective action may be required); (3) its results may be inconclusive and not provide the environmental professional with sufficient information to confirm the presence of contamination exceeding regulatory action levels; or (4) the environmental professional may conclude that recognized environmental conditions have not impacted the site.

In the latter case, additional investigation may be necessary to evaluate the potential for contamination to exist that might impose significant environmental liabilities. In such situations, because the results of the Phase II ESA are not reportable to the regulatory agency, the user's environmental policy, level of acceptable risk, and business strategy dictate the client's response.

The environmental industry standard of care and accepted practices currently dictates the scope of Phase II ESA. The ASTM has prepared a Phase II guidance document outlining the scope, activities, and presentation of an intrusive investigation; it was scheduled for release during 1997. As with the transaction screen and the Phase I ESA standards, ASTM's contribution of the Phase II ESA standard certainly will be incorporated into the environmental industry standard of care.

Components of the Phase II ESA

The following are primary components of the Phase II ESA:

1. **Work plan and project scope.** The environmental professional and client determine a mutually acceptable scope of work for the investigation. A sampling plan, analytical plan, quality assurance plan, and health and safety plan are constituents of the work plan.
2. **Fieldwork and sample acquisition.** Careful execution and documentation of the actual sample collection process, field screening methodologies, and appropriate sample-handling techniques are critical components of the field activities.
3. **Data analysis.** Presentation and analysis of the data collected during the fieldwork phase is an integral portion of the Phase II ESA.
4. **Report preparation and recommendations.** An accurate and concise narrative along with maps, geologic cross sections, data tables, and diagrams should be prepared to support the conclusions of the environmental professional.

Phase II ESA—Work Plan and Project Scoping

A variety of elements should be considered when defining the relationship and responsibilities of the parties involved in completing the Phase II ESA. The implications and limitations of an intrusive investigation must be understood by the property owner, potential buyer, lender, and environmental professional. The following are examples of initial considerations that must be understood to reduce the potential for litigation:

1. The current property owner must provide written permission to access the site, collect samples, and prepare the Phase II ESA.
2. The client's acceptable level of risk and the potential for uncertainty must be clearly defined.
3. Potential errors and/or omissions that may affect the results or recommendations include laboratory error, sampling error, and unidentified environmental concerns. These must be discussed.

4. Use of the report must be restricted to the parties concerned and have a specified lifetime before confirmation sampling is required.
5. The current owner must understand the legal reporting requirements should contamination be encountered.

The work plan should carefully detail all recognized environmental conditions associated with the property and specify which tasks are intended to address each particular condition. The work plan also must present a summary of the site conditions, the investigation's objectives, and an inventory of the contaminants that may be present at the site. Additional technical details such as proposed analytical work, health and safety of on-site personnel, quality assurance, and sampling techniques should also be included.

An extensive review of the site conditions and the behavior of the hazardous substances or petroleum products that may be present is essential to understand how the contaminant may be transported in the subsurface. The environmental professional relies upon a knowledge of fate and transport processes, and subsurface conditions at the site to prepare the sampling plan. The sampling plan is a component of the work plan that provides the rationale for sample acquisition associated with each recognized environmental condition. It presents the locations of proposed soil borings or test pits, the number and types of samples to be collected, the sampling methodology, and equipment scheduled to facilitate the collection of representative samples. The objective of the sampling plan is to ascertain that the soil and groundwater samples are properly collected from locations most likely to exhibit contamination from the recognized environmental condition being evaluated. Procedures to prevent cross-contamination of the samples and decontamination of equipment should be included in the sampling plan and the quality assurance plan. A description of the testing techniques should be presented in the sampling plan to indicate whether grab or composite samples will be collected for direct or indirect testing techniques. (Geophysical techniques or tightness testing of a UST are examples of indirect testing methods.) The sampling plan also may provide contingencies to expand the drilling program in an effort to delineate the extent of impacted soil or groundwater if contamination is encountered.

The work plan also must contain an analytical plan indicating the laboratory techniques to be utilized to detect targeted contaminants. The environmental professional must ensure that the laboratory is qualified and approved by the state or the EPA to conduct the analyses requested and attain meaningful quantification limits for the purpose of the investigation. Table 11–3 includes a partial sampling of the materials commonly encountered at an automobile service station and the associated regulated substances that may be generated (Shineldecker, 1992).

Table 11–3 indicates that the typical Phase II ESA sampling plan could not reasonably test for all the potential contaminants that might be found at an automobile service station. The environmental professional is responsible for targeting those compounds most likely to be present given the specific recognized environmental conditions of the property. The actual EPA analytical method used to evaluate the presence of a particular contaminant depends on the media sampled (drinking water, surface water, waste water, air, soil, or solid waste), detection limit required, and the budget for sample analysis.

▶ **TABLE 11–3**
Sample of service station materials and associated regulated substances that may be generated.*

Material	Associated Regulated Substance
Abrasives	Aluminum, antimony, carborundum, zirconium
Adhesives	Benzene, carbon disulfide, carbon tetrachloride, dichloroethane, ethyl-benzene, isocyanates, ketones, sulfuric acid, xylenes
Antifreeze	Ethyl alcohol, ethylene glycol, isopropyl alcohol, methyl alcohol
Batteries	Antimony, cadmium, cobalt, lead, mercury, picric acid, sulfuric acid, zinc
Brake fluid	Ethylene glycol, ketones, PCBs
Brake linings	Acetone, asbestos, benzene, graphite, phenols, tetrachloroethylene, trichloroethane, trichloroethylene
Diesel fuel	Benzene, ethylbenzene, polynuclear aromatic hydrocarbons (PAHs), toluene, xylenes
Gasoline	Benzene, dichloroethane, ethylbenzene, lead, methyl tertiary butyl ether, toluene, xylenes
Hydraulic fluids	Ethylene glycol, ketones, PCBs, PAHs
Lubricants	Arsenic, graphite, ketones, hydrocarbons, lead
Paint and paint thinners	Arsenic, benzene, chromium, cobalt, copper, ethylbenzene, ethylene glycol, ketones, lead, mercury, methylene chloride, PAHs, phenols, tetrachloroethylene, toluene, trichloroethylene, xylenes
Waste oil	Benzene, ethylbenzene, PAHs, toluene, xylenes

*These materials are commonly found in an automotive service station. They may generate the regulated substances specified here.

The hazard communication requirements of the Occupational Safety and Health Administration (OSHA) dictate that a site-specific health and safety plan be prepared for sites subject to intrusive sampling programs and that have the potential for worker exposure to chemical hazards. The health and safety plan must identify the possible contaminants present, exposure symptoms, emergency procedures, site control, decontamination procedures, personal protective equipment, and other issues of concern.

The quality assurance plan specifies procedures for evaluating laboratory accuracy, precision, and potential sources of cross-contamination. A predetermined number of blank, spike, and duplicate samples should be submitted along with the actual samples to evaluate laboratory performance and the quality of data obtained. Other quality assurance concerns include proper sampling techniques, sample preservation, handling, and custody protocol. The level of quality control depends on the needs of the client and the substances being analyzed.

The work plan is submitted for client approval to ensure that the specific needs of the investigation are satisfied. Once the client and/or regulatory agency approves the work plan, fieldwork can be scheduled.

Phase II ESA—Fieldwork and Sample Acquisition

Prior to the intrusive sampling site activities, all utilities in the vicinity of the borings must be accurately located. Samples are collected according to the procedures and techniques specified in the work plan. It is imperative that accurate maps, lithologic logs, photographs, and field observations be recorded for presentation in the Phase II ESA final report.

Field screening techniques represent one of the most important aspects of the fieldwork phase. The instrumentation used for field screening measurements must be subjected to calibration and quality assurance procedures to obtain reproducible and reliable data. Several instruments that can be used to rapidly indicate the presence of contamination are available. Laboratory samples are commonly selected based on the results of the field screening data.

Phase II ESA—Data Analysis

An analysis of the data collected should determine whether the appropriate chemical analyses were performed and the samples were collected from the proper locations to adequately evaluate the potential for contamination from the recognized environmental conditions. If contamination was not detected, the environmental professional also must determine whether the results of the quality control procedures were acceptable. The laboratory data should generally correlate with the field screening results and boring log observations.

If the laboratory did detect contamination the environmental professional must evaluate the quality assurance data and determine whether human health or the environment could be affected by the release. If a federal or state standard exists for the contaminant detected, the environmental professional must determine whether that concentration was exceeded by the samples collected during the intrusive investigation. Generally, the maximum contaminant level (MCL) as established by the Clean Water Act is the enforceable standard for commonly encountered groundwater or surface water contaminants. A range of state guidelines and corrective action levels have been established for various contaminant concentrations in soil. Additional site-specific soil contamination guidelines may be formulated based on the contaminant concentration in soil that may present a leaching hazard to groundwater quality. The environmental professional also must determine whether the contaminant detected might occur naturally at the concentrations observed.

Phase II ESA—Report Preparation and Recommendations

The environmental professional must prepare a site characterization report, compile field data, site maps, geologic cross sections, data tables, and diagrams to support the conclusions of the investigation and to assist the user in making an informed business decision regarding the presence or absence of environmental risks associated with the property.

The results of the Phase II ESA investigation could indicate that no contamination was encountered. The completion of the Phase II ESA would meet the due diligence requirements for the specific environmental conditions and might be relied upon to establish the innocent landowner defense if the other components of the defense are satisfied.

Another scenario is that the Phase II ESA might encounter contamination in excess of federal or state standards. The responsibility to report the contamination generally resides with the current property owner. If the contamination presents an immediate threat to human health or the environment, the environmental professional and the owner may be obligated by law to report the release to the appropriate federal or state agency. Further characterization, abatement actions, or remediation may then be required under the direction of the regulatory agency.

Phase II ESA—Process Summary

The completion of the Phase II ESA should provide the environmental professional with the data necessary to conclude whether the recognized environmental conditions have negatively impacted the environment, or if contamination exceeding regulatory action levels exists and that a certain degree of environmental liability may be associated with the property. Any of these conclusions should aid the purchaser in making an informed decision regarding the acquisition of the property.

SUMMARY

- ▶ Since 1980, and the advent of CERCLA liability, a series of environmental assessment techniques has evolved. The techniques were intended to assist potential property buyers in evaluating potential environmental hazards associated with a site. Originally, the scope of work and the final report for an environmental assessment exhibited significant variability and inconsistent recommendations, depending on the project type and the skill of the preparer. Not until the early 1990s did ASTM release a set of unifying standards for the transaction screen process and the Phase I ESA. A standard for the Phase II ESA was scheduled for release in 1997 and should again result in a more consistent, accurate, and meaningful evaluation of a property.

- ▶ The primary reason for conducting a preacquisition environmental site assessment is to provide the purchaser with one component of the innocent landowner defense to avoid liability associated with the cleanup of a contaminated site and to evaluate the environmental integrity of a property.

- ▶ Parties that are potentially liable for the cleanup of a contaminated site include the present property owner, the property owner at the time of the chemical release, a person who arranged for the improper disposal or treatment of hazardous substances, or the transporter who selects the treatment or disposal site.

- ▶ The transaction screen process allows for the rapid evaluation of a property not likely to be impacted by hazardous substances or petroleum products, such as commercial property or undeveloped land. The transaction screen process may be completed by anyone who is able to record accurate responses to the questions presented in ASTM's transaction screen questionnaire. The person completing the questionnaire need not be as extensively trained as the environmental professional because there is little opportunity for interpretation and additional investigation within the scope of this assessment. Any affirmative response could potentially result in an intrusive sampling program or a comprehensive Phase I ESA to evaluate the condition of the site.

- ▶ The Phase I ESA is significantly more detailed than the transaction screen and allows increased flexibility to resolve specific concerns that arise during the investigation. A variety of professionals have the training and skills necessary to be considered "environmental professionals" and are, therefore, qualified to prepare, or supervise, the Phase I ESA and subsequent investigations.

▶ Upon completion of the Phase I ESA, or a transaction screen, the environmental professional should be able to prepare a work plan for the Phase II ESA that will serve to confirm or deny environmental impacts resulting from each of the recognized environmental conditions specified during the preliminary investigations.

▶ The work plan is a critical component to the Phase II ESA that details the objectives, strategy, and scope of activities for the intrusive investigation. The following are the most significant components of the work plan:

1. **The sampling plan** is prepared to ensure that the most appropriate samples are collected as dictated by law and the objectives of the investigation.
2. **The analytical plan** ensures that the proper laboratory procedures are applied to the samples and that the desired level of detection is obtained.
3. **The quality assurance plan** specifies the statistical techniques and the sample handling protocol to minimize the potential for cross-contamination or erroneous results.
4. **The health and safety plan** mandated by OSHA requires the employer to convey all potential site dangers to those persons potentially exposed to chemical hazards.

▶ The goal of the Phase II ESA is to provide the purchaser with sufficient information to make an informed business decision regarding property acquisition based on the desired level of risk. The goal of the environmental professional is to confirm, or dispute, that the property has not been impacted or that contamination exceeding applicable cleanup standards was encountered. Frequently, the results of the Phase II ESA may not be crystal clear; the environmental professional must then make conclusions and recommendations based on the data obtained, experience, and the purchaser's level of risk.

QUESTIONS FOR REVIEW

1. What are the purpose and fundamental difference among each of the following?
   a. Transaction screen
   b. Phase I environmental site assessment
   c. Phase II environmental site assessment
   d. Environmental audit
2. Who is generally recognized as being capable of conducting a transaction screen on a property?
3. What are the professional qualifications necessary to conduct Phase I environmental site assessments?
4. What does the statement "CERCLA liability is strict, joint, and several" mean?
5. Whom may the EPA name as a potentially responsible party for the cleanup of a SuperFund site?
6. What statutory defenses are available to protect property owners from CERCLA liability?
7. What conditions must be satisfied to establish the innocent landowner defense?
8. What are the similarities and differences between the fundamental components of the transaction screen and the Phase I environmental site assessment?
9. Who is exempt from liability for the cleanup of a contaminated site under CERCLA?
10. What are the four components of the Phase II environmental site assessment work plan? Describe each.

ACTIVITIES

1. Submit a Freedom of Information Act request to your state department of environmental quality for a list of federal and state CERCLA, leaking underground storage tank, abandoned landfill, and emergency spill response sites for the county or state in which you live. Allow approximately four to six weeks for a response. Prepare a map showing the location of the environmental sites within a 1-mile radius from the center of the town. Write a brief description indicating the name and address of each site, nature of the environmental hazard, and the status of the cleanup (if known).

2. For any of the sites identified in the previous activity, conduct a historical review of the property to identify the nature, source location, and age of the release. You may use a reverse city directory, aerial photographs, fire insurance maps, topographic maps, or newspaper clippings available at your local public library. You may also interview people who are familiar with the property or have knowledge of the environmental hazards associated with it.

REFERENCES

American Society for Testing and Materials. May 1993. *Standard Practice For Environmental Site Assessments: Transaction Screen Process,* ASTM Designation E1528-93, 30 pp.

American Society for Testing and Materials. June 1994. *Standard Practice For Environmental Site Assessments: Phase I Environmental Site Assessment Process,* ASTM Designation E1527-94, 24 pp.

Brown, Johnine J. 1994. "Federal Environmental Laws Effecting Business and Real Estate Transactions," *Conference on Economic Redevelopment and Environmental Risk Management in Business and Real Estate Transactions.* San Francisco, CA: RTM Communications, Inc., 13 pp.

Fleer, James E. June/July 1993. "ASTM Releases New Assessment Standard, Soils," pp. 6–11.

McHugh, Laura D. 1992. *Due Diligence—Environmental Assessment Standards.* North Vancouver, BC: Specialty Technical Publishers, pp. 3-1 through 5-28.

National Ground Water Association. September 1992. *Guidance to Environmental Site Assessments.* Dublin, OH: NEWA, 72 p.

Prescott, Michael K. 1990. *The Environmental Handbook for Property Transfer and Financing.* Chelsea, MI: Lewis Publishers, 125 pp.

Schoenbaum, Thomas J. 1985. Environmental Policy Law. Mineola, NY: Foundation Press, 1011 pp.

Shineldecker, Chris L. 1992. *Handbook of Environmental Contaminants: A Guide For Site Assessment.* Chelsea, MI: Lewis Publishers, 371 pp.

12

Remedial Technologies

R. Ryan Dupont

Upon completion of this chapter, you will be able to meet the following objectives:

▸ Provide an overview of the Corrective Action process related to contamination at facilities permitted under the Resource Conservation and Recovery Act (RCRA).

▸ Highlight relevant site, soil, and waste characteristics that limit the feasibility and expected effectiveness of given remedial technologies.

▸ Provide a summary of the general characteristics of chemical, biological, and thermal treatment techniques that might be carried out at a contaminated site.

OVERVIEW

The purpose of hazardous waste management is to minimize the dangers to workers, the public, and the environment during the use, storage, and transport of hazardous materials. When the inevitable release does occur from an operating facility or a hazardous waste treatment, storage, or disposal facility (TSDF), the responsible party must take action to minimize the release and initiate cleanup activities that will reach the goals of protecting public health and the environment. A wide variety of techniques is available to carry out this cleanup action, ranging from simple excavation of contaminated soil and off-site disposal at an RCRA–permitted facility, to complex *in situ* (in place) treatment of contaminated soil and groundwater. This chapter details the information necessary to determine what techniques are appropriate for a given situation and summarizes the general characteristics of each technique that might be used at a given site. The emphasis is on the discussion of site, soil, and waste characteristics that limit the feasibility and expected effectiveness of given remedial technologies.

THE CORRECTIVE ACTION PROCESS

Ongoing activities that generate hazardous waste or that result in the transport, treatment, storage, and/or disposal of hazardous wastes are regulated under RCRA. When a release of hazardous chemicals and wastes from such facilities occurs, RCRA regulations require initiation of spill and containment measures. If environmental contamination is evident (a fish kill, contaminated drinking water supply, or contaminated monitoring wells, for example), a *Corrective Action program* must be initiated. This Corrective Action program is initiated by the owner/operator and is conducted under the supervision of state or regional EPA personnel responsible for the RCRA program in the jurisdiction where the release occurred. The intent of EPA regulation is to streamline the corrective action process so that the cleanup of minor releases or those releases requiring self-evident remedial alternatives (i.e., overexcavation, drumming, and disposal) can be carried out rapidly. In addition, once the corrective action is accomplished, the RCRA facility is expected to return to normal operating status.

Table 12–1 summarizes the basic components of a Corrective Action process. The process generally consists of an assessment of the magnitude of the potential problem at the facility, called an *RCRA Facility Assessment (RFA);* the investigation of the nature and extent of contamination at the site, called an *RCRA Facility Investigation (RFI);* an evaluation of the remedial options that could be used for the cleanup of the site and the selection of the best option for a given contamination problem, called a *Corrective Measures Study (CMS);* and implementation of the best option for

▶ **TABLE 12–1**
The Corrective Action process for cleanup of hazardous waste/materials releases from RCRA facilities.

Activity	Responsible Party	Purpose
RCRA facility assessment (RFA)	EPA	Evaluation of releases or potential releases of hazardous wastes and/or constituents threatening public health and the environment.
Corrective action plan (CAP) RCRA facility investigation (RFI)	Owner/Operator	Site characterization to identify the nature and extent of contamination at the site for help in selection of remedial alternatives.
Corrective measures study (CMS)	Owner/Operator	Detailed feasibility study to identify cleanup alternatives that could work at the site based on RFI results for site/soil/waste characteristics.
		The result of the CMS is the selection of the best alternative for cleanup at a given site.
Corrective measures implementation (CMI)	Owner/Operator	The design and construction of technology options selected as the best cleanup approach for a given site based on RFI and CMS results. This phase also includes performance monitoring to ensure that the technology is performing as designed.

site cleanup, called a *Corrective Measures Implementation (CMI)*. The RFI, CMS, and CMI activities are conducted as part of what the EPA defines as the *Corrective Action Plan (CAP)*.

If the RFA and/or the owner/operator find that the hazardous waste or hazardous material release represents an immediate threat to human health and/or the environment, *interim measures* may be required before the actual CAP takes place. Examples of the types of interim measures possible are summarized in Table 12–2. They include such physical control activities as segregating nonleaking containers from leaking ones, pumping liquid out of a leaking storage tank or an impoundment to reduce the leak rate of hazardous materials/waste, placing temporary covers on contaminated soils and/or waste piles, fencing off contaminated areas to prevent direct contact, or

▶ **TABLE 12–2**
Examples of interim measures that can be used prior to developing the corrective action plan for the protection of public health and the environment from imminent threats from hazardous waste/materials releases at RCRA facilities (from U.S. EPA, 1988a).

Source/Receptor	Possible Interim Measures
Waste/Product containers	Overpack/Redrum Move to new storage area Segregate from nonleaking containers Temporary cover
Tanks	Secondary containment Partial or complete removal of liquid
Surface impoundments	Partial or complete removal of liquid Temporary cover Run-on, run-off controls Sampling to verify contamination levels
Landfills	Run-on, run-off controls Pump out leachate collection system liquids Temporary cap
Waste piles and soils	Run-on, run-off controls Temporary cover Waste removal and disposal
Gas migration	Gas barriers Active gas collection and treatment system
Groundwater	Interceptor trench Pump and treat for groundwater flow control Sampling upgradient from contamination to identify source
Surface water	Run-on, run-off controls Collection or diversion of source of hazardous waste/material
Other actions	Fencing to prevent direct contact exposure Provide alternative water supplies to replace contaminated drinking water Temporary relocation of exposed population

simple sampling and analysis to verify actual levels of release and exposure before implementing other interim actions (U.S. EPA, 1988a).

THE CORRECTIVE ACTION PLAN

The RFI Process

As indicated previously, the first step in the corrective action process is to identify the nature and extent of the release so that a cost-effective cleanup can take place. This assessment is the first step in the problem definition phase of the CAP, called by the EPA the *RFI process.* Its components are summarized in Table 12–3. As indicated in Table 12–3, the RFI process is designed to provide information necessary for the rational selection of containment and treatment options necessary to control and clean up a release at a given site. It is important to note that the correction action process is flexible,

▶ TABLE 12–3
Components of the RFI effort.

Components	Purpose
Collection of historical data on past practices	To identify possible sources, types, and quantities of release possible based on past practices at the facility. From this, specific compounds of concern and the environmental media that they may contaminate can be determined.
Screening of containment and recovery technologies	Based on an identification of possible contaminants and their properties, identification of technologies that could be used to halt their migration in a given medium can be identified.
Field investigation to identify migration	Collection of data from the site to verify migration and to quantify the nature and level of contamination expected.
Screening of treatment technologies	Based on an identification of possible contaminants and their properties, identification of technologies that could be used for contaminant removal/destruction in a given medium can be identified.
Field investigation to characterize site	Collection of site and soil data necessary to fully evaluate the ability of a given technology to remove/destroy the hazardous contaminant(s) of concern. Usually conducted in conjunction with migration potential field investigation (above).
Bench- and/or pilot-scale studies	Collection of technology-specific performance data affected by site and soil characteristics to verify on a small scale that a specific technology will remove/destroy the contaminant(s) of concern.

allowing different containment and treatment options for different portions of the site, based on findings of the RFI. It also encourages the use of a combination of containment/treatment technologies (treatment trains) to optimize site cleanup activities.

Preinvestigation Evaluation

Because owners/operators should know the hazardous materials being used in their facility and the waste being generated, they should understand the general nature of the release, such as the release being a chlorinated solvent used in degreasing operations rather than being a hydrocarbon with aromatic and straight-chained compound content derived from a specific processing unit. Knowing the specific process streams that are involved in the release, the owners/operators can determine the potential pathways for hazardous contaminant movement in the environment and the remedial technologies needed to control them. For example, the release of a non-chlorinated solvent that contains volatile and aromatic compounds in addition to a nonvolatile, hydrocarbon carrier liquid would impact the air medium from volatilization of some components, impact ground or surface waters due to aromatic compound solubility in water, and also impact the soil due to the adsorption of the nonvolatile hydrocarbons by the soil surfaces. This problem is then a multimedia one, requiring the monitoring and remediation of air, soil, and surface or groundwater that might have received the released hazardous contaminants.

Screening of Candidate Containment/Recovery Options

Once the general nature of the release and the media being impacted are known, the need for containment and recovery and the type of system that can control the migration of hazardous contaminants can be identified. If the properties of the chemicals expected in the release are volatile and toxic to humans, the containment and collection of vapors are necessary. They may be a part of any interim action activities carried out soon after the release. If the contaminants of concern are water soluble and may move into the groundwater and contaminate a drinking water supply, a reasonable containment option is to cap the soil where the release took place to prevent the compounds from leaching out of the soil due to rainfall infiltration. Table 12–4 summarizes the containment and recovery technologies that are appropriate for each medium that can be impacted, along with the technology selection criteria that should be considered based on RFI information for a given release event. It is important to note that the selection of a given containment/recovery system is based not only on site characteristics (such as flat terrain), but also on soil (low sorption potential, etc.) and waste, (e.g., highly volatile) properties that affect the movement of hazardous chemicals in the environment.

Examples of typical containment and recovery systems for controlling groundwater contaminant migration are shown in Figures 12–1 through 12–3. Active hydraulic controls for source protection (see Figure 12–1) using groundwater extraction require the treatment of the extracted water to remove/destroy the hazardous contaminants it contains. Water-injection systems do not require water treatment, nor do they result in contaminant removal. Instead, they merely divert the plume to prevent drinking-well contamination. Passive hydraulic control systems using subsurface trenches are limited to shallow (approximately 30 feet) contamination situations and require collected groundwater treatment, as do extraction well systems.

▶ **TABLE 12–4**
Containment and recovery options to be considered and selection criteria for various impacted media (from U.S. EPA, 1989).

Media	Option	Selection Criteria
Soils	Excavation	Small aerial extent of release and soil contamination Shallow vertical extent due to confining soil layer beneath release
	Vacuum extraction	Volatile contaminants in released chemical(s) Low sorption potential for contaminant on soil Deep groundwater and permeable soil Little to no soil layering below release
	Cap/Cover	Flat terrain
	No action	Low risk to human health and the environment
Groundwater	Groundwater pumping and hydraulic barriers	Permeable groundwater aquifer Little to no soil layering in groundwater aquifer Soluble plume that can be controlled by groundwater pumping
	Subsurface drains and low permeability barriers	Shallow groundwater table depth Soluble plume that can be controlled by groundwater controls
	No action	Low risk to human health and the environment
Surface water	Run-off/Run-on controls and diversion/collection	Flat terrain with access for control structure placement Nature of impacted surface water Magnitude of precipitation and run-off in area
	No action	Low risk to human health and the environment
Air	Cap/Cover	Flat terrain Shallow vertical extent so deep gas migration minimal
	Gas collection	Low sorption potential for contaminant on soil Deep contamination, deep groundwater, and permeable soil Little to no soil layering below release
	No action	Low risk to human health and the environment

The major advantage of passive systems is that they do not require energy to pump groundwater out of the aquifer. They simply use a drainage trench to intercept contaminated groundwater flowing into it.

The use of low-permeability barrier walls is depicted in Figure 12–3. Contaminant migration is prevented not by recovery of contaminated groundwater, but by isolating the contaminated site from flowing groundwater, preventing any contamination in the first place. These walls are generally composed of mixtures of soil or cement and bentonite, or sheet piles. They require a great deal of earth moving, use

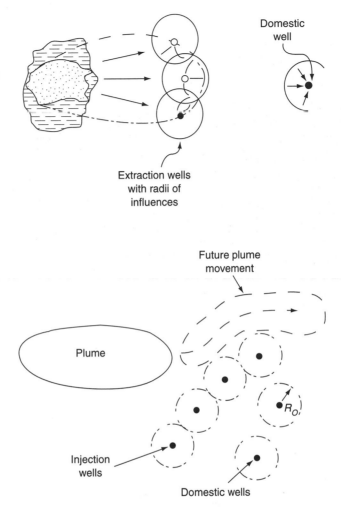

▶ **FIGURE 12–1**
Active groundwater contain-
ment/recovery systems used
to prevent contaminated
groundwater migration (from
U.S. EPA, 1985).

significant amounts of the site's land area, and are relatively expensive to construct. These barrier systems are limited to placement in areas with shallow depths of contamination, because conventional trenching techniques are normally used for their placement. They can be installed without disturbing the waste material itself, protecting the surrounding environment from additional releases that can result from waste excavation and handling. Because of long-term deterioration and leakage of groundwater under and through these barriers, however, some form of groundwater control within the barrier is generally required, as shown in Figure 12–3.

Figure 12–4 provides examples of passive and active gas migration control systems that can be applied for vapor phase contaminant migration from an RCRA facility release site. As with groundwater control systems, passive systems are intended to intercept vapor migration, while active systems utilize an active control system to collect vapor from a much wider area of influence than is possible using passive controls. Once vapors are collected, gas treatment is generally required prior to discharge of the collected gas to ensure that no hazardous contaminants are released into the environment.

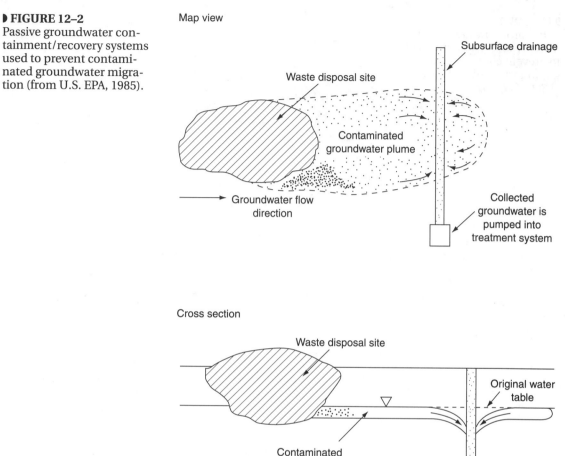

Details of design and installation of specific containment and recovery technologies are provided in U.S. EPA (1989) documents and Spooner et al. (1984), referenced at the end of this chapter. These containment/recovery systems are normally part of remedial technologies for the in-place treatment of soil and groundwater. The need for their use as an interim action to protect public health and the environment prior to beginning long-term corrective action depends, however, on determing the risk that uncontrolled contaminant migration poses to the exposed population. This can be found in part from the data collected in the field investigation phase of the RFI.

Field Investigation to Identify Migration and for Site Characterization

Based on an understanding of the potential contaminants involved in a release from a given facility, field investigations should be conducted soon after the release to verify the nature and extent of contaminant release and contaminant migration potential. If, for example, it is known that volatile, nonchlorinated solvents were involved in a release to shallow soils, field investigations should be carried out to attempt to determine the extent of the release both in terms of the area and depth of the spill and to quantify the concentration of volatile compounds both within the soil and being released into the atmosphere.

▶ **FIGURE 12–3**
Low permeability barrier
containment/recovery sys-
tems used for prevention of
contaminated groundwater
migration (from Spooner et
al., 1984).

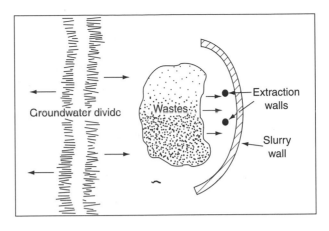

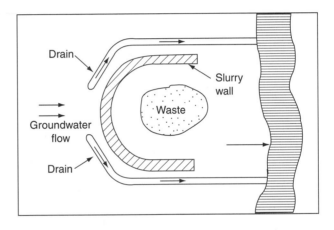

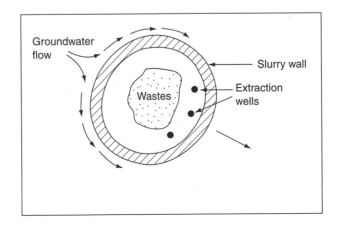

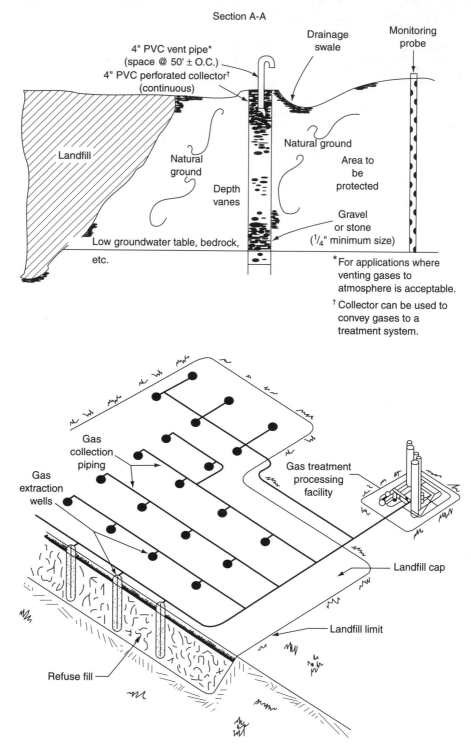

Section A-A

4" PVC vent pipe*
(space @ 50' ± O.C.)
4" PVC perforated collector†
(continuous)

Drainage swale

Monitoring probe

Landfill

Natural ground

Natural ground

Depth vanes

Area to be protected

Gravel or stone ($\frac{1}{4}$" minimum size)

Low groundwater table, bedrock, etc.

*For applications where venting gases to atmosphere is acceptable.

†Collector can be used to convey gases to a treatment system.

Gas collection piping

Gas extraction wells

Gas treatment processing facility

Landfill cap

Landfill limit

Refuse fill

▶ **FIGURE 12–4**
Passive and active containment/recovery systems used to prevent contaminated gas migration (from U.S. EPA, 1985).

196

Field investigation techniques were described earlier in this book. For corrective action sites they would typically involve soil gas surveys, shallow soil sampling, and headspace gas measurements that are designed to provide rapid information regarding the relative severity of contamination at locations where samples are collected. From these initial screening results, decisions can be made on whether interim measures are needed to contain and recover released materials. In addition, site areas can be identified that require further detailed investigation to provide information required for treatment technology selection.

Screening of Candidate Treatment Technologies

Based on the identification of possible contaminants released at a facility and their physical/chemical properties, and upon verifying the nature and extent of the release from field investigation results, the initial screening of applicable treatment technologies can take place. The ultimate objective of treatment efforts in the corrective action process is to make hazardous waste or hazardous material nonhazardous. As indicated in Table 12–5, if hazard reduction itself is not possible, mobility reduction to reduce possible exposure is preferred. Volume reduction to reduce the magnitude of the release to which the public may be exposed is the least desirable option to reduce the overall risk of the hazardous waste or hazardous material release.

Table 12–5 lists four general treatment technology classes that can be used for site remediation. These include physical, chemical, biological, and thermal treatment

▶ **TABLE 12–5**
Treatment approaches for RCRA facility remediation.

		Possible Technology Classes	Data Needs
	Hazard reduction	Biological Thermal Chemical	Which chemicals? What concentrations? Where are they? What volume is contaminated?
Most Desirable			
↓	Mobility reduction	Physical • containment • solidification	What are the short-term risks? What are the long-term risks? What are the migration pathways? What volume is contaminated?
Least Desirable			
	Volume reduction	Physical • separation	Which chemicals? What concentrations? Where are they? What volume is contaminated?

technologies. In addition, a number of these technologies can be carried out in place *(in situ)* or using extracted material *(ex situ)* methods. Physical treatment technologies generally focus on mobility and volume reduction, and do not alter the nature of the hazardous wastes/materials of concern. The selection of a particular treatment technology class that does result in hazard reduction depends to a large extent on the characteristics of the hazardous waste or hazardous material that was released. The selection of the particular treatment system within a given technology class and the means of operation (i.e., in situ or ex situ) depends not only on the hazardous waste/hazardous material, but also on the extent of the release, soil conditions, and site characteristics where the release occurred. Table 12–6 provides a summary of the technology classes that may be applicable for different waste types based on their characteristics and physical form along with the information necessary to make a final selection of technologies warranting further investigation.

Table 12–7 summarizes the general criteria that affect in situ versus ex situ process selection. The primary criteria are the extent of the release—that is, whether cost-effective overexcavation and treatment above ground are possible, and whether the delivery and recovery of material is possible at the site. If the release is widespread and involves soil and groundwater contamination, if surface disturbances cannot be allowed, and if the soil beneath the site is uniform in its structure and texture, in situ treatment systems are generally applicable. If the release is small in extent, if the subsurface is nonuniform, and if excavation from the surface is possible, then ex situ processes are more desirable.

Bench- and/or Pilot-Scale Treatability Studies

The use of *bench-* and/or *pilot-scale treatability studies* may be necessary to avoid technology misapplication at a full-scale release site. If the selected technology has a long performance history for wastes and site conditions found at the release site being remediated, treatability studies will not be necessary and full-scale remedial system design can begin. However, if the technology being considered is new or unproven, or if an old technology has not been used for the specific waste type or under the specific site and/or soil conditions, bench- and/or pilot-scale treatability studies may be necessary.

Bench-scale treatability studies are designed to screen a large selection of possible technologies to determine whether soil and waste conditions specific to the release site limit the efficiency of hazard reduction under ideal laboratory conditions. This large selection of technologies can be screened for $10,000 to $50,000 within a four- to six-week period (U.S. EPA, 1989), eliminating from further investigation those technologies that prove incapable of providing the hazard reduction necessary for the release. Once bench-scale screening is complete, *pilot-scale treatability testing* at the actual field site may be necessary to collect further site-specific data for the final design of the full-scale remedial technology system. The decision to carry out pilot-scale studies, which cost from $25,000 to $1 million or more and last from twelve weeks to twelve months, is driven by the cost of replacing the technology if it fails at the field-scale. Again, pilot-scale studies generally are not necessary if the technology is not new or if the short-term risk of the release is low so that if necessary, design and implementation of other technologies can take place without compromising the health and safety of the public.

▶ TABLE 12–6
Summary of treatment technology classes and information needs for technology selection as a function of the characteristics and physical form of the hazardous waste/ hazardous material release (U.S. EPA, 1989).

Waste Type	Form	Technology Options	Preliminary Technology Data Needs
Corrosive	Liquid	Chemical	pH Constituent analysis Oil and grease content Total suspended solids Total dissolved solids
Ignitable	Liquid	Chemical	pH Constituent analysis
		Biological	Gross organic components (BOD, TOC) Dissolved oxygen Nutrients analysis (NH_3, PO_4, NO_3) pH Priority pollutant analysis ORP
		Thermal	Heat content Ash content Halogen content Moisture content Heavy metal content Volatile matter content
	Gas	Thermal	Heat content Halogen content
		Pretreat to get to treatable form by chemical and biological means	
	Solid	Biological	Gross organic components (BOD, TOC) Nutrient analysis (NH_3, PO_4, NO_3) Priority pollutant analysis pH
		Thermal	Heat content Volatile matter content Ash characteristics Ash content Halogen content Moisture content Heavy metals content

▶ TABLE 12–6
(continued)

Waste Type	Form	Technology Options	Preliminary Technology Data Needs
Reactive	Liquid	Chemical	pH Constituent analysis
		Biological	Gross organic components (BOD, TOC) Dissolved oxygen Nutrients analysis (NH_3, PO_4, NO_3) pH Priority pollutant analysis ORP
		Thermal	Heat content Ash content Halogen content Heavy metal content Volatile matter content
	Gas	Thermal	Heat content Halogen content
		Pretreat to get to treatable form by chemical and biological means	
	Solid	Biological	Gross organic components (BOD, TOC) Nutrient analysis (NH_3, PO_4, NO_3) Priority pollutant analysis pH
		Thermal	Heat content Volatile matter content Ash characteristics Ash content Halogen content Moisture content Heavy metals content
Toxic-Inorganic	Liquid	Chemical	pH Constituent analysis Oil and grease content Total suspended solids Total dissolved solids
	Gas	Pretreatment required to get to liquid phase for chemical treatment or solid phase for solidification/stabilization	Solubility (in H_2O, organic) Solvents, oils, etc. Size distribution Constituent analysis

Waste Type	Form	Technology Options	Preliminary Technology Data Needs
Organics	Liquid	Chemical	pH Constituent analysis Halogen content Total suspended solids heavy metal content
		Biological	Gross organic components (BOD, TOC) Dissolved oxygen Nutrient analysis (NH_3, PO_4, NO_3) pH Priority pollutant analysis ORP
		Thermal	Heat content Ash content Halogen content Heavy metals content Volatile matter content
	Gas	Thermal	Heat content Halogen content
		Pretreat to get to treatable form by chemical and biological means	
	Solid	Biological	Gross organic components (BOD, TOC) Nutrient analysis (NH_3, PO_4, NO_3) Priority pollutant analysis pH
		Thermal	Heat content Volatile matter content Ash characteristics Ash content Halogen content Moisture content Heavy metals content

The Corrective Measures Study and Corrective Measures Implementation

Once the RFI process has been completed, an engineering review and analysis of RFI findings takes place. This corrective measures study (CMS) provides a final review of the treatment technologies identified by the RFI as usable for remediating

▶ TABLE 12–7
Applicability of in situ versus ex situ technologies for RCRA site corrective actions.

Technology	Site Characteristic	Applicability of Technology
Ex situ	Extent of contamination	Small aerial and vertical extent favoring excavation
	Subsurface soil	Highly variable, making migration pathways difficult to determine
	Surface structures	No structures so that surface excavation is possible
	Proximity of receptors	No close receptors so releases during excavation pose little risk
	Remedial objectives	Rapid cleanup rate required, forcing excavation and ex situ treatment
	Cost constraints	No cost constraints, so costly excavation is possible
In situ	Extent of contamination	Large aerial and/or vertical extent that makes excavation costly
	Subsurface soil	Uniform, making migration pathways definable
	Surface structures	Structures that cannot be disturbed, making excavation impossible
	Proximity of receptors	Close receptors so releases during excavation pose high risk
	Remedial objectives	Rapid cleanup rate not required
	Cost constraints	Cost constraints, so costly excavation is traded for longer-term monitoring costs

the site-specific release. It also provides the opportunity to conduct an engineering evaluation of the likelihood of success of these treatment options in meeting cleanup goals. Corrective measures alternatives are evaluated based on the following criteria:

▶ **Technical**. Expected level of performance, expected reliability, ease of construction of the alternative, and safety of the process.
▶ **Environmental**. Adverse effects on environmentally sensitive areas.
▶ **Human health.** Reduction in short- and long-term health risks.
▶ **Institutional**. Governing local, state, federal regulations, and community relations aspects of each alternative.
▶ **Cost**. Capital (construction, equipment, land) and operation/maintenance costs and indirect costs (engineering, legal, and license fees, and start-up costs) of each alternative.

From the findings of the CMS, the best technology option is selected for the remediation of a release based on waste, site, soil, and community-specific characteristics. This selection generally requires a trade-off to be made between human health and environmental risks and other important technical, community, and cost considerations. It is important to note, however, that the EPA does not consider cost to be an overriding consideration when selecting a remedial technology. Cost becomes relevant, based on EPA policy, only when one remedial technology must be selected from several providing equivalent protective levels (U.S. EPA, 1989).

REMEDIAL TECHNOLOGY TREATMENT OPTIONS FOR HAZARD REDUCTION

A summary of various treatment technologies applicable for the remediation of releases from RCRA facilities is discussed next, focusing on summarizing their applicability and limitations related to the corrective action process. This summary provides only an introduction to the range of common treatment technologies used in corrective action applications. If you are interested in further information on the subject, refer to U.S. EPA (1988b, 1989, and 1993) and Powell (1994) for more details on conventional and innovative technology applications and expected technology cost and performance levels.

Chemical Treatment Processes

Chemical treatment consists of processes that, through the addition of specific chemicals and energy, result in the chemical transformation of the hazardous waste or hazardous material to a less-hazardous form. It can be used either as a complete treatment method or as a pretreatment step prior to further waste hazard reduction using other treatment. One notable characteristic of the chemical treatment process is the production of chemical sludge that must also be processed before it can be disposed of without further environmental damage. Figure 12–5 summarizes the general types of chemical treatment processes available for use in RCRA corrective actions.

Applicability of Chemical Treatment Options

Chemical treatment is generally applicable to a broad range of organic and inorganic wastes and, as indicated in Figure 12–5, can be used for reactions as different as the destruction of organic compounds through oxidation/reduction, or *lysis,* to the separation of inorganic compounds via chemical precipitation. Chemical treatment tends to be very compound and waste specific, often requiring a series of reaction steps to treat a complex waste mixture. Finally, the efficiency of chemical reactions is affected by the level of mixing between the added chemical reactant and the waste material. Consequently, chemical treatment processes are applicable to hazardous wastes or hazardous materials in the aqueous or slurry form and are not generally applicable to *in situ* treatment applications.

Adjustment of pH

Adjustment of pH is used either to neutralize a highly alkaline or acidic waste to reduce its reactivity or corrosiveness or to cause the precipitation of metals from a liquid waste stream by raising the pH of the waste stream so that the metal's solubility in the liquid is very low. Typical neutralizing chemicals include the following: $Ca(OH)_2$, CaO, $NaOH$, $NaHCO_3$, Na_2CO_3, and $Mg(OH)_2$ for acid waste streams; and H_2SO_4, HCl, and HNO_3 for basic waste streams. The neutralization process is designed to change the pH of the waste stream to a neutral pH 7 range; doing so may

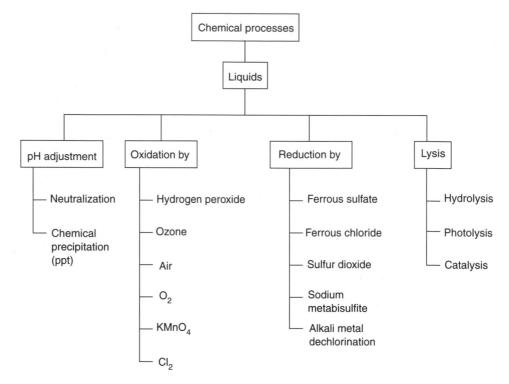

result in chemical precipitation. Figure 12–6 shows the solubility of a variety of metal
hydroxide compounds as a function of pH and indicates that, particularly for both
acid wastes, as their pHs move toward pH 8, their solubility decreases significantly.
The intent of the chemical precipation process then is to reach not simply a neutral
pH level by pH adjustment but also the minimum solubility point for the metal(s) of
interest so that the waste-stream metal concentration can be minimized.

Neutralization results in an increase of dissolved inorganic salt concentrations
in the waste stream and, perhaps, some chemical precipitation due to chemical addition. The *chemical precipitation process* results in potentially large volumes of
heavy metal sludge that require further treatment prior to disposal. In addition, because chemical precipitation often requires waste pH levels to reach 9 or higher before
effective chemical precipitation can take place, the waste stream may have to be
brought back to neutral conditions via acid addition prior to its discharge.

Chemical Oxidation/Reduction Processes

Chemical oxidation/reduction reactions are used to change the oxidation state of a
hazardous waste or hazardous material to reduce its toxicity, solubility, or stability,
thus reducing its overall hazard in the environment. Typical oxidation/reduction
agents used for waste treatment are listed in Figure 12–5. Those waste contaminants
for which chemical oxidation and reduction reactions are particularly applicable are
summarized in Table 12–8. Oxidation/reduction reactions are generally nonselective,

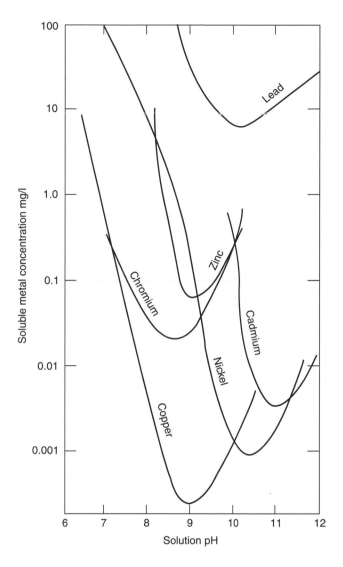

▶ **FIGURE 12–6**
Solubility of various metal
hydroxides as a function of
solution pH (from U.S. EPA,
1989).

so the waste composition must be well known to ensure that the unexpected production of more hazardous reaction intermediates does not occur. These processes may be applicable for compounds that are not easily treatable with other treatment techniques (alkali metal dechlorination of PCBs and dioxins, for example), but they may also result in violent, explosive reactions and may produce undesirable air emissions and odors during treatment.

Hydrolysis and Photolysis Processes

The purpose of these chemical reactions is to facilitate chemical bond breakage so that further treatment of the hazardous waste or hazardous material is possible using other treatment processes. *Hydrolysis reactions* utilize water to break chemical bonds; *photolysis* employs ultraviolet (UV) radiation to produce the desired effect. The effectiveness of a hydrolysis or photolysis reaction is very compound

▶ TABLE 12–8
Waste contaminants for which chemical oxidation/reduction reactions are applicable
(from U.S. EPA, 1989).

Reaction	Waste Stream (in order of reactivity)
Oxidation	Phenols, aldehydes, aromatics, amines, some sulfur compounds
	Alcohols, ketones, organic acids, esters, alkyl-substituted aromatics, nitro-substituted aromatics, carbohydrates
	Halogenated hydrocarbons, saturated aliphatics, benzene
	Cyanide wastes
Reduction	Hexavalent chromium waste
	Mercury waste
	Hexavalent selenium waste
	Organic lead compounds
	Chelated metal-bearing wastes
Alkali metal	PCBs
Dechlorination	Dioxins
	Chlorinated solvent waste
	Organochlorine pesticides

specific because not all organic and inorganic compounds react to water or UV light. Hydrolysis is usually carried out in the presence of acids or bases and is appropriate for the pretreatment of phenols and some chlorinated organics. Photolysis reactions can be used for the chemical transformation of dioxins, chlorinated organics, and pesticides, along with some nitrated waste streams. Photolysis may be severely limited, however, by dissolved and suspended materials in the waste that are not hazardous but that absorb the UV light being used to destroy the hazardous compounds. Consequently, pretreatment of the waste stream to remove dissolved and suspended materials may be necessary before photolysis can be effectively carried out.

Biological Treatment Processes

Biological treatment processes destroy or transform hazardous wastes/materials using microoorganisms in either ex situ or in situ reactor systems (Figure 12–7). The conditions within these systems are maintained so that the microorganisms contained within them are encouraged to utilize the hazardous waste or hazardous material as a source of energy to grow on, resulting in the conversion of this waste material to oxidized end products (CO_2, H_2O, and other oxidized inorganic compounds) and more microorganisms.

Applicability of Biological Treatment Options

Biological processes are applicable to waste existing as a liquid, slurry, or contaminated soil, as long as the waste is biodegradable and is not toxic to the microorganisms being used to carry out the biodegradation process. Table 12–9 summarizes the range of RCRA–related compounds that are known to be degradable and/or transformable using biological systems.

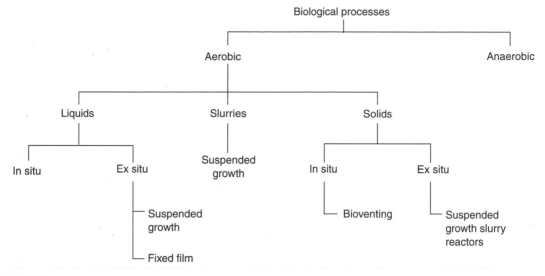

Biological processes

Aerobic — Anaerobic

Liquids — Slurries — Solids

In situ — Ex situ

Suspended growth (Slurries)

In situ — Ex situ

Suspended growth / Fixed film

Bioventing

Suspended growth slurry reactors

▶ FIGURE 12–7

Biological treatment processes that can be used for RCRA facility remediation (from U.S. EPA, 1989).

Once the waste's biodegradability has been determined, the treatment system variables listed in Table 12–10 must be controlled to ensure that biodegradation will not be inhibited. *Aerobic processes* (those that operate in the presence of oxygen) are the most common of the biological treatment systems used because they can degrade a wide variety of organic compounds (see Table 12–9), they are the most operationally stable, and they result in the conversion of complex organic compounds to simple oxidized end products. *Anaerobic systems* are used to dechlorinate alkyl halides and some nitrogen-substituted organics that will not degrade aerobically, but because of their instability and low reaction rates, they are not widely used in RCRA corrective actions.

Suspended Growth Systems

Suspended growth systems are designed to provide intimate contact between the hazardous waste or hazardous material stream and the microorganisms used for its biodegradation by suspending the microorganisms in this waste material and mixing them with it. These systems are capable of treating both liquids and solid slurries. As indicated in Figure 12–8, these systems are typically composed of a reaction or aeration tank to which oxygen is added and actual biodegradation and growth of microorganisms occur, plus a liquids/solids separation tank. This liquids/solids separation tank, or *clarifier,* allows the treated liquid to be separated from the microbial solids so that it can be discharged and provides a means for recycling the necessary microbial solids (called *activated sludge*) back into the reactor for use in treating additional hazardous waste or hazardous material entering the process. Excess biomass is processed further (by stabilization and drying) before its disposal.

These suspended growth systems are able to treat waste streams with moderate (1,000 mg/L BOD) to high (10,000 mg/L BOD) organic content and are generally able to resist variable loadings of influent waste materials due to the dilution effect of the aeration tank and the continual supply of active microorganisms entering the system

▶ TABLE 12–9

Degradable and/or transformable RCRA–related compounds using biological treatment processes (U.S. EPA, 1985).

Substrate Compounds	Respiration		
	Aerobic	Anaerobic	Fermentation
Straight chain alkanes	+	+	+
Branched alkanes	+	+	+
Saturated alkyl halides		+	
Unsaturated alkyl halides		+	
Esters, glycols, epoxides	+	+	+
Alcohols	+	+	
Aldenhydes, ketones	+	+	
Carboxylic acids	+	+	
Amides	+	+	
Esters	+	+	
Nitriles	+	+	
Amines	+	+	
Phthalate esters	+	+	
Nitrosamines		+	
Thiols			
Cyclic alkanes	+		+
Unhalogenated aromatics	+	+	
Halogenated aromatics	+	+	
Simple aromatic nitro compounds	+	+	
Aromatic nitro compounds with other functional groups	+	+	
Phenols	+	+	+
Halogenated side chain aromatics	+		+
Fused ring hydroxy compounds	+		
Nitrophenols		+	
Halophenols	+		
Phenols—dihydrides polyhydrides	+		
Two- & three-ring fused polycyclic hydrocarbons	+		
Biphenyls	+		
Chlorinated biphenyls	+		
Four-ring fused polysyclic hydrocarbons	+		
Five-ring fused polycyclic hydrocarbons	+		
Fused polycyclic hydrocarbons	+		
Organophosphates	+	+	
Pesticides and herbicides	+	+	

with the recycle stream. Mixing intensity must be adequate to keep the waste material in suspension, which is a particular concern when treating high solid-content slurries, and to provide an adequate supply of oxygen to the microorganisms. Finally, if the hazardous waste or hazardous material being treated contains high levels of volatile compounds, the aeration tank may have to be covered to capture and further treat these released vapors so they do not pose an unacceptable exposure risk.

▶ TABLE 12–10
Biological treatment system variables that must be controlled to ensure uninhibited biodegradation rates.

Variable	Desirable Range	Effect of Variable
Toxicity	Nontoxic	Results in death of microorganisms, resulting in no reactants in system to carry out the biodegradation reaction.
Oxygen	>2 mg/L in water >2 vol% in soil gas	For aerobic processes, oxygen is the primary reactant necessary for the destruction of organic compounds. When oxygen falls below these levels the system turns anaerobic and efficiency drops.
	<1 mg/L in water <2 vol% in soil gas	For anaerobic processes necessary for some hazardous compounds (see Table 12–9). Many anaerobic microorganisms cannot survive in the presence of oxygen and efficiency drops.
pH	6 to 8	Optimal treatment usually occurs at neutral pH. Extreme pH values (<4 or >10) tend to be toxic to microorganisms and efficiency drops.
Nutrients	C:N:P =20:50:1	Primary nutrients required for microorganism growth are nitrogen and phosphorous. They are required on a C:N:P basis of 20:5:1 expressed on a weight basis. If N or P are inadequate relative to the carbon in the waste stream, efficiency drops.
Temperature	15 to 45°C	Biodegradation rate varies by a factor of 2 for every 10°C temperature change, i.e., goes up by 2 with 10°C temperature increase and down by 2 with 10°C temperature drop.

Fixed-film Systems

Fixed-film systems are applicable for treating low organic-content waste material and for complex organic waste streams that require slow growing, specialized microbial populations for their efficient degradation. The reactor portions of these systems contain inert packing material on which the microorganisms (biomass) grow and are concentrated. Unlike a suspended growth system, biomass in a fixed-film system is retained for long periods of time by the packing material and degrades the contaminants as the waste stream passes through the reactor. Because of the concentrated biomass within the reactor, reactor volume can be small, making the system cost effective. A schematic of a typical packed tower fixed-film system, showing the packed bed reactor and clarifier used for liquid solids separation, is presented in Figure 12–9. It should be noted as indicated in Figure 12–9 that fixed-film systems typically recirculate treated water rather than settled solids that are circulated in suspended

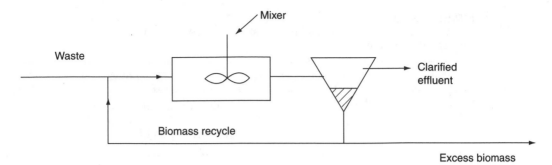

▶ FIGURE 12–8
Schematic of a suspended growth biological treatment system (from U.S. EPA, 1989).

growth systems. The purpose of the liquid recycle stream in fixed-film systems is to provide the following:

1. Dilution of the influent waste stream to reduce the impact of high incoming concentrations.
2. Hydraulic shearing so that excess biomass within the packing material does not clog the media.
3. Increased contact between the waste stream and the biomass on the media by returning the wastewater to the packed bed several times before it leaves the treatment system.

Because of their internal packing, fixed-film systems can be used only for liquid waste streams. However, the in situ treatment of contaminated soils is possible using the soil media as a fixed-film reactor.

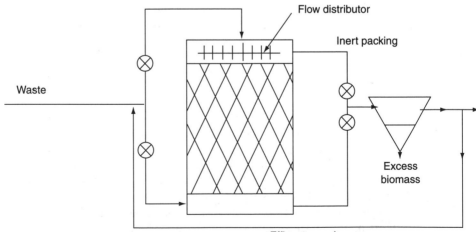

▶ FIGURE 12–9
Schematic of a typical fixed-film biological treatment system applied to RCRA site remediation (from U.S. EPA, 1989).

In Situ Biological Treatment Systems

As indicated earlier, in situ treatment systems are desirable if the extent of contamination is large, and if extraction and/or excavation activities are limited by site conditions. In situ biological treatment systems can be applied for the biodegradation of hazardous waste or hazardous material both above and below the groundwater table.

In situ treatment of contamination above the groundwater table is called *bioventing*. It incorporates the concepts of vacuum extraction for active gas migration control (Table 12–4 and Figure 12–4) with those of biological treatment using fixed-film systems. Figure 12–10 shows a schematic of a bioventing system in which air is moved through the soil above the groundwater table to provide oxygen to the microorganisms, encouraging aerobic degradation of soil contaminants. The system is designed to supply oxygen at a rate that is needed to keep the soil gas between 2 and 21 percent oxygen by volume. These systems are operated at low vapor-flow rates—that is, 1 to 10 cubic feet per minute (cfm)—and if operated properly, contain only carbon dioxide in the collected vapor so it may be discharged to the atmosphere without treatment.

One limitation to effective treatment that can occur in bioventing systems is low water content in the soil due to drying. The soil water content should be maintained above 50 percent field capacity (the soil water content after the soil has been drained by gravity) to ensure that adequate water exists in the soil to maintain an active microbial population. If drying does occur, water may be added to the system either by surface irrigation or by humidifying the air being pulled into the system by the vacuum extraction system. A 55-gallon drum of water connected to passive injection wells can serve as an inexpensive and effective "bubbler" for this purpose.

In situ treatment of contamination below the groundwater table is carried out using the soil below the groundwater table as a fixed-film reactor, with infiltration

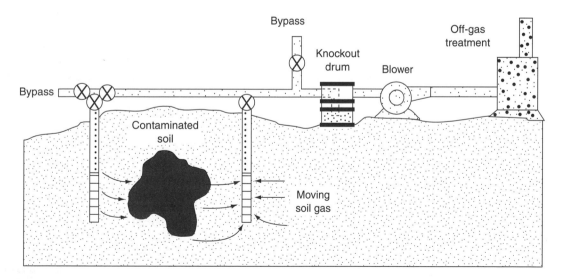

▶ **FIGURE 12–10**
Schematic of a typical bioventing system applied to contaminated soil above the groundwater table at an RCRA site.

galleries and injection wells supplying required nutrients and oxygen to the subsurface. Water content is not a concern in this type of system as it is with a bioventing system. What is a concern, however, is the difficulty in transporting nutrients and oxygen to the area of contamination using this injected water. The flow rate of water injection is typically low, and degradation rates are so rapid that nutrients, particularly oxygen, do not travel far from the point of injection before they are completely utilized. In addition, these systems tend to clog rapidly around the injection areas due either to stimulation of microbial growth or oxidation and precipitation of dissolved iron in the groundwater. Care must be taken in using and applying in situ groundwater treatment because of these limitations. The current state of the practice uses pumping to control the migration of contaminants in the groundwater, with the treatment of this pumped water ex situ in above-ground reactors described previously.

Thermal-Treatment Processes

Thermal-treatment processes can be operated at either low temperature as a pretreatment option for soils and sludges or at high temperature for the thermal destruction of hazardous organic wastes or hazardous material under oxygen-rich conditions. Low-temperature systems called *thermal stripping systems* are generally followed by a high-temperature process, *incineration*. Low-temperature thermal stripping systems are used to thermally separate combustible organics from noncombustible, high ash-content solids, producing a combustible gas stream and a treated solid residue.

Applicability of Incineration Treatment Options

Incineration systems are typically used for biologically toxic or highly concentrated organic waste streams and wastes that are not easily degradable using other treatment technologies, such as those containing PCBs, dioxins, and more than 1,000 parts-per-million chlorine. The heat content of the waste is also a consideration, as indicated in Figure 12–11, with wastes having a heat content of between 2,500 and 8,500+ BTU per pound being combustible in incineration systems. Incinerator systems are available for the destruction of all waste forms including liquids, solids, and gases. Table 12–11 summarizes the various types of incinerator systems applicable for each of these waste types. Note that air pollution control systems, which are required for the removal of particulates and acid gases from the incinerator flue gas, are an integral part of incineration systems.

Liquid Injection Systems

A schematic of a typical *liquid-injection incineration* system is shown in Figure 12–12. Liquid hazardous waste or hazardous material is introduced into the incinerator combustion chamber through injection nozzles that produce fine droplets that mix with injected air and fuel in the combustion zone. These fine droplets combust rapidly and completely, producing flame temperatures of 1,200 to 1,300°F for reaction times of 0.1 to 2 seconds. These incinerators can be used for all pumpable organic wastes but cannot handle waste with significant particulate content because of the potential for nozzle clogging. Heat recovery is possible from these systems as indicated in Figure 12–12; both acid gas scrubbing with caustic solution and dry or wet systems for particulate removal may be necessary to meet emission control standards for incineration systems.

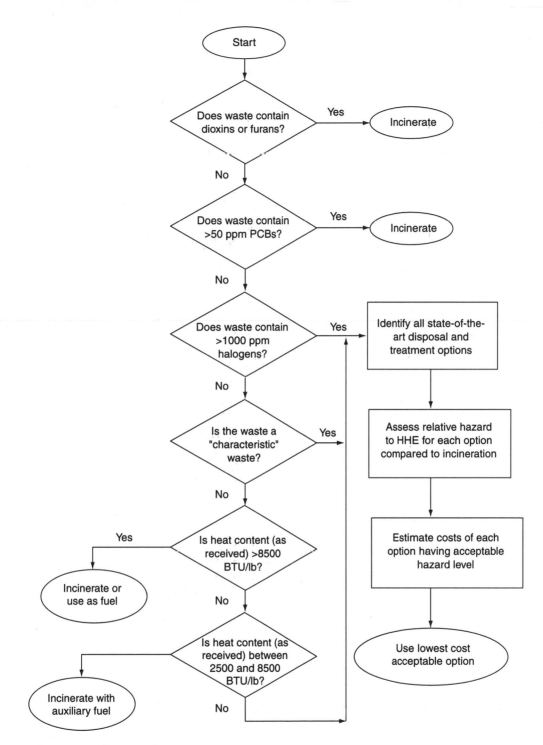

▶ **FIGURE 12–11**
Decision tree for determining the applicability of incineration for a given hazardous waste/material released at an RCRA site (from U.S. EPA, 1989).

213

▶ TABLE 12–11
Incinerator types applicable for various waste materials.

Incinerator	Liquids	Solids/Sludges	Soil	Gases
Liquid injection	X			X
Plasma arc	X	X		X
Rotary kiln	X	X	X	X
Fluidized bed	X	X		X
Circulating bed	X	X	X	X

Plasma Arc Systems

The *plasma arc system* is actually a pyrolytic system, providing high-temperature destruction of hazardous wastes or hazardous materials in the absence of oxygen. A high-temperature (9,000°F) plasma gas is developed within the system by passing an electrical charge between a plasma burner and the bottom of the plasma reactor (Figure 12–13). This plasma gas is ionized and when the waste comes in contact with it, the waste rapidly decomposes, primarily into H_2, CO, and HCl gases. As indicated in Figure 12–13, these decomposition products are either removed by the caustic scrubber (acid gases) or are completely combusted in a flare before the flue gas is released to the environment. Plasma arc systems are particularly applicable for de-

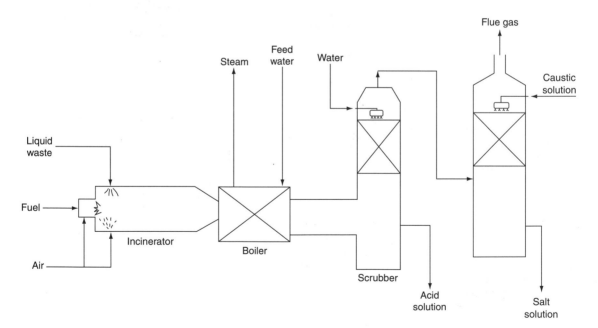

▶ FIGURE 12–12
Schematic of a typical liquid-injection system showing energy recovery and required flue gas treatment units (from U.S. EPA, 1989).

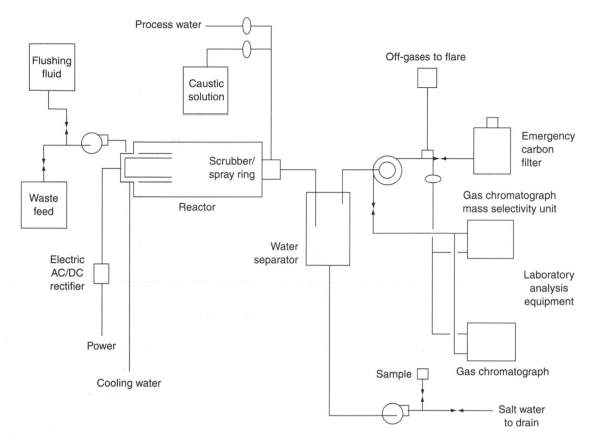

▶ **FIGURE 12–13**
Schematic of a plasma arc system showing required flue gas treatment units (from U.S. EPA, 1989).

stroying of high-chlorine content wastes, such as those containing chlorinated pesticides, PCBs, or dioxins, and can treat both liquids and sludges that can be fluidized by adding liquid so as to not clog the waste-injection system.

Rotary Kiln Systems

Rotary kiln systems (Figure 12–14) consist of a long, inclined, slowly rotating, refractory-lined tube in which solids, sludges, soils, and/or liquids are placed. Auxiliary fuel is provided at the inlet end of the kiln, and as the kiln slowly rotates, the waste mixture in the kiln is tumbled, providing the turbulence necessary to effectively mix and heat the waste. The kiln itself is designed to raise the temperature of the waste material so that combustible waste components are volatilized and brought to approximately 500 to 600°F before entering the secondary combustion chamber. Additional liquid waste and/or auxiliary fuel is added to the secondary combustor, where the final destruction of the gaseous waste products take place at 1,200 to 1,500°F. Ash residue from the combusted solids exits the outlet end of the kiln, and additional heat recovery and acid gas/particulate removal are provided as indicated in Figure 12–14.

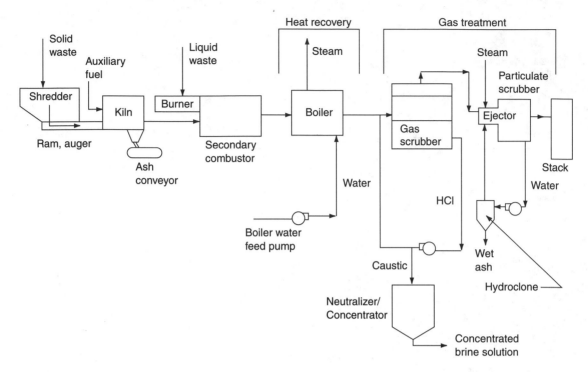

▶ **FIGURE 12–14**

A typical rotary kiln incineration system showing heat recovery and required flue gas treatment units (from U.S. EPA, 1989).

Fluidized Bed and Circulating Bed Systems

Fluidized bed incinerators (Figure 12–15) utilize a bed of fine, inert packing material (usually sand) to provide turbulence and heat transfer to the waste stream being incinerated by preheating and fluidizing the bed. The bed is fluidized by passing the combustion air at a flow rate high enough up through the bed to expand it by 25 to 30 percent of its original volume. Once the bed reaches the desired combustion temperature, normally 750 to 1,000°F, the waste feed is initiated at multiple points throughout the fluidized bed. The system is effective for the incineration of slurries and sludges, but cannot treat soils or viscous wastes.

SUMMARY

▶ The determination of the need for immediate interim measures to minimize the migration of hazardous waste/material following a release at an RCRA-permitted facility and the selection of long-term corrective measures depend to a large extent on the nature of the waste itself and on the conditions of the site and soil beneath the facility.

▶ The RCRA facility investigation is designed to identify the specific nature and extent of contamination and to aid in the selection of remedial alternatives best suited to actual waste/site/soil conditions once a release has been detected.

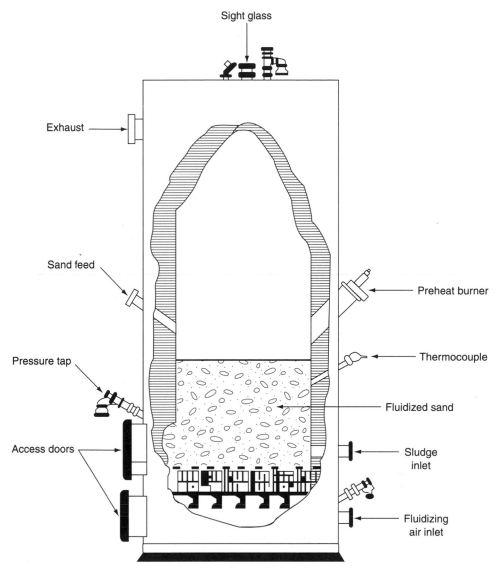

Sight glass

Exhaust

Sand feed

Preheat burner

Thermocouple

Pressure tap

Fluidized sand

Access doors

Sludge
inlet

Fluidizing
air inlet

▶ **FIGURE 12–15**
Schematic of a typical fluidized bed incineration system (from U.S. EPA, 1989).

▶ The corrective measures study leads to the selection of the best technology for site cleanup as affected by technical as well as environmental, human health, institutional, and cost constraints.

▶ The selection of a particular remedial option for overall hazard reduction can come from one of three possible treatment technology classes: chemical, biological, or thermal systems.

▶ Physical treatment options are focused not on hazard reduction, but on the reduction of the volume and/or mobility of the hazardous contaminants.

▶ Treatment options are selected based on the type of hazardous waste/material and the site and soil constraints that may exist for a given release scenario.

▶ With adequate knowledge of waste, site, and soil conditions, a cost-effective corrective action approach can be developed that will reach the ultimate goal of protection of human health and the environment.

QUESTIONS FOR REVIEW

1. Define the four parts of the Corrective Action process, identify the purpose of each, and identify who is responsible for carrying them out.
2. What is a CAP? How is it developed and what is it used for?
3. Describe the role that interim measures play in the overall Corrective Action process.
4. What is to be accomplished in the RFI process?
5. What is the desirable hierarchy for remediation of a release at an RCRA facility, and what is the purpose of each level in the hierarchy?
6. Define the term *in situ* and describe under what circumstances this type of treatment system might be used versus an *ex situ* system.

7. What containment and recovery option should be considered for a site with the following characteristics: groundwater contamination, permeable aquifer, deep water table, and high risk to human health and the environment?
8. Explain when a biological treatment option would be selected for Corrective Action at a given RCRA facility and when it would not.
9. For what waste types are thermal processes particularly useful? Why?
10. Would a waste containing no PCBs, 500 ppm chlorine, with an energy content of 1,000 BTU/lb, be considered a candidate for incineration? Why or why not?

ACTIVITIES

1. Contact a local manufacturing company and request that it provide you with information regarding a waste produced there. Using the information provided regarding the composition and concentration of individual compounds in the waste and general waste characteristics (pH, water content, BOD), determine what type of technology would be appropriate for the remediation of contaminated soil and groundwater if a release of this material occurred at the facility.
2. Obtain a soil survey for the location where you are taking this course. From the soil survey, determine what the relevant properties of the soil in this location are related to the feasibility of implementing containment and recovery options for the following types of releases:
   a. Small quantity of nonvolatile oily waste material at the surface

   b. Moderate quantity of volatile, high-risk waste material from a pipeline below the surface
   c. Large quantity of water-soluble, high-risk contaminant released to the surface

   First identify what media might be impacted from each of these types of releases. Next, based on the soil properties taken from the soil survey, identify which if any techniques listed in Table 12-4 might be appropriate for each type of release. Finally, identify any additional information you might need before making a final selection of appropriate containment and recovery actions needed at your site.

REFERENCES

Powell, D.M. 1994. *Selecting Innovative Cleanup Technologies: EPA Resources,* Chemical Engineering Progress (5):33–41.

Spooner, P.A., et. al., *Slurry Trench Construction for Pollution Migration Control,* EPA/540/2–84/001.

U.S. EPA. 1985. Handbook—Remedial Action at Waste Disposal Sites (Revised), EPA/625/6–85/006, Washington, D.C.: Office of Emergency and Remedial Response.

U.S. EPA. 1988a. RCRA Corrective Action Interim Measures Guidance—Interim Final, EPA/530–SW–88–029, OSWER 9902.4, Washington, D.C.: Office of Solid Waste and Emergency Response.

U.S. EPA. 1988b. Cleanup of Releases from Petroleum USTs: Selected Technologies, EPA/530/UST–88/001, Washington, D.C.: Office of Underground Storage Tanks.

U.S. EPA. 1989. Seminar Publication. Corrective Action: Technologies and Applications. EPA/625/4–89/020, Cincinnati, OH: Center for Environmental Research Information.

U.S. EPA. 1993. Remediation Technologies: Screening Matrix and Reference Guide, EPA/542–B–93–005, Washington, D.C.: Office of Solid Waste and Emergency Response, OS–110W, Tyndall AFB, FL: U.S. Air Force Armstrong Laboratory.

Appendix A

State Pollution Prevention Offices

Alabama
Department of Environmental Mgmt.
1751 Congressman William L. Dickinson
    Drive
Montgomery, AL 36130
(205) 260-2779

Alaska
Alaska Health Project
1818 West Northern Lights Boulevard
Suite 103
Anchorage, AK 99517
(907) 276-2864

Department of Environmental
    Conservation
Pollution Prevention Office
3601 C Street Suite 1334
Anchorage, AK 99503-1795
(907) 563-6529

Arizona
Department of Environmental Quality
Pollution Prevention Unit
3033 North Central Avenue, Room 558
Phoenix, AZ 85012
(602) 207-4210

Arkansas
Department of Pollution Prevention and
    Ecology
Hazardous Waste Division
P.O. Box 8913
Little Rock, AR 72219-8913
(501) 570-2861

California
California Integrated Waste Management
    Board
8800 Cal Center Drive
Sacramento, CA 95826
(916) 255-2289

California Local Government Commission
909 12th Street
Suite 205
Sacramento, CA 95826
(916) 448-1198

Department of Toxic Substances Control
Pollution Prevention, Public & Regulatory
    Assistance Division
P.O. Box 806
Sacramento, CA 95812-0806
(916) 322-3670

U.S. Environmental Protection Agency
Office of Pacific Island/Native
American Programs E-4
75 Hawthorne Street
San Francisco, CA 94105
(415) 744-1599

Colorado
Department of Health
Pollution Prevention Waste Reduction Program
4300 Cherry Creek Drive S
Denver, CO 80220
(303) 692-3003

Connecticut
Connecticut Hazardous Waste Management
  Service
Connecticut Technical Assistance Program
  (ConnTAP)
900 Asylum Avenue
Suite 360
Hartford, CT 06105-1904
(203) 241-0777

Department of Environmental Protection
Bureau of Waste Management
165 Capitol Avenue
Hartford, CT 06106
(203) 566-5217

Delaware
Department of Natural Resources &
  Environmental Control
Pollution Prevention Program
Kings Highway
P.O. Box 1401
Dover, DE 19903
(302) 739-5071

University of Delaware
Department of Civil Engineering
Newark, DE 19716
(302) 451-8522

Florida
Department of Environmental Regulation
Waste Reduction Assistance Program
2600 Blair Stone Road
Tallahassee, FL 32399-2400
(904) 488-0300

Georgia
Department of Natural Resources
Environmental Protection Division
4244 International Parkway
Suite 104
Atlanta, GA 30334
(404) 362-2537

Hawaii
Department of Health
Office of Solid Waste
5 Waterfront Plaza
Suite 250
500 Ala Moana Blvd.
Honolulu, HI 96813
(808) 586-4373

Idaho
Department of Health and Welfare
1410 North Hilton Street
Boise, ID 83720-9000
(208) 334-5860

Illinois
Department of Energy and Natural Resources
Hazardous Waste Research & Information
  Center
One East Hazelwood Drive
Champaign, IL 61820-7465
(217) 333-8569

Environmental Protection Agency
Office of Pollution Prevention
2200 Churchill Road
P.O. Box 19276
Springfield, IL 62794-9276
(217) 785-0533

Indiana
Department of Environmental Mgmt.
Office of Pollution Prevention and
Technical Assistance
P.O. Box 6015
105 South Meridian Street
Indianapolis, IN 46225
(317) 232-8172

Purdue University
Environmental Management and
Education Program
2129 Civil Engineering Bldg.
West Lafayette, IN 47907-i284
(317) 494-5038

Iowa
Department of Natural Resources
Waste Management Authority Div.
Wallace State Office Building
Des Moines, IA 50319
(515) 281-8941

University of Northern Iowa
Iowa Waste Reduction Center
75 Biology Research Complex
Cedar Falls, IA 50614-0185
(319) 273-2079

Kansas
Department of Health and
Environment
State Technical Action Plan
Forbes Field, Building 740
Topeka, KS 66620
(913) 296-1603

Kansas State University
Engineering Extension Programs
133 Ward Hall
Manhattan, KS 66506-2508
(913) 532-6026

Kentucky
University of Louisville
Kentucky Partners—State Waste
Reduction Center
Ernst Hall, Room 312
Louisville, KY 40292
(502) 588-7260

Louisiana
Department of Environmental Quality
P.O. Box 82263
Baton Rouge, LA 70884-2263
(504) 765-0720

Maine
Department of Environmental Protection
State House Station #17
Augusta, ME 04333
(207) 287-2811

Maine Waste Management Agency
State House Station 154
Augusta, ME 04333
(207) 287-5300

Maryland
Department of the Environment
Hazardous Waste Program
2500 Broening Highway, Building 40
Baltimore, MD 21224
(410) 631-3344

Maryland Environmental Services
2020 Industrial Drive
Annapolis, MD 21401
(301) 974-7281

University of Maryland
Technical Extension Service
Engineering Research Center
College Park, MD 20742
(301) 454-1941

Massachusetts
Department of Environment
Office of Technical Assistance
100 Cambridge Street
Boston, MA 02202
(617) 727-3260

Department of Environmental Protection
Toxics Use Reduction Act Implementation
1 Winter Street
Boston, MA 02108
(617) 292-5870

University of Massachusetts—Lowell
Toxics Use Reduction Institute
1 University Ave.
Lowell, MA 01854
(508) 934-3262

Michigan
Department of Commerce and Natural Resources
Office of Waste Reduction Services
Environmental Services Division
P.O. Box 30004
116 West Allegan Street
Lansing, MI 48909-1178
(517) 335-2142

Michigan Technological University
Waste Reduction and Management Program
1400 Townsend Drive
Houghton, MI 49931
(906) 487-2098

Minnesota
Minnesota Office of Waste Management
1350 Energy Lane
Suite 201
St. Paul, MN 55108-5272
(612) 649-5744

Minnesota Pollution Control Agency
Environmental Assessment Office
520 Lafayette Road
St. Paul, MN 55155
(612) 296-8643

University of Minnesota
Technical Assistance Program
School of Public Health
1313 5th Street, S.E., Suite 207
Minneapolis, MN 55414
(612) 627-4646

Mississippi
Department of Environmental Quality
Waste Reduction/Waste Minimization Program
P.O. Box 10385
Jackson, MS 39289-0385
(601) 961-5171

Technical Assistance Program and Solid Waste
Reduction Assistance
P.O. Drawer CN
Mississippi State, MS 39762
(601) 325-8454

Missouri
Department of Natural Resources
Division of Environmental Quality
205 Jefferson Street
P.O. Box 176
Jefferson City, MO 65102
(314) 751-3176

Environmental Improvement and Energy
Resources Authority
225 Madison Street
P.O. Box 744
Jefferson City, MO 65102
(314) 751-4919

Montana
Department of Health and Environmental
Sciences
Water Quality Bureau
Room A-206, Cogswell Building
Helena, MT 59620
(406) 444-2406

Montana State U. Extension Service
807 Leon Johnson Hall
Bozeman, MT 59717-0312
(406) 994-5683

Nebraska
Department of Environmental Control
Hazardous Waste Section
301 Centennial Mall South
P.O. Box 98922
Lincoln, NE 68509
(402) 471-421 7

Nevada
Bureau of Waste Management
Division of Environmental Protection
123 West Nye Lane
Carson City, NV 89710
(702) 687-5872

Nevada Energy Conservation
Program
Office of Community Services—
  Capitol Complex
201 South Fall Street
Carson City, NV 89710
(702) 885-4420

University of Nevada—Reno
Business Environmental Program
Nevada Small Business Development Center
Reno, NV 89557-0100
(702) 784-1717

New Hampshire
Department of Environmental Services
Waste Management Division
New Hampshire P2 Program
6 Hazen Drive
Concord, NH 03301-6509
(603) 271-2912

New Hampshire Business and Industry
  Association
New Hampshire Waste Cap
122 North Main Street
Concord, NH 00301
(603) 224-5388

New Jersey
Department of Environmental Protection
Office of Pollution Prevention
CN-402
401 East State Street
Trenton, NJ 08623
(609) 777-0518

New Jersey Institute of Technology
Technical Assistance Program
Hazardous Substance Management
Research Center
323 Martin Luther King Blvd.
Newark, NJ 07102
(609) 292-8341

New Mexico
New Mexico Environmental Department
Municipal Water Pollution Prevention Program
1190 St. Francis Drive
P.O. Box 26110
Santa Fe, NM 87502
(505) 827-2804

New York
Clarkson University
Hazardous Waste/Toxic Substance Re-
  search/Management Center
Rowley Laboratories
Potsdam, NY 13699
(315) 268-6542

Department of Environmental Conservation
Bureau of Pollution Prevention
50 Wolf Road
Albany, NY 12233-7253
(518) 457-7276

New York Pollution Prevention Program
Erie County Office of Pollution Prevention
Erie Co. Office Bldg.
95 Franklin St.
Buffalo, NY 14202
(716) 858-6231

New York State Environmental Facilities Corp.
Technical Advisory Services Division
50 Wolf Road
Albany, NY 12205
(518) 457-4138

North Carolina

Department of Environmental, Health, and
    Natural Resources
Pollution Prevention Program
P.O. Box 27687
Raleigh, NC 27611-7687
(919) 571-4100

North Dakota

Department of Health and Consolidated
    Laboratories
Environmental Health Section
1200 Missouri Avenue
Rm. 201
P.O. Box 5520
Bismarck, ND 58502
(701) 221-5150

Ohio

Department of Development Technology
    Transfer Organization
77 South High Street, 26th Floor
Columbus, OH 43255-0330
(614) 644-4286

Department of Natural Resources
Division of Litter Prevention and Recycling
Fountain Square Court—Building F2
Columbus, OH 43224-1387
(614) 265-6333

Environmental Protection Agency
Pollution Prevention Section
Division of Hazardous Waste Management
P.O. Box 1049
Columbus, OH 43266-0149
(614) 644-3969

Ohio's Thomas Edison Program
77 South High Street, 26th Floor
Columbus, OH 4321
(614) 466-3887

Oklahoma

Department of Health
Environmental Quality Council
Environmental Health Adm.—0200
1000 N.E.10th Street
Oklahoma City, OK 73117-1299
(405) 271-7353

Department of Health
Pollution Prevention Technical Assistance
    Program
Hazardous Waste Management
1000 N.E.10th Street
Oklahoma City, OK 73117-1299
(405) 271-7047

Oregon

Department of Environmental Quality
Hazardous and Solid Waste Division
811 S.W. 6th Avenue
Portland, OR 97204
(503) 229-5458

Pennsylvania

Department of Environmental Resources
Office of Air and Waste Management
P.O. Box 2063
Harrisburg, PA 17105-8472
(717) 787-7382

Penn State University
Pennsylvania Tech. Assistance Program
110 Barbara Bldg. II
810 North University Drive
University Park, PA 16802
(814) 865-0427

University of Pittsburgh
Center for Hazardous Materials Research
Applied Research Center
320 William Pitt Way
Pittsburgh, PA 15238
(412) 826-5320

Rhode Island
Department of Environmental Mgmt.
Hazardous Waste Reduction Program
83 Park Street
Providence, Ri 02903-1037
(401) 277-3434

South Carolina
Clemson University
Continuing Engineering Education
Hazardous Waste Management
Research Fund
P.O. Drawer 1607
Clemson, SC 29633
(803) 656-3308

Department of Health and Environmental
    Control
Center for Waste Minimization
2600 Bull Street
Columbia, SC 29201
(802) 734-4715

South Dakota
Department of Environment and Natural
    Resources
Waste Management Program
Joe Foss Building
523 East Capitol Avenue
Pierre, SD 57501-3181
(605) 773-4216

Tennessee
Department of Health and Environment
Bureau of Environment
14th Floor, L & C Building
401 Church Street
Nashville, TN 37213-0455
(605) 741-3657

Tennessee Valley Authority
Waste Reduction & Management
Section
311 Board Street
Chattanooga, TN 37406
(615) 7;1-4574

University of Tennessee
Waste Reduction Assistance Program
Center for Industrial Services
226 Capitol Blvd. Building
Nashville, TN 37219-1804
(615) 242-2456

Texas
Texas Tech University
Center for Hazardous and Toxic Waste Studies
P.O. Box 4679
Lubbock, TX 79409-3121
(806) 742-1413

Texas Water Commission
Office of Pollution Prevention and
Conservation
P.O. Box 13087
Capitol Station
Austin, TX 78711-3087
(512) 463-7869

Utah
Department of Environmental Quality
Office of Executive Director
168 N 1950 West Street
P.O. Box 144810
Salt Lake City, UT 84114-4810
(801) 536-4477

Vermont
Department of Environmental Conservation
Pollution Prevention Division
103 South Main Street
Waterbury, VT 05676
(802) 563-8702

Department of Environmental Conservation
Recycling and Resource Conservation Section
103 South Main Street
Waterbury, VT 05676
(802) 244-8702

Virginia
Department of Waste Management
Waste Minimization Program
Monroe Bldg., 11th Floor
101 North 14th Street
Richmond, VA 23219
(804) 371-8716

Virginia Polytechnic Institute and
    State University
University Center for Environment
and Hazardous Materials Studies
Blacksburg, VA 24061-0113
(703) 231-7508

Washington
Washington Department of Ecology
Toxics Reduction Section
Mail Stop PV-11
Olympia, WA 98504-8711
(206) 407-6723

West Virginia
Division of Natural Resources
Waste Management Section
Pollution Prevention and Open Dump
    Program
1356 Hansford Street.
Charleston, WV 25301
(304) 558-4000

Wisconsin
Department of Development
Hazardous Pollution Prevention Audit
Grant Program
P.O. Box 7979
132 West Washington Ave.
Madison, Wl 53707
(608) 266-3075

Department of Natural Resources
Bureau of Solid and Hazardous Waste
    Management
P.O. Box 7921
Madison, Wl 53707-7921
(608) 267-3763

Wyoming
Department of Environmental Quality
Solid Waste Management Program
122 W. 25th Street—Herschler Building
Cheyenne, WY 82002
(307) 777-7752

Appendix B

The New Toxicity Characteristic Rule: Information and Tips for Generators

United States
Environmental Protection
Agency

Solid Waste and
Emergency Response
(OS-305)

EPA/530-SW-90-028
January 1990

⊛EPA

The New Toxicity Characteristic Rule

Information and Tips for Generators

TC
REPLACES
EP
TOXICITY

LQG
GUIDELINES

WASTE
MINIMIZATION

Waste Minimization
Should Be the Key
Component of Your
Company's Hazardous
Waste Management
Program

Printed on recycled paper

The Toxicity Characteristic rule will bring a significant volume of previously unregulated waste into the federal hazardous waste management system. Businesses that take steps to reduce generation of these wastes may realize substantial savings in waste management costs. Tips on waste minimization practices are provided below, along with contacts for more information.

If you produce 1,000 kilograms or more of hazardous waste per month, or think you might, this publication is for you. If you produce less waste, please order our publication for small quantity generators, *Does Your Business Produce Hazardous Waste? Many Small Businesses Do* (EPA/530-SW-90-027).

The final Toxicity Characteristic rule replaces the existing Extraction Procedure Toxicity Characteristic (EPTC) by adding 25 chemicals to the eight metals and six pesticides on the EPTC list of constituents. The rule also replaces the Extraction Procedure (EP) leach test with the Toxicity Characteristic Leaching Procedure (TCLP). Regulatory levels for the 25 new constituents are determined by the product of a health-related concentration threshold and a dilution/attenuation factor (DAF). The DAF is derived using a new ground-water transport model.

What the TC Rule Means to Generators

Facilities that produce 1,000 kilograms (2,200 pounds) or more of hazardous waste within a calendar month are classified by EPA as hazardous waste generators. For the sake of discussion, we will call this group "Large Quantity Generators," or LQGs. LQGs are subject to more stringent storage requirements under federal law than their small quantity generator (SQG) counterparts, who produce less than 1,000 kilograms, but more than 100 kilograms of hazardous waste in a month.

Under the new rule, generators are required to use the Toxicity Characteristic criteria in place of the EPTC when determining whether any hazardous waste is produced by their industrial processes. **Some facilities previously regulated as small quantity generators may be brought into the LQG category** with this rule, since the volume of "TC waste" generated must now be added to monthly hazardous waste volume calculations. Newly regulated LQGs are subject to all generator requirements as outlined in Title 40, Part 262 of the Code of Federal Regulations (CFR).

The new rule might also affect some businesses previously not subject to hazardous waste regulation. These businesses are now required to determine whether TC constituents are present in their waste stream at or above the regulated levels. If so, they must comply with appropriate generator requirements.

Compliance Guidelines

Under the new rule, waste generators must determine whether one or more of the TC constituents are present in their waste at or above the regulated levels. If so, LQGs must take the following steps to comply with the new rule:

1. Notify EPA and obtain an EPA Identification Number (using EPA Form 8700-12) **within six months** of the rule's publication. Businesses that already have an EPA ID Number for their facility are not required to renotify EPA.

2. Comply with all on-site waste storage requirements, including time constraints (maximum 90 days) and satellite storage regulations. Facilities that have permits to treat or dispose of their waste on site may require permit modifications under this rule. Contact your EPA Regional office for more information.

3. Ensure that a Uniform Hazardous Waste Manifest (EPA Form 8700-22) accompanies any waste that is transported offsite. **Generators are legally responsible for the proper disposal of their hazardous waste.** TC waste is subject to all disposal requirements found in Part 265 of the CFR; generators must ensure that their waste is disposed of only at properly permitted facilities.

LQGs must be in full compliance with these requirements by September 25, 1990. Generators who fail to take these steps are subject to fines and other EPA enforcement actions.

(continued on back panel)

Reducing Management Costs

Waste minimization means reducing the amount of waste your company generates. You can reduce your waste volume in several ways, including:

- Separating hazardous waste from nonhazardous waste.
- Recycling (also called recovery and reuse).
- Substitution of nonhazardous for hazardous raw materials.
- Changing manufacturing processes so that less hazardous waste is produced.
- Redesigning or reformulating end-products to be less hazardous.

The payoffs to your company for waste minimization can be great. Waste minimization can save money through more efficient use of resources and reduced waste treatment and disposal costs. It can also reduce your hazardous waste-related financial liabilities. Finally, waste minimization can pay off when local residents are confident that your industry is making every effort to handle wastes responsibly.

For More Information...

For more information about the TC rule, how it affects you and your business, and waste minimization technologies, contact the following sources:

- Your trade association
- Regional EPA office
- State hazardous waste program office
- RCRA/Superfund Hotline at 800-424-9346 or TDD 800-553-7672 for the hearing impaired
- Pollution Prevent Information Clearinghouse (access through the RCRA/Superfund Hotline)

If you have not already received the following materials, they might be available through your trade association—or you can order them through the RCRA/Superfund Hotline:

- Industry-specific Sheets. These provide waste stream information for the following 18 industries potentially affected by the TC rule [EPA/530-SW-90-027a-r].

  a. Vehicle Maintenance
  b. Drycleaning and Laundry
  c. Furniture/Wood Refinishing
  d. Equipment Repair
  e. Textile Manufacturing
  f. Wood Preserving
  g. Printing and Allied Industry
  h. Chemical Manufacturing
  i. Pesticide End-users
  j. Construction
  k. Railroad Transport
  l. Educational/Vocational
  m. Laboratories
  n. Metal Manufacturing
  o. Pulp and Paper Industry
  p. Formulators
  q. Cleaning and Cosmetics
  r. Leather/Leather Products

- *Waste Minimization: Environmental Quality and Economic Benefits* [EPA/530-SW-87-026]
- *The EPA Manual for Waste Minimization Opportunity Assessments* [EPA-600/2-88-025]
- *Modifying RCRA Permits* [EPA/530-SW-89-050]
- *How to Set Up a Local Program to Recycle Used Oil* [EPA/530-SW-89-039A] (Pamphlets: 039B-D).

The Following Constituents Are Now Regulated under the TC Rule:

Old EP Constituents
Arsenic
Barium
Cadmium
Chromium
Lead
Mercury
Selenium
Silver
Endrin
Lindane
Methoxychlor
Toxaphene
2,4-Dichlorophenoxycetic acid
2,4,5-Trichlorophenoxypropionic acid

New Organic Constituents
Benzene
Carbon Tetrachloride
Chlordane
Chlorobenzene
Chloroform
m-Cresol
o-Cresol
p-Cresol
Cresol
1,4-Dichlorobenzene
1,2-Dichloroethane
1,1-Dichloroethylene
2,4-Dinitrotoluene
Heptachlor (and its hydroxide)
Hexachloro-1,3-butadiene
Hexachlorobenzene
Hexachloroethane
Methyl ethyl ketone
Nitrobenzene
Pentachlorophenol
Pyridine
Tetrachloroethylene
Trichloroethylene
2,4,5-Trichlorophenol
2,4,6-Trichlorophenol
Vinyl chloride

Appendix C

Notification of Regulated Waste Activity (EPA Form 8700-12) and Instructions

Please print or type with ELITE type (12 characters per inch) in the unshaded areas only

Form Approved, OMB No. 2050-0028 Expires 9-30-96
GSA No. 0246-EPA-OT

Please refer to the *Instructions for Filing Notification* before completing this form. The information requested here is required by law (Section 3010 of the *Resource Conservation and Recovery Act*).	**⊕EPA** **Notification of Regulated Waste Activity** United States Environmental Protection Agency	**Date Received** **(For Official Use Only)**

I. Installation's EPA ID Number *(Mark 'X' in the appropriate box)*

☐ **A. First Notification**	☐ **B. Subsequent Notification** *(Complete item C)*	**C. Installation's EPA ID Number**

II. Name of Installation *(Include company and specific site name)*

III. Location of Installation *(Physical address not P.O. Box or Route Number)*

Street

Street *(Continued)*

City or Town	**State**	**Zip Code**

County Code	**County Name**

IV. Installation Mailing Address *(See Instructions)*

Street or P.O. Box

City or Town	**State**	**Zip Code**

V. Installation Contact *(Person to be contacted regarding waste activities at site)*

Name *(Last)*	*(First)*

Job Title	**Phone Number** *(Area Code and Number)*

VI. Installation Contact Address *(See Instructions)*

A. Contract Address Location Mailing Other	**B. Street or P.O. Box**

City or Town	**State**	**Zip Code**

VII. Ownership *(See Instructions)*

A. Name of Installation's Legal Owner

Street, P.O. Box, of Route Number

City or Town	**State**	**Zip Code**

Phone Number *(Area Code and Number)*	**B. Land Type**	**C. Owner Type**	**D. Change of Owner Indicator** Yes No	**(Date Changed)** Month Day Year

EPA Form 8700-12 (Rev. 11-30-93) Previous edition is obsolete.

Continued on Reverse

Please print or type with ELITE type (12 characters per inch) in the unshaded areas only

Form Approved, OMB No. 2050-0028 Expires 9-30-96
GSA No. 0246-EPA-OT

ID - For Official Use Only

IX. Description of Regulated Wastes *(Additional Sheet)*

B. Listed Hazardous Wastes. *(See 40 CFR 261.31 - 33; Use this page only if you need to list more than 12 waste codes.)*

13	14	15	16	17	18
19	20	21	22	23	24
25	26	27	28	29	30
31	32	33	34	35	36
37	38	39	40	41	42
43	44	45	46	47	48
49	50	51	52	53	54
55	56	57	58	59	60
61	62	63	64	65	66
67	68	69	70	71	72
73	74	75	76	77	78
79	80	81	82	83	84
85	86	87	88	89	90
91	92	93	94	95	96
97	98	99	100	101	102
103	104	105	106	107	108
109	110	111	112	113	114
115	116	117	118	119	120

EPA Form 8700-12 (Rev. 11-30-93) **Previous edition is obsolete.**

V. Line-by-Line Instructions for Completing EPA Form 8700-12

Type or print in black ink all items except Item X, "Signature," leaving a blank box between words. The boxes are spaced at 1/4" intervals which accommodate elite type (12 characters per inch). When typing, hit the space bar twice between characters. If you print, place each character in a box. Abbreviate if necessary to stay within the number of boxes allowed for each Item. If you must use additional sheets, indicate clearly the number of the Item on the form to which the information on the separate sheet applies.

Note: When submitting a **subsequent notification** form, notifiers must complete in their entirety Items I, II, IV, VI, VII, VIV and X. Other sections that are being added to (i.e., newly regulated activities) or altered (i.e., installation contact) must also be completed. All other sections may be left blank.

Item I -- Installations EPA ID Number:

Place an "X" in the appropriate box to indicate whether this is your first or a subsequent notification *for this site.* If you have filed a previous notification, enter the EPA Identification Number assigned to this site in the boxes provided. Leave EPA ID Number blank if this is your first notification *for this site.*

Note: When the owner of an installation changes, the new owner must notify U.S. EPA of the change, even if the previous owner already received a U.S. EPA Identification Number. Because the U.S. EPA ID Number is site-specific, the new owner will keep the existing ID number. If the installation moves to another location, the owner/operator must notify EPA of this change. In this instance a new U.S. EPA Identification Number will be assigned, since the installation has changed locations.

Items II and IV -- Name and Location of Installation:

Complete Items II and IV. Please note that the address you give for Item IV, Location of Installation, must be a physical address, *not a post office box or route number.*

Notification of Regulated Waste Activity

County Code and Name: Give the county code, if known. If you do not know the county code, enter the county name, from which EPA can automatically generate the county code. If the county name is unknown contact the local Post Office. To obtain a list of county codes, contact the National Technical Information Service U.S. Department of Commerce, Springfield, Virginia, 22161 or at (703) 487-4650. The list of codes is contained in the Federal Information Processing Standards Publication (FIPS PUB) number 6-3.

Item IV -- Installation Mailing Address:

Please enter the Installation Mailing Address. If the Mailing Address and the Location of Installation (Item IV) are the same, you can print "Same" in the box for Item IV.

Item V -- Installation Contact:

Enter the name, title, and business telephone number of the person who should be contacted regarding information submitted on this form.

Item VI -- Installation Contact Address:

A. **Code:** If the contact address is the same as the location of installation address listed in Item IV or the installation mailing address listed in Item IV, place an "X" in the appropriate box to indicate where the contact may be reached. If the location of installation address, the installation mailing address, and the installation contact address are all the same, mark the "Location" box. If the contact address is *not* the same as those addresses in either Item III or IV, place an "X" in the "Other" box and complete Item VI.B. If an "X" is entered in either the location or mailing box, Item VI.B. should be left blank.

B. **Address:** Enter the contact address *only* if the contact address is different from either the location of installation address (Item IV) or the installation mailing address (Item IV), and if Item VI.A. was marked "Other."

Item VII - Ownership:

A. **Name:** Enter the name of the legal owner(s) of the installation, including the property owner. Also enter the address and phone number where this individual can be reached. Use the comment section in XI or additional sheets if necessary to list more than one owner.

B. **Land Type:** Using the codes listed below, indicate in VII.B. the code which *best describes* the current legal status of the land on which the installation is located:

F = Federal
S = State
I   = Indian
P = Private
C = County
M = Municipal*
D = District
O = Other

Note: If the Owner Type is **best described as Indian, County or District, please use those codes. Otherwise, use Municipal.*

C. **Owner Type:** Using the codes listed below, indicate in VII. C. the code which *best describes* the legal status of the current owner of the installation:

F = Federal
S = State
I   = Indian
P = Private
C = County
M = Municipal*
D = District
O = Other

Note: If the Owner Type is **best described as Indian, County or District, please use those codes. Otherwise, use Municipal.*

D. **Change of Owner Indicator:** *(If this is your installations first notification, leave Item VII.D. blank and skip to Item VIII. If this is a subsequent notification, complete Item VII.D. as directed below.)*

If the owner of this installation has changed since the installation's original notification, place an "X" in the box marked "Yes" and enter the date the owner changed.

If the owner of this installation has not changed since the installation's original notification, place an "X" in the box marked "No" and skip to Item VIII.

If an additional owner(s) has been added or replaced since the installation's original notification, place an "X" in the box marked "Yes." Use the comment section in XI to list any additional owners, the dates they became owners, and which owner(s) (if any) they replaced. If necessary attach a separate sheet of paper.

Item VIII -- Type of Regulated Waste Activity:

A. **Hazardous Waste Activity:** Mark an "X" in the appropriate box(es) to show which hazardous waste activities are going on **at this installation.**

    1. **Generator:** If you generate a hazardous waste that is identified by characteristic or listed in 40 CFR Part 261, mark an "X" in the appropriate box for the quantity of non-acutely hazardous waste that is generated per calendar month. If you generate acutely hazardous waste please refer to 40 CFR Part 262 for further information.

    2. **Transporter:** If you transport hazardous waste, indicate if it is your own waste, for commercial purposes, or mark both boxes if both classifications apply. Mark an "X" in each appropriate box to indicate the method(s) of transportation you use. Transporters do not have to complete Item IX of this form, but must sign the certification in Item X. The Federal regulations for hazardous waste transporters are found in 40 CFR Part 263.

    3. **Treater/Storer/Disposer:** If you treat, store or dispose of regulated hazardous waste, then mark an "X" in this box. You are reminded to contact the appropriate addressee listed for your State in Section III.C. of this package to request Part A of the RCRA Permit Application. The Federal regulations for hazardous waste installation owners/operators are found in 40 CFR Parts 264 and 265.

    4. **Hazardous Waste Fuel:** If you market hazardous waste fuel, place an "X" in the appropriate box(es). If you burn hazardous waste fuel on-site, place an "X" in the appropriate box and indicate the type(s) of combustion devices in which hazardous waste fuel is burned. (Refer to the definition section for complete descriptions of each device).

Note: Generators are required to notify for waste-as-fuel activities only if they market directly to the burner.

"Other Marketer" is defined as any person, other than a generator marketing hazardous waste, who markets hazardous waste fuel.

5. **Underground Injection Control:** If you generate and/or treat, store or dispose of hazardous waste, place an "X" in the box if an injection well is located at your installation. "Underground Injection" means the subsurface emplacement of fluids through a bored, drilled or driven well; or through a dug well, where the depth of the dug well is greater than the largest surface dimension.

B. **Used Oil Recycling Activities:** Mark an "X" in the appropriate box(es) to indicate which used oil recycling activities are taking place at this installation.

1. **Used Oil Fuel Marketer:** If you market off-specification used oil, mark an "X" in box 1a. If you are the first to claim the used oil meets the used oil specification established in 40 CFR Part 279.11, mark an "X" in box 1b. If either of these boxes are marked, you must also notify (or have previously notified) as a used oil transporter, off-specification used oil fuel burner, or used oil processor/re-refiner, unless you are a used oil generator. (Used oil generators are not required to notify.)

2. **Used Oil Burner:** If you burn off-specification used oil fuel, place an "X" in the box(es) to indicate the type(s) of combustion device(s) in which off-specification used oil fuel is burned. (Refer to the definition section for complete descriptions of each device.)

3. **Used Oil Transporter:** If you transport used oil and/or own/operate a used oil transfer facility, place an "X" in the appropriate box(es) to indicate this used oil recycling activity.

4.   **Used Oil Processor/Re-refiner:** If you process and/or re-refine used oil, place an "X" in the appropriate box(es) to indicate this used oil activity.

Item IX -- Description of Hazardous Wastes:

Note: Only persons involved in hazardous waste activity (Item VIII.A.) need to complete this item. Transporters requesting a U.S. EPA Identification Number do not need to complete this item, but must sign the "Certification" in Item X.

You will need to refer to 40 CFR Part 261 (enclosed as Section VII) in order to complete this section. Part 261 identifies those wastes that EPA defines as hazardous. If you need help completing this section, please contact the appropriate addressee for your State as listed in Section IV. C. of this package.

A.   **Characteristics of Nonlisted Hazardous Wastes:** If you handle hazardous wastes which are not listed in 40 CFR Part 261, Subpart D, but do exhibit a characteristic of hazardous waste as defined in 40 CFR Part 261, Subpart C, you should describe these wastes by the EPA hazardous waste number for the characteristic. Place an "X" in the box next to the characteristic of the wastes that you handle. If you mark "4. Toxicity Characteristic," please list the specific EPA hazardous waste number(s) for the specific contaminant(s) in the box(es) provided. Refer to Section VIV to determine the appropriate hazardous waste number(s).

B.   **Listed Hazardous Wastes:** If you handle hazardous wastes that are listed in 40 CFR Part 261, Subpart D, enter the appropriate 4-digit numbers in the boxes provided.

   Note: If you handle more than 12 listed hazardous wastes, please continue listing the waste codes on the extra sheet provided at the end of this booklet. If it is used, attach the additional page to the rest of the form before mailing it to the appropriate EPA Regional or State Office.

C.   **Other Wastes:** If you handle other wastes or State regulated wastes that have a waste code, enter the appropriate code number in the boxes provided.

Item X -- Certification:

This certification must be signed by the owner, operator, or an authorized representative of your installation. An "authorized representative" is a person responsible for the overall operation of the installation (i.e., a plant manger or superintendent, or a person of equal responsibility). *All notifications must include this certification to be complete.*

Item XI -- Comments:

Use this space for any additional comments.

VI. Definitions

The following definitions are included to help you to understand and complete the Notification Form:

Act or RCRA means the Solid Waste Disposal Act, as amended by the Resource Conservation and Recovery Act of 1976, as amended by the Hazardous and Solid Waste Amendments of 1984, 42 U.S.C. Section 6901 *et seq.*

Authorized Representative means the person responsible for the overall operation of the installation or an operational unit (i.e., part of a installation), e.g., superintendent or plant manager, or person of equivalent responsibility.

Boiler means an enclosed device using controlled flame combustion and having the following characteristics:

1.   The unit has physical provisions for recovering and exporting energy in the form of steam, heated fluids, or heated gases;

2.   The units combustion chamber and primary energy recovery section(s) are of integral design (i.e., they are physically formed into one manufactured or assembled unit);

3.   The unit continuously maintains an energy recovery efficiency of at least 60 percent, calculated in terms of the recovered energy compared with the thermal value of the fuel;

Notification of Regulated Waste Activity

4. The unit exports and utilizes at least 75 percent of the recovered energy, calculated on an annual basis (excluding recovered heat used internally in the same unit, for example, to preheat fuel or combustion air or drive fans or feedwater pumps); and

5. The unit is one which the Regional Administrator has determined on a case-by-case basis, to be a boiler after considering the standards in 40 CFR 260.32.

Burner means the owner or operator of any boiler or industrial furnace that burns hazardous waste fuel for energy recovery and that is not regulated as a RCRA hazardous waste incinerator.

Disposal means the discharge, deposit, injection, dumping, spilling, leaking, or placing of any solid waste or hazardous waste into or on any land or water so that such solid waste or hazardous waste or any constituent thereof may enter the environment or be emitted into the air or discharged into any waters, including ground waters.

Disposal Installation means a installation or part of a installation at which hazardous waste is intentionally placed into or on any land or water, and at which waste will remain after closure.

EPA Identification (I.D.) Number means the number assigned by EPA to each generator, transporter, and treatment, storage, or disposal installation.

Generator means any person, by site, whose act or process produces hazardous waste identified or listed in 40 CFR Part 261.

Hazardous Waste means a hazardous waste as defined in 40 CFR 261.3.

Hazardous Waste Fuel means hazardous waste and any fuel that contains hazardous waste that is burned for energy recovery in a boiler or industrial furnace that is not subject to regulation as a RCRA hazardous waste incinerator. However, the following hazardous waste fuels are subject to regulation as used oil fuels:

1. Used oil fuel burned for energy recovery that is also a hazardous waste solely because it exhibits a characteristic of hazardous waste identified in Subpart C of 40 CFR Part 261; and

Notification of Regulated Waste Activity

Table 1
Alphabetized State Listing of Contacts for Obtaining and Submitting the Notification Form

Alabama

Obtain information or forms from, and mail completed forms to:
Land Division
Alabama Department of
  Environmental Management
1751 Cong. Wm. L. Dickinson Drive
Montgomery, Alabama 36130
(205) 271-7730

Alaska

Obtain information or forms from:
Department of Environmental Conservation
410 Willoughby Avenue, Suite 105
Juneau, Alaska 99801-1795
(907) 465-5150
Mail completed forms to:
U.S. EPA Region 10
Waste Management Branch, HW-105
1200 Sixth Avenue
Seattle, Washington 98101
(206) 553-0151

American Samoa

Obtain information from:
Environmental Quality Commission
Government of American Samoa
Pago Pago, American Samoa 96799
Overseas Operator Commercial call (684) Country
Code 663-2304
Obtain forms from and mail completed forms to:
U.S. EPA Region 9
Hazardous Waste Management Division
75 Hawthorne Street, H-2-1
San Francisco, CA 94105
(415) 744-2074

Arizona

Obtain information or forms from:
Office of Waste Programs
Arizona Department of
  Environmental Quality
3033 N. Central Avenue
Phoenix, Arizona 85012
(602) 207-4108

Mail completed forms to:
U.S. EPA Region 9
Hazardous Waste Management Division
75 Hawthorne Street, H-2-1
San Francisco, CA 94105
(415) 744-2074

Arkansas

Obtain information or forms from, and mail completed forms to:
Arkansas Department of Pollution
  Control and Ecology
8001 National Drive
P.O. Box 8913
Little Rock, Arkansas 72219-8913
(501) 570-2872

California

Obtain information or forms from, and mail completed forms to:
U.S. EPA Region 9
Hazardous Waste Management Division
75 Hawthorne Street, H-2-1
San Francisco, CA 94105
(415) 744-2074

Colorado

Obtain information or forms from, and mail completed forms to:
Colorado Department of Health
4300 Cherry Creek Drive, South
HMWMD-HWC-B2
Denver, Colorado 80222-1530
(303) 692-3300

Connecticut

Obtain information or forms from, and mail completed forms to:
Hazardous Materials Management Unit
Department of Environmental Protection
State Office Building
165 Capitol Avenue
Hartford, Connecticut 06106
(203) 566-4869

Notification of Regulated Waste Activity

Table 1 (continued)

Delaware
Obtain information or forms from, and mail completed forms to:
Delaware Department of Natural Resources
& Environmental Control
Division of Air and Waste Management
Hazardous Waste Management Branch
P.O. Box 1401, 89 Kings Highway
Dover, Delaware 19903
(302) 739-3689
(302) 739-3672

District of Columbia
Obtain information or forms from, and mail completed forms to:
Department of Consumer and
   Regulatory Affairs
Environmental Regulation Administration
Hazardous Waste Branch
2100 Martin Luther King Jr. Ave., S.E.
Washington, D.C. 20020
(202) 404-1167

Florida
Obtain information or forms from, and mail completed forms to:
Hazardous Waste Regulation Section
Department of Environmental Regulation
Twin Towers Office Building
2600 Blair Stone Road
Tallahassee, Florida 32399-2400
(904) 488-0300

Georgia
Obtain information or forms from, and mail completed forms to:
Land Protection Branch
Industrial and Hazardous
Waste Management Program
1154 East Tower
205 Butler Street, S.E.
Atlanta, Georgia 30334
(404) 656-7802

Guam
Obtain information from:
Guam Environmental Protection Agency
Harmon Plaza
Complex Unit B-107
103 Orjas Street
Harmon, Guam 96911

Obtain forms from and mail completed forms to:
U.S. EPA Region 9
Hazardous Waste Management Division
75 Hawthorne Street, H-2-1
San Francisco, CA 94105
(415) 744-2074

Hawaii
Obtain information from:
Hawaii Department of Health
Solid and Hazardous Waste Branch
Five Waterfront Plaza, Suite 250
500 Ala Moana Boulevard
Honolulu, Hawaii 96813
*Obtain forms from and mail
completed forms to:*
U.S. EPA Region 9
Hazardous Waste Management Division
75 Hawthorne Street, H-2-1
San Francisco, California 94105
(415) 744-2074

Idaho
Obtain information or forms from, and mail completed forms to:
Department of Environmental Quality
1410 N. Hilton, Third Floor
Boise, Idaho 83706
(208) 334-5898

Illinois
Obtain information or forms from, and mail completed forms to:
Illinois Environmental Protection Agency
Division of Land Pollution Control
2200 Churchill Road
Springfield, Illinois 62706
(217) 785-8452

Indiana
Obtain information or forms from:
Indiana Department of
   Environmental Management
105 S. Meridian Street
P.O. Box 6015
Indianapolis, Indiana 46225
(317) 232-8925

Notification of Regulated Waste Activity

Table 1 (continued)

Mail completed forms to:
U.S. EPA Region 5
RCRA Activities
P.O. Box A3587
Chicago, IL 60690
(312) 886-4001

Iowa
Obtain information or forms from, and mail completed forms to:
U.S. EPA Region 7
RCRA Branch
Attn: RCRA/IOWA
726 Minnesota Avenue
Kansas City, Kansas 66101
(913) 551-7861

Kansas
Obtain information or forms from, and mail completed forms to:
Department of Health and Environment
Attn: Hazardous Waste Section
Forbes Field, Building 740
Topeka, Kansas 66620
(913) 296-1600

Kentucky
Obtain information or forms from, and mail completed forms to:
Division of Waste Management
Department of Environmental Protection
Cabinet for Natural Resources
& Environmental Protection
Fort Boone Plaza, Bldg. #2
14 Reilly Road
Frankfort, Kentucky 40601
(502) 564-6716

Louisiana
Obtain information or forms from, and mail completed forms to:
Louisiana Department of
  Environmental Quality
Department of Solid and Hazardous Waste
P.O. Box 82178
Baton Rouge, Louisiana 70884-2178
(504) 765-0332

Maine
Obtain information or forms from, and mail completed forms to:
Bureau of Oil and
  Hazardous Materials Control
Department of Environmental Protection
State House Station #17
Augusta, Maine 04333
(207) 289-2651

Maryland
Obtain information or forms from, and mail completed forms to:
Maryland Department of the Environment
Waste Management Administration
Hazardous Waste Program
2500 Broening Highway
Baltimore, Maryland 21224
(410) 631-3343
(410) 631-3344

Massachusetts
Obtain information or forms from, and mail completed forms to:
Division of Solid and Hazardous Waste
Massachusetts Department of
  Environmental Protection
One Winter Street, 5th Floor
Boston, Massachusetts 02108
(617) 292-5851

Michigan
Obtain information or forms from:
Waste Management Division
Michigan Department of Natural Resources
Box 30241
Lansing, Michigan 48909
(517) 373-2730

Notification of Regulated Waste Activity

Table 1 (continued)

Mail completed forms to:
U.S. EPA Region 5
RCRA Activities
P.O. Box A3587
Chicago, IL 60604
(312) 886-4001

Minnesota
Obtain information or forms from:
Solid and Hazardous Waste Division
Minnesota Pollution Control Agency
520 Lafayette Road, North
St. Paul, Minnesota 55155
(612) 297-8330
Mail completed forms to:
U.S. EPA Region 5
RCRA Activities
P.O. Box A3587
Chicago, IL 60604
(312) 886-4001

Mississippi
Obtain information or forms from, and mail completed forms to:
Department of Environmental Quality
Attn: Jerry Banks
P.O. Box 10385
Jackson, Mississippi 39289-0385
(601) 961-5171

Missouri
Obtain information or forms from, and mail completed forms to:
Waste Management Program
Department of Natural Resources
Jefferson Building
205 Jefferson Street (13/14 floor)
P.O. Box 176
Jefferson City, Missouri 65102
(314) 751-3176

Montana
Obtain information or forms from, and mail completed forms to:
Solid and Hazardous Waste Bureau
Department of Health and
   Environmental Sciences
Cogswell Building
Helena, Montana 59620
(406) 444-1430

Nebraska
Obtain information or forms from, and mail completed forms to:
Hazardous Waste Management Section
Department of Environmental Quality
State House Station
P.O. Box 98922
Lincoln, Nebraska 68509-8922
(402) 471-2186

Nevada
Obtain information or forms from:
Waste Management Program
Division of Environmental Protection
The Bureau of Waste
   Management Permits Branch
333 West Nye Lane
Carson City, NV 89710
(702) 687-5872
Mail completed forms to:
U.S. EPA Region 9
Hazardous Waste Management Division
75 Hawthorne Street, H-2-1
San Francisco, CA 94105
(415) 744-2074

New Hampshire
Obtain information or forms from, and mail completed forms to:
Division of Public Health Services
Office of Waste Management
Bureau of Hazardous Waste
   Classification & Manifests
Department of Health and Welfare
Health and Welfare Building
6 Hazen Drive
Concord, New Hampshire 03301
(603) 271-2900

New Jersey
Obtain information from:
New Jersey Department of Environmental
   Protection
Hazardous Waste Regulation Program
Bureau of Advisement and Manifests
401 East State Street, CN-421
Trenton, New Jersey 08625-0421
(609) 292-8341

Notification of Regulated Waste Activity

Table 1 (continued)

*Obtain forms from and mail
completed forms to:*
U.S. EPA Region 2
Air and Waste Management Division
Attn: RCRA Notifications
26 Federal Plaza, Room 505
New York, NY 10278
(212) 264-9883

New Mexico
Obtain information and forms from:
Hazardous Waste Bureau
525 Camino De Loss Marquez
Sante Fe, New Mexico 87501
(505) 827-4358
Mail completed forms to:
U.S. EPA Region 6
Hazardous Waste Management Division
First Interstate Bank Tower
1445 Ross Avenue, Suite 1200
Dallas, Texas 75202-2733
(214) 655-6750

New York
Obtain information from:
New York Department of
  Environmental Conservation
Solid and Hazardous Waste
  Manifest Section
50 Wolfe Road
Albany, New York 12212
(518) 457-6858
*Obtain forms from and mail
completed forms to:*
U.S. EPA Region 2
Air and Waste Management Division
Attn: RCRA Notifications
26 Federal Plaza, Room 505
New York, NY 10278
(212) 264-9883

North Carolina
*Obtain information or forms from, and mail
completed forms to:*
Hazardous Waste Management Branch
Division of Solid Waste Management
Environment, Health & Natural Resources
P.O. Box 27687
Raleigh, North Carolina 27611-7687
(919) 733-2178

North Dakota
*Obtain information or forms from, and mail
completed forms to:*
Division of Waste Management
Department of Health and
  Consolidated Laboratories
1200 Missouri Avenue
P.O. Box 5520
Bismarck, North Dakota 58506-5520
(701) 328-5166

Northern Mariana Islands
Obtain information from:
Department of Public Health and
  Environmental Services
Division of Environmental Quality
Saipan, Mariana Islands 96950
Overseas Operator: (676) 234-6984
Cable Address: Gov. NMI Saipan
*Obtain forms from and mail
completed forms to:*
U.S. EPA Region 9
Hazardous Waste Management Division
75 Hawthorne Street, H-2-1
San Francisco, California 94105
(415) 744-2074

Ohio
*Obtain information or forms from, and mail
completed forms to:*
Ohio Environmental Protection Agency
1800 WaterMark Drive
Columbus, Ohio 43215
(614) 644-2977

Oklahoma
Obtain information or forms from:
Department of Environmental Quality
Hazardous Waste Quality
  Management Service
1000 Northeast 10th Street
Oklahoma City, Oklahoma 73117-1212
(405) 271-5338
Mail completed forms to:
U.S. EPA Region 6
Hazardous Waste Management Division
First Interstate Bank Tower
1445 Ross Avenue, Suite 1200
Dallas, Texas 75202-2733
(214) 655-6750

Notification of Regulated Waste Activity

Table 1 (continued)

Oregon
Obtain information or forms from, and mail completed forms to:
Oregon Department of
  Environmental Quality
Hazardous Waste Operations
811 Southwest 6th Avenue
Portland, Oregon 97204
(503) 229-5913

Pennsylvania
Obtain information from:
Pennsylvania Department of
  Environmental Resources
Bureau of Waste Management
Market Street State Office Building
400 Market Street, 14th Floor
Harrisburg, Pennsylvania 17105-8471
(717) 787-6239
Obtain forms from and mail completed forms to:
U.S. EPA Region 3
RCRA Programs Branch
Pennsylvania Section (3 HW51)
841 Chestnut Street
Philadelphia, PA 19107
(215) 597-1230

Puerto Rico
Obtain information from:
Puerto Rico Environmental Quality Board
Land Pollution Control Area
Inspection, Monitoring and Surveillance
P.O. Box 11488
Santurce, Puerto Rico 00910-1488
(809) 722-0439
Obtain forms from and mail completed forms to:
U.S. EPA Region 2
Air and Waste Management Divison
Attn: RCRA Notifications
26 Federal Plaza, Room 505
New York, New York 10278
(212) 264-9883

Rhode Island
Obtain information or forms from, and mail completed forms to:
Solid Waste Management Program
Department of Environmental Management
204 Canon Building 75 Davis Street
Providence, Rhode Island 02908
(401) 277-2797

South Carolina
Obtain information or forms from, and mail completed forms to:
Bureau of Solid and Hazardous
  Waste Management
Department of Health and
  Environmental Control
2600 Bull Street
Columbia, South Carolina 29201
(803) 734-5214

South Dakota
Obtain information or forms from, and mail completed forms to:
Department of Environment and
  Natural Resources
Office of Waste Management
319 Coteau
c/o 500 E. Capital Avenue
Pierre, South Dakota 57501-5070
(605) 773-3153

Tennessee
Obtain information or forms from, and mail completed forms to:
Division of Solid Waste Management
Tennessee Department of Public Health
401 Church Street
LNC Tower, 5th Floor
Nashville, Tennessee 37243-1535
(615) 532-0780

Texas
Obtain information or forms from:
Industrial and Hazardous Waste Division
Waste Evaluation Section
P.O. Box 13087, Capitol Station
Austin, Texas 78711-3087
(512) 908-6832

Notification of Regulated Waste Activity

Table 1 (continued)

Mail completed forms to:
U.S. EPA Region 6
Hazardous Waste Management Division
First Interstate Bank Tower
1445 Ross Avenue, Suite 1200
Dallas, Texas 75202-2733
(214) 655-6750

Utah
Obtain information or forms from, and mail completed forms to:
Division of Solid and Hazardous Waste
Department of Environmental Quality
P.O. Box 144880
Salt Lake City, Utah 84114-4880
(801) 538-6170

Vermont
Obtain information or forms from, and mail completed forms to:
Waste Management Division
Agency of Environmental Conservation
103 South Main Street
Waterbury, Vermont 05676
(802) 241-3888

Virgin Islands
Obtain information from:
Virgin Islands Department of Planning & Natural Resources
Division of Environmental Protection
179 Altona and Welgunst
St. Thomas, Virgin Islands 00801
(809) 774-3320
Obtain forms from and mail completed forms to:
U.S. EPA Region 2
Air and Waste Management Division
Attn: RCRA Notifications
26 Federal Plaza, Room 505
New York, New York 10278
(212) 264-9883

Virginia
Obtain information or forms from, and mail completed forms to:
Virginia Department of Waste Management
Monroe Building, 11th Floor
101 North 14th Street
Richmond, Virginia 23219
(804) 225-3863

Washington
Obtain information or forms from, and mail completed forms to:
Department of Ecology
P.O. Box 47658
Olympia, Washington 98504-7658
(206) 459-6316

West Virginia
Obtain information or forms from, and mail completed forms to:
Department of Commerce, Labor and Environmental Protection
Division of Environmental Protection
Office of Waste Management
1356 Hansford Street
Charleston, West Virginia 25301
(304) 558-5393

Wisconsin
Obtain information or forms from:
Bureau of Solid Waste
Department of Natural Resources
P.O. Box 7921
Madison, Wisconsin 53707
(608) 266-1327
Mail completed forms to:
U.S. EPA Region 5
RCRA Activities
P.O Box A3587
Chicago, IL 60690
(312) 886-4001

Wyoming
Obtain information or forms from, and mail completed forms to:
U.S. EPA Region 8
Hazardous Waste Management Division (8HWM-ON)
999 18th Street, Suite 500
Denver, Colorado 80202-2405
(303) 294-1361

Notification of Regulated Waste Activity

Table 2
U.S. EPA Regional Contacts for the Notification Form

U.S. EPA Region 1
RCRA Support Section
JFK Federal Building
Boston, MA 02203-2211
(617) 573-5750
> *Connecticut, Maine,*
> *Massachusetts, New Hampshire,*
> *Rhode Island, Vermont*

U.S. EPA Region 2
Air and Waste Management Division
Attn: RCRA Notifications
26 Federal Plaza, Room 505
New York, NY 10278
(212) 264-9883
> *New Jersey, New York, Puerto*
> *Rico, Virgin Islands*

U.S. EPA Region 3
RCRA Programs Branch (3 HW50)
841 Chestnut Street
Philadelphia, PA 19107
(215) 597-1230 (PA, DC)
(215) 597-3884 (VA, WV, DE, MD)
> *Delaware, District of Columbia,*
> *Maryland, Pennsylvania, Virginia,*
> *West Virginia*

U.S. EPA Region 4
Hazardous Waste Management Division
RCRA Permitting Section
345 Courtland Street, NE
Atlanta, GA 30365
(404) 347-3433
> *Alabama, Florida, Georgia,*
> *Kentucky, Mississippi, North*
> *Carolina, South Carolina,*
> *Tennessee*

U.S. EPA Region 5
RCRA Activities
P.O. Box A3587
Chicago, IL 60690
(312) 886-4001
> *Illinois, Indiana, Michigan,*
> *Minnesota, Ohio, Wisconsin*

U.S. EPA Region 6
Hazardous Waste Management Division
First Interstate Bank Tower
1445 Ross Avenue, Suite 1200
Dallas, TX 75202-2733
(214) 655-6750
> *Arkansas, Louisiana, New Mexico,*
> *Oklahoma, Texas*

U.S. EPA Region 7
RCRA Branch, Permitting Section
726 Minnesota Avenue
Attn: WSTIN/RCRA/PRMT
Kansas City, KS 66101
(913) 551-7654
> *Iowa, Kansas, Missouri, Nebraska*

U.S. EPA Region 8
Hazardous Waste Management Division
999 18th Street, Suite 500
Denver, CO 80202-2405
(303) 294-1361
> *Colorado, Montana, North Dakota,*
> *South Dakota, Utah, Wyoming*

U.S. EPA Region 9
Hazardous Waste Management Division
75 Hawthorne Street, H-3-4
San Francisco, CA 94105
(415) 744-2074
> *Arizona, California, Hawaii,*
> *Nevada, American Samoa, Guam,*
> *Northern Mariana Islands*

U.S. EPA Region 10
Waste Management Branch, HW-105
1200 Sixth Avenue
Seattle, WA 98101
(206) 553-0151
> *Alaska, Idaho, Oregon,*
> *Washington*

Appendix D

Uniform Hazardous Waste Manifest (EPA Form 8700-22) and Instructions

Please print or type.   (Form designed for use on elite (12-pitch) typewriter.)   Form Approved. OMB No. 2050-0039. Expires 9-30-96

UNIFORM HAZARDOUS WASTE MANIFEST	1. Generator's US EPA ID No.	Manifest Document No.	2. Page 1 of	Information in the shaded areas is not required by Federal law.

3. Generator's Name and Mailing Address

A. State Manifest Document Number

B. State Generator's ID

4. Generator's Phone (   )

5. Transporter 1 Company Name	6.	US EPA ID Number	C. State Transporter's ID
			D. Transporter's Phone
7. Transporter 2 Company Name	8.	US EPA ID Number	E. State Transporter's ID
			F. Transporter's Phone
9. Designated Facility Name and Site Address	10.	US EPA ID Number	G. State Facility's ID
			H. Facility's Phone

11. US DOT Description (Including Proper Shipping Name, Hazard Class and ID Number)		12. Containers		13. Total Quantity	14. Unit Wt/Vol	I. Waste No.
	HM	No.	Type			
a.						
b.						
c.						
d.						

J. Additional Descriptions for Materials Listed Above

K. Handling Codes for Wastes Listed Above

15. Special Handling Instructions and Additional Information

16. GENERATOR'S CERTIFICATION: I hereby declare that the contents of this consignment are fully and accurately described above by proper shipping name and are classified, packed, marked, and labeled, and are in all respects in proper condition for transport by highway according to applicable international and national government regulations.

If I am a large quantity generator, I certify that I have a program in place to reduce the volume and toxicity of waste generated to the degree I have determined to be economically practicable and that I have selected the practicable method of treatment, storage, or disposal currently available to me which minimizes the present and future threat to human health and the environment; OR, if I am a small quantity generator, I have made a good faith effort to minimize my waste generation and select the best waste management method that is available to me and that I can afford.

Printed/Typed Name	Signature	Month	Day	Year

17. Transporter 1 Acknowledgement of Receipt of Materials

Printed/Typed Name	Signature	Month	Day	Year

18. Transporter 2 Acknowledgement of Receipt of Materials

Printed/Typed Name	Signature	Month	Day	Year

19. Discrepancy Indication Space

20. Facility Owner or Operator: Certification of receipt of hazardous materials covered by this manifest except as noted in Item 19.

Printed/Typed Name	Signature	Month	Day	Year

Style F15 REV-6 Labelmaster, An American Labelmark Co., Chicago, IL 60646 (800)621-5808   EPA Form 8700-22 (Rev. 9-88) Previous editions are obsolete.

Please print or type.   (Form designed for use on elite (12-pitch) typewriter.)   *Form Approved. OMB No. 2050-0039. Expires 9-30-96*

UNIFORM HAZARDOUS WASTE MANIFEST *(Continuation Sheet)*	21. Generator's US EPA ID No.	Manifest Document No.	22. Page	Information in the shaded areas is not required by Federal law.

23. Generator's Name	L. State Manifest Document Number
	M. State Generator's ID

24. Transporter ____ Company Name	25. US EPA ID Number	N. State Transporter's ID
		O. Transporter's Phone
26. Transporter ____ Company Name	27. US EPA ID Number	P. State Transporter's ID
		Q. Transporter's Phone

28. US DOT Description *(Including Proper Shipping Name, Hazard Class, and ID Number)*		29. Containers		30. Total Quantity	31. Unit Wt/Vol	R. Waste No.
HM		No.	Type			
a.						
b.						
c.						
d.						
e.						
f.						
g.						
h.						
i.						

S. Additional Descriptions for Materials Listed Above	T. Handling Codes for Wastes Listed Above

32. Special Handling Instructions and Additional Information

33. Transporter ____ Acknowledgement of Receipt of Materials		Date
Printed/Typed Name	Signature	Month Day Year
34. Transporter ____ Acknowledgement of Receipt of Materials		Date
Printed/Typed Name	Signature	Month Day Year

35. Discrepancy Indication Space

Style F16 REV-6 LABELMASTER, An AMERICAN LABELMARK CO., CHICAGO, IL 60646 800-621-5808   EPA Form 8700-22A (Rev. 9-88) Previous editions are obsolete.

Letters down left margin: GENERATOR, TRANSPORTER, FACILITY

INSTRUCTIONS FOR COMPLETION OF UNIFORM HAZARDOUS WASTE MANIFEST

U.S. EPA Form 8700-22

General Information

Utah State regulation requires proper completion of all information on a
manifest. Omissions, false coding, or illegibility is considered a
violation. All generators are responsible under Utah State and Federal Law
for the proper identification, labeling, manifesting, and ultimate disposal of
all hazardous wastes they generate. The manifest system is designed to track
hazardous waste from the point of generation until its final disposal (cradle
to grave). In order to accomplish this goal, it is essential that all items
on a manifest be properly completed.

Read all instructions before completing this form.

This form has been designed for use on a 12-pitch (elite) typewriter; a firm
point pen may also be used-press down hard.

Utah and Federal regulations require generators and transporters of hazardous
waste and owners or operators of hazardous waste treatment, storage, and
disposal facilities to use this form (8700-22) and, if necessary, the
continuation sheet (Form 8700-22A) for both inter and intrastate
transportation.

Utah and Federal regulations also require generators and transporters of
hazardous waste and owners or operators of hazardous waste treatment, storage
and disposal facilities to complete the following information.

GENERATORS

Item 1.
Generator's U.S. EPA ID Number-Manifest Document Number

Enter the generator's U.S. EPA twelve digit identification number and the
unique five digit number assigned to this Manifest (e.g., 00001) by the
generator.

Item 2.
Page 1 of _____

Enter the total number of pages used to complete this Manifest, i.e., the
first page (EPA Form 8700-22) plus the number of Continuation Sheets (EPA Form
8700-22A), if any.

Item 3.
Generator's Name and Mailing Address

Enter the name and mailing address of the generator. The address should be
the location that will manage the returned Manifest forms.
Item 4.
Generator's Phone Number

Enter a telephone number where an authorized agent of the generator may be
reached in the event of an emergency.

Item 5.
Transporter 1 Company Name

Enter the company name of the first transporter who will transport the waste.

Item 6.
U.S. EPA ID Number

Enter the U.S. EPA twelve digit identification number of the first transporter
identified in item 5.

Item 7.
Transporter 2 Company Name

If applicable, enter the company name of the second transporter who will
transport the waste. If more than two transporters are used to transport the
waste, use a Continuation Sheet(s) (EPA Form 8700-22A) and list the
transporters in the order they will be transporting the waste.

Item 8.
U.S. EPA ID Number

If applicable, enter the U.S. EPA twelve digit identification number of the
second transporter identified in item 7.

Note.--If more than two transporters are used, enter each additional
transporter's company name and U.S. EPA twelve digit identification number in
items 24-27 on the Continuation Sheet (EPA Form 8700-22A). Each Continuation
Sheet has space to record two additional transporters. Every transporter used
between the generator and the designated facility must be listed.

Item 9.
Designated Facility Name and Site Address

Enter the company name and site address of the facility designated to receive
the waste listed on this Manifest. The address must be the site address,
which may differ from the company mailing address.

Item 10.
U.S. EPA ID Number

Enter the U.S. EPA twelve digit identification number of the designated
facility identified in item 9.

Item 11.
U.S. DOT Description [Including Proper Shipping Name, Hazard Class, and ID
Number (UN/NA)]

Enter the U.S. DOT Proper Shipping Name, Hazard Class, and ID Number (UN/NA)
for each waste as identified in 49 CFR 171 through 177.

Note.--If additional space is needed for waste descriptions, enter these
additional descriptions in item 28 on the Continuation Sheet (EPA Form
8700-22A).

Item 12.
Containers (No. and Type)

Enter the number of containers for each waste and the appropriate abbreviation
from Table 1 (below) for the type of container.

Table 1--Types of Containers

DM=Metal drums barrels, kegs
DW=Wooden drums, barrels, kegs
DF=Fiberboard or plastic drums, barrels kegs
TP=Tanks portable
TT=Cargo tanks (tank trucks)
TC=Tank cars
DT-Dump truck
CY=Cylinders
CM=Metal boxes, cartons, cases (including roll-offs)
CW=Wooden boxes, cartons, cases
CF=Fiber or plastic boxes, cartons, cases
BA=Burlap, cloth, paper or plastic bags

Item 13.
Total Quantity

Enter the total quantity of waste described on each line.

Item 14.
Unit (Wt./Vol.)

Enter the appropriate abbreviation from Table II (below) for the unit of measure.

Table II--Units of Measure

G=Gallons (liquids only) *[8.3 pounds]
P=Pounds
T=Tons (2000 lbs)
Y=Cubic yards *[.85 tons]
L=Liters (Liquids only) *[2.2 pounds]
K=Kilograms [2.2 pounds]
M=Metric tons (1000 kg) [2,200 pounds]
N=Cubic meters *[2,200 pounds]

*Specific gravity - 1.00 unless indicated in Box J

Item 15.
Special Handling Instructions and Additional Information

Generators may use this space to indicate special transportation, treatment, storage, or disposal information or Bill of Lading information. States may not require additional, new or different information in this space. For international shipments, generators must enter in this space the point of departure (City and State) for those shipments destined for treatment, storage, or disposal outside the jurisdiction of the United States.

Item 16.
Generator's Certification

The generator must read, sign (by hand), and date the certification statement. If a mode other than highway is used, the word "highway" should be lined out and the appropriate mode (rail, water, or air) inserted in the space below. If another mode in addition to the highway mode is used, enter the appropriate additional mode, (e.g., and rail) in the space below.

In signing the waste minimization certification statement, those generators who have not been exempted by State or Federal statute or regulation from the duty to make a waste minimization certification are also certifying that they have complied with the waste minimization requirements.

Note.--All of the above information except the handwritten signature require in item 16 may be preprinted.

TRANSPORTERS

Item 17.
Transporters 1 Acknowledgement of Receipt of Materials

Enter the name of the person accepting the waste on behalf of the first
transporter. That person must acknowledge acceptance of the waste described
on the Manifest by signing and entering the date of receipt.

Item 18.
Transporter 2 Acknowledgement of Receipt of Materials

Enter, if applicable, the name of the person accepting the waste on behalf of
the second transporter. That person must acknowledge acceptance of the waste
described on the Manifest by signing and entering the date of receipt.

Note.--International Shipment--Transporter Responsibilities.

Exports--Transporters must sign and enter the date of the waste left the
United States in item 15 of Form 8700-22.

Imports--Shipments of hazardous waste regulated by RCRA and transported into
the United States from another country must upon entry be accompanied by the
U.S. EPA Uniform Hazardous Waste Manifest. Transporters who transport
hazardous waste into the United States from another country are responsible
for completing the Manifest 6.1.3(a) UHWMR

Owners and Operators of Treatment, Storage, or Disposal Facilities.

Item 19.
Discrepancy Indication Space

The authorized representative of the designated (or alternate) facility's
owner or operator must note in this space any significant discrepancy between
the waste described on the Manifest and the waste actually received at the
facility.

Owners and operators of the facilities located in unauthorized States (i.e.,
the U.S. EPA administers the hazardous waste management the hazardous waste
management program) who cannot resolve significant discrepancies within 15
days or receiving the waste must submit to their Regional Administrator (see
list below) a letter with a copy of the Manifest at issue describing the
discrepancy and attempts to reconcile it (40 CFR 264.72 and 265.72).

Owners and operators of facilities located in authorized States (i.e., those States that have received authorization from the U.S. EPA to administer the hazardous waste program) should contact their State agency for information on State Discrepancy Report requirements.

EPA Regional Administrators

Regional Administrator, U.S. EPA
Region I, J.F. Kennedy Fed. Bldg.
Boston, MA 02203

Regional Administrator, U.S. EPA
Region II, 26 Federal Plaza
New York, NY 10278

Regional Administrator, U.S. EPA
Region III, 6th and Walnut Sts.
Philadelphia, PA 19106

Regional Administrator, U.S. EPA
Region IV, 345 Courtland St., NE.
Atlanta, GA 30365

Regional Administrator, U.S. EPA
Region V, 230 S. Dearborn St.
Chicago, IL 60604

Regional Administrator, U.S. EPA
Region VI, 1201 Elm Street
Dallas, TX 75270

Regional Administrator, U.S. EPA
Region VII, 324 East 11th Street
Kansas City, MO 64106

Regional Administrator, U.S. EPA
Region VIII, 1860 Lincoln Street
Denver, CO 80295

Regional Administrator, U.S. EPA
Region IX, 215 Freemont Street
San Francisco, CA 94105

Regional Administrator, U.S. EPA
Regional X, 1200 Sixth Avenue
Seattle, WA 98101

Item 20.
Facility Owner or Operator Certification of Receipt of Hazardous Materials Covered by This Manifest Except as Noted in Item 19.

Print or type the name of the person accepting the waste on behalf of the owner or operator of the facility. That person must acknowledge acceptance of the waste described on the Manifest by signing and entering the date of receipt.

Items A-K are not required by Federal regulations for intra-or interstate transportation. However, States may require generators and owners or operators of treatment, storage, or disposal facilities to complete some or all of items A-K as part of State manifest reporting requirements. Generators and owners and operators of treatment, storage, or disposal facilities are advised to contact State officials for guidance on completing the shaded areas of the Manifest.

GENERATORS

Item 21.
Generator's U.S. EPA ID Number--Manifest Document Number

Enter the generator's U.S. EPA twelve digit identification number and the unique five digit number assigned to this Manifest (e.g., 00001) as it appears in item 1 on the first page of the Manifest.

Item 22.
Page _____

Enter the page number of this Continuation Sheet.

Item 23.
Generator's name

Enter the generator's name as it appears in item 3 on the first page of the Manifest.

Item 24.
Transporter _____ Company Name

If additional transporters are used to transport the waste described on this Manifest, enter the company name of each additional transporter in the order in which they will transport the waste. Enter after the word "Transporter" the order of the transporter. For example, Transporter 3 Company name. Each Continuation Sheet will record the names of two additional transporters.

Item 25.
U.S. EPA ID Number

Enter the U.S. EPA twelve digit identification number of the transporter described in item 24.

Item 26.
Transporter _____ Company Name

If additional transporters are used to transport the waste described on this Manifest, enter the company name of each additional transporter in the order in which they will transport the waste. Enter after the word "Transporter" the order of the transporter. For example, Transporter 4 Company Name. Each Continuation Sheet will record the names of two additional transporters.

Item 27.
U.S. EPA ID Number

Enter the U.S. EPA twelve digit identification number of the transporter described in item 26.

Item 28.
U.S. DOT Description Including Proper Shipping Name, Hazardous Class, and ID Number (UN/NA)

Refer to item 11.

Item 29.
Containers (No. and Type)

Refer to item 12.

Item 30.
Total Quantity

Refer to item 13.

Item 31.
Unit (Wt./Vol.)

Refer to item 14.

Item 32.
Special Handling Instructions

Generators may use this space to indicate special transportation, treatment, storage, or disposal information or Bill of Lading information. States are not authorized to require additional, new, or different information in this space.

TRANSPORTERS

Item 33.
Transporter--Acknowledgement of Receipt of Materials

Enter the same number of the Transporter as identified in item 24. Enter also
the name of the person accepting the waste on behalf of the Transporter
(Company Name) identified in item 24. That person must acknowledge acceptance
of the waste described on the Manifest by signing and entering the date of
receipt.

Item 34.
Transporter--Acknowledgement of Receipt of Materials.

Enter the same number as identified in item 26. Enter also the name of the
person accepting the waste on behalf of the Transporter (Company Name)
identified in item 26. That person must acknowledge acceptance of the waste
described on the Manifest by signing and entering the date of receipt.

OWNERS AND OPERATORS OF TREATMENT, STORAGE, OR DISPOSAL FACILITIES

Item 35.
Discrepancy Indication Space

Refer to item 19.

Items L-R are not required by Federal regulations for intra- or interstate
transportation. However, States may require generators and owners or
operators of treatment, storage, or disposal facilities to complete some of
all of items L-R as part of State manifest reporting requirements. Generators
and owners and operators of treatment, storage, or disposal facilities are
advised to contact State officials for guidance on completing the shaded areas
of the manifest.

Appendix E

1995 Hazardous Waste Report (EPA Form 8700-13A/B) and Instructions

OMB#: 2050-0024 Expires 8/31/96

BEFORE COPYING FORM, ATTACH SITE IDENTIFICATION LABEL OR ENTER:

SITE NAME: _____

EPA ID NO: ⌞_⌟⌞_⌟⌞_⌟ ⌞_⌟⌞_⌟⌞_⌟ ⌞_⌟⌞_⌟⌞_⌟ ⌞_⌟⌞_⌟

U.S. ENVIRONMENTAL PROTECTION AGENCY

1995 Hazardous Waste Report

FORM IC

IDENTIFICATION AND CERTIFICATION

INSTRUCTIONS: Read the detailed instructions beginning on page 9 of the 1995 Hazardous Waste Report booklet before completing this form.

Sec. I Site name and location address. Complete A through H. Check the box ☐ in items A, C, E, F, G, and H if same as label; if different, enter corrections. If label is absent, enter information. Instruction page 10.

A. EPA ID No.
Same as label ☐ or → ⌞_⌟⌞_⌟⌞_⌟ ⌞_⌟⌞_⌟⌞_⌟ ⌞_⌟⌞_⌟⌞_⌟ ⌞_⌟⌞_⌟

B. County

C. Site/company name
Same as label ☐ or →

D. Has the site name associated with this EPA ID changed since 1993? ☐ 1 Yes ☐ 2 No

E. Street name and number. If not applicable, enter industrial park, building name, or other physical location description.
Same as label ☐ or →

F. City, town, village, etc.
Same as label ☐ or →

G. State
Same as label ⌞_⌟⌞_⌟

H. Zip Code
Same as label ⌞_⌟⌞_⌟⌞_⌟⌞_⌟⌞_⌟ - ⌞_⌟⌞_⌟⌞_⌟⌞_⌟

Sec. II Mailing address of site. Instruction page 10.

A. Is the mailing address the same as the location address? ☐ 1 Yes (SKIP TO SEC. III) ☐ 2 No (GO TO BOX B)

B. Number and street name of mailing address

C. City, town, village, etc.

D. State ⌞_⌟⌞_⌟

E. Zip Code ⌞_⌟⌞_⌟⌞_⌟⌞_⌟⌞_⌟ - ⌞_⌟⌞_⌟⌞_⌟⌞_⌟

Sec. III Name, title, and telephone number of the person who should be contacted if questions arise regarding this report. Instruction page 10.

A. Please print: Last Name First name M.I.

B. Title

C. Telephone ⌞_⌟⌞_⌟⌞_⌟ ⌞_⌟⌞_⌟⌞_⌟ - ⌞_⌟⌞_⌟⌞_⌟⌞_⌟
Extension ⌞_⌟⌞_⌟⌞_⌟⌞_⌟

Sec. IV "I certify under penalty of law that this document and all attachments were prepared under my direction or supervision in accordance with a system designed to assure that qualified personnel properly gather and evaluate the information submitted. Based on my inquiry of the person or persons who manage the system, or those persons directly responsible for gathering the information, the information submitted is, to the best of my knowledge and belief, true, accurate and complete. I am aware that there are significant penalties under Section 3008 of the Resource Conservation and Recovery Act for submitting false information, including the possibility of fine and imprisonment for knowing violations."

A. Please print: Last Name First name M.I.

B. Title

C. Signature

D. Date of signature ⌞_⌟⌞_⌟ ⌞_⌟⌞_⌟ ⌞_⌟⌞_⌟
MO. DAY YR.

Page 1 of ___

FORM IC

EPA ID NO: ⌊_⌋_⌋ ⌊_⌋_⌋ ⌊_⌋_⌋ ⌊_⌋_⌋

Sec.V - Generator Status.   Instruction pages 10, 12.

A. 1995 RCRA generator status	B. Reason for not generating
(CHECK ONE BOX BELOW)	(CHECK ALL THAT APPLY)
□ 1 LQG □ 2 SQG SKIP to SEC. VI □ 3 CESQG □ 4 Non generator (Continue to Box B)	□ 1 Never generated □ 5 Periodic or occasional generator □ 2 Out of business □ 6 Waste minimization activity □ 3 Only excluded or delisted waste □ 7 Other (SPECIFY COMMENTS IN BOX BELOW) □ 4 Only non-hazardous waste

---

Sec.VI - On-Site Waste Management Status.   Instruction pages 13, 14.

A. Storage subject to RCRA permitting requirements	B. Treatment, disposal, or recycling subject to RCRA permitting requirements	C. RCRA-exempt treatment, disposal, or recycling

---

Sec.VII - Waste Minimization Activity during 1994 or 1995.   Instruction pages 14, 15.

A. Did this site begin or expand a <u>source reduction</u> activity during 1994 or 1995?	B. Did this site begin or expand a <u>recycling</u> activity during 1994 or 1995?	C. Did this site systematically investigate opportunities for <u>source reduction or recycling</u> during 1994 or 1995?
□ 1 Yes □ 2 No	□ 1 Yes □ 2 No	□ 1 Yes □ 2 No

D. Did any of the factors listed below delay or limit this site's ability to initiate new or additional <u>source reduction</u> activities in 1994 or 1995?
(CHECK YES OR NO FOR EACH ITEM)

Yes	No		
□ 1	□ 2	a.	Insufficient capital to install new source reduction equipment or implement new source reduction practices
□ 1	□ 2	b.	Lack of technical information on source reduction techniques applicable to the specific production processes
□ 1	□ 2	c.	Source reduction is not economically feasible: cost savings in waste management or production will not recover the capital investment
□ 1	□ 2	d.	Concern that product quality may decline as a result of source reduction
□ 1	□ 2	e.	Technical limitations of the production processes
□ 1	□ 2	f.	Permitting burdens
□ 1	□ 2	g.	Source reduction previously implemented - additional reduction does not appear to be technically feasible
□ 1	□ 2	h.	Source reduction previously implemented - additional reduction does not appear to be economically feasible
□ 1	□ 2	i.	Source reduction previously implemented - additional reduction does not appear to be feasible due to permitting requirements
□ 1	□ 2	j.	Other (SPECIFY COMMENTS IN BOX BELOW)

E. Did any of the factors listed below delay or limit the site's ability to initiate new or additional on-site or off-site <u>recycling</u> activities during 1994 or 1995?
(CHECK YES OR NO FOR EACH ITEM)

Yes	No			Yes	No		
□ 1	□ 2	a.	Insufficient capital to install new recycling equipment or implement new recycling practice	□ 1	□ 2	g.	Technical limitations of production processes inhibit shipments off-site for recycling
□ 1	□ 2	b.	Lack of technical information on recycling techniques applicable to this site's specific production process	□ 1	□ 2	h.	Technical limitations of production processes inhibit on-site recycling
				□ 1	□ 2	i.	Permitting burdens inhibit recycling
□ 1	□ 2	c.	Recycling is not economically feasible: cost savings in waste management will not recover the capital investment	□ 1	□ 2	j.	Lack of permitted off-site recycling facilities
				□ 1	□ 2	k.	Unable to identify a market for recycled materials
□ 1	□ 2	d.	Concern that product quality may decline as a result of recycling	□ 1	□ 2	l.	Recycling previously implemented - additional recycling does not appear to be technically feasible
□ 1	□ 2	e.	Requirements to manifest wastes inhibit shipments of off-site for recycling	□ 1	□ 2	m.	Recycling previously implemented - additional recycling does not appear to be economically feasible
□ 1	□ 2	f.	Financial liability provisions inhibit shipments off-site for recycling	□ 1	□ 2	n.	Recycling previously implemented - additional recycling does not appear to be feasible due to permitting requirements
				□ 1	□ 2	o.	Other (SPECIFY COMMENTS IN BOX BELOW)

Comments:

FORM IC

INSTRUCTIONS FOR FILLING OUT

FORM IC - IDENTIFICATION AND CERTIFICATION

WHO MUST SUBMIT THIS FORM?

All sites required to file the 1995 Hazardous Waste Report must submit Form IC.

PURPOSE OF THIS FORM

Form IC is divided into seven sections. Sections I through III identify the site. Section IV certifies the information reported throughout is truthful, accurate, and complete. Sections V and VI update the site's EPA notification of hazardous waste activities. Finally, Section VII records information on waste minimization activities during 1994 and 1995.

HOW TO FILL OUT THIS FORM

You should fill out all seven sections. Please print or type (12 pitch) all information. Throughout the form, enter "NA" if the information requested is not applicable. Use the Comments section at the end of the form to clarify or continue any entry. Preceding the comment, reference the section number and box letter to which it refers.

Please note the following list of information you must provide if you are required to submit the Form IC.

Section I

	Block A	EPA ID No.
	Block C	Site/company name
	Block E	Street name and number
	Block F	City, town, village, etc.
	Block G	State
	Block H	Zip Code

Section II

	Block B	Number and street name of mailing address
	Block C	City, town, village, etc.
	Block D	State
	Block E	Zip Code

Section III

	Block A	Last Name, First Name, and M.I.
	Block B	Title
	Block C	Telephone number and extension

Section IV

	Block A	Last Name, First Name, and M.I.
	Block B	Title
	Block C	Signature
	Block D	Date of signature

Section V

	Block A	1995 RCRA Generator Status

FORM IC

Section VI

Block A	Storage subject to RCRA Permitting requirements
Block B	Treatment, disposal, or recycling subject to RCRA permitting requirements

Section VII

Block A	Began <u>source reduction activity</u> during 1994 or 1995 (Y/N)
Block B	Began or expanded a <u>recycling</u> activity during 1994 or 1995 (Y/N)
Block C	Investigated opportunities for <u>source reduction or recycling</u> during 1994 or 1995 (Y/N)

ITEM-BY-ITEM INSTRUCTIONS

Section I: <u>Site name and location address</u>
Fill out Boxes A through H. In Box B, enter the county, borough, or parish in which the site is located, unless that information is present and correct on any label provided. Check the box "Same as label" if the address information provided on a pre-printed label is correct. In Box D, check "Yes" or "No" to indicate whether the site/company name associated with this EPA Identification Number has changed since 1993. The EPA Identification Number is address specific and cannot be transferred to a new location. Blocks A, C, E, F, G, and H are required fields.

Section II: <u>Mailing address of site</u>
Check "Yes" or "No" to indicate if the site's mailing address is the same as the location address listed in Section I. If you checked "No", enter the site's mailing address in Boxes B through E. Blocks B, C, D, and E are required fields.

> ☞ **Skip to Section III,** if you checked "Yes".
> **Continue to Box B,** if you checked "No".

Section III: <u>Contact information</u>
Enter the full name, title, and phone number of the person who should be contacted if questions arise regarding the information provided in the 1995 Hazardous Waste Report submitted by your site. Blocks A, B, and C are required fields.

Section IV: <u>Certification</u>
Do not fill out Section IV until all forms required for submission are present, complete, and accurate. The 1995 EPA Hazardous Waste Report Submission Checklist at the back of this booklet is provided to assist you. After you have filled out all required forms, enter your full name and title, and the date. Read the certification statement, and sign the form. Refer to the list beginning on page vi for the mailing address for your Report. Blocks A, B, C, and D are required fields.

Section V: <u>Generator Status</u>
Complete Box A and follow the instructions to fill out Box B or skip to Section VI. Block A is a required field.

Box A:	<u>1995 RCRA generator status</u>

Check one box to indicate the site's RCRA hazardous waste generation status in 1995. (See NOTE box on page 11.)

Report your site's Federal RCRA generator status in Box A even if your State defines generator status categories differently.

If the site did generate any RCRA hazardous waste during 1995, review the definitions of LQG, SQG, and CESQG (see below) to determine your generator status. Then check the appropriate box.

If your site did not generate RCRA hazardous waste during 1995, check "4 Non generator" and proceed to Box B.

☞ | **NOTE:** | A site that generates solid waste must determine if that waste is a RCRA hazardous waste, or if it is excluded from regulation under 40 CFR 261.4(b). RCRA hazardous waste managed solely in units exempt from RCRA permitting requirements are not to be counted in determining if a site is a large quantity generator. However, if a site is required to file the 1995 Hazardous Waste Report, EPA requests that RCRA hazardous waste treated solely in exempt units be reported. If a waste is excluded, or if it is regulated only by your State, its quantity should not be counted in determining RCRA generator status.

<u>Code</u>   <u>1995 RCRA generator status</u>

1   <u>LQG: Large Quantity Generator</u>
           This site is a Large Quantity Generator if, in 1995, it met **any** of the following criteria:

   a)   The site generated in one or more months, during 1995, 1,000 kg (2,200 lbs) or more of RCRA hazardous waste; **or**

   b)   The site generated in one or more months, during 1995, or accumulated at any time, 1 kg (2.2 lbs) of RCRA acute hazardous waste; **or**

   c)   The site generated or accumulated at any time more than 100 kg (220 lbs) of spill cleanup material contaminated with RCRA acute hazardous waste.

2   <u>SQG: Small Quantity Generator</u>
           This site is a Small Quantity Generator if, in 1995, it met **all** the following criteria:

   a)   In one or more months the site generated more than 100 kg (220 lbs) of hazardous waste, <u>but in no month did the site</u>: (1) generate 1,000 kg (2,200 lbs) or more of hazardous waste; or (2) generate 1 kg (2.2 lbs) or more of acute hazardous waste; or (3) generate 100 kg (220 lbs) or more of material from the cleanup of a spillage of acute hazardous waste; **and**

   b)   The site accumulated no more than 1 kg (2.2 lbs) of acute hazardous waste **and** no more than 100 kg (220 lbs) of material from the cleanup of a spillage of acute hazardous waste; **and**

   c)   The site stored its wastes in tanks or containers in a manner consistent with regulatory provisions.

           OR, the site is a Small Quantity Generator if, in 1995,

   a)   The site met all other criteria for a Conditionally Exempt Small Quantity Generator (CESQG), but

   b)   The site accumulated 1,000 kg (2,200 lbs) or more of hazardous waste.

3   <u>CESQG: Conditionally Exempt Small Quantity Generator</u>
           This site's hazardous waste activities met the definition of a RCRA CESQG <u>every month</u> during 1995. A RCRA CESQG is defined by the following criteria:

   a)   The site generated no more than 100 kg (220 lbs) of hazardous waste, **and** no more than 1 kg (2.2 lbs) of acute hazardous waste, **and** no more than 100 kg (220 lbs) of material from the cleanup of a spillage of acute hazardous waste; **and**

FORM IC

    b)   The site accumulated no more than 1,000 kg (2,200 lbs) of hazardous waste, **and** no more than 1 kg (2.2 lbs) of acute hazardous waste, **and** no more than 100 kg (220 lbs) of material from the cleanup of a spillage of acute hazardous waste; **and**

    c)   The site treated or disposed of the hazardous waste in a manner consistent with regulatory provisions (40 CFR 261.5(f)(3) and 261.5(g)(3)).

4   <u>Non generator</u>
This site did not generate RCRA hazardous waste during 1995.

☞   **Continue to Box B,** if you checked 4.
    Skip to Section VI, if you checked 1, 2, or 3.

Box B: <u>Reason for not generating</u>
    If the site did not generate RCRA hazardous waste during 1995, check as many boxes as necessary to explain the reason. The alternatives are:

<u>Code</u>   <u>Reason for not generating</u>

1   <u>Never generated</u>: The site has never generated RCRA hazardous waste and did not do so during 1995.

2   <u>Out of business</u>: The site has gone out of business and did not generate hazardous waste at this location during 1995.

3   <u>Only excluded or delisted waste</u>: The site generated only excluded or delisted wastes not subject to RCRA regulation during 1995. Wastes not subject to RCRA regulation are delisted wastes and excluded wastes. A list of excluded wastes is provided beginning on page 57.

4   <u>Only non-hazardous waste</u>: The site generated no wastes subject to RCRA regulation. This includes wastes regulated only by your State and RCRA Subtitle D wastes (non-hazardous).

5   <u>Periodic or occasional generator</u>: This site generates RCRA hazardous waste only occasionally, and generated none during 1995.

6   <u>Waste minimization activity</u>: This site was previously a generator of RCRA hazardous waste, but did not generate any during 1995 due to an effective waste minimization program. (See the definition of Waste Minimization on page 55.)

7   <u>Other</u>: This site had other reasons for not generating in 1995. Specify in the Comments box at the bottom of the form and reference Section V, Box B.

📖   Excluded Wastes, page 57.

Section VI: <u>On-Site Waste Management Status</u>
Blocks A and B are required fields.

Box A: <u>Storage subject to RCRA permitting requirements</u>
    Did the site have any storage subject to RCRA permitting requirements on site during 1995? Select one code from the list below and record in the response space in Box A.

FORM IC

NOTE:	Short-term accumulation under the 90, 180, or 270 day rules is exempt from RCRA permitting requirements. If the ONLY type of storage at your site was accumulation of wastes under these rules prior to shipment, answer "1-No storage subject to RCRA permitting requirements."

Code   Storage subject to RCRA permitting requirements

1   No storage subject to RCRA permitting requirements

2   Tanks

3   Containers

4   Tanks and containers

5   Other

8   Don't know

Box B: Treatment, disposal, or recycling subject to RCRA permitting requirements
During 1995, was treatment, disposal, or recycling of RCRA hazardous waste conducted on site in units subject to RCRA permitting requirements? Select one code from the list below and record in the response space in Box B.

Code   Treatment, disposal, or recycling subject to RCRA permitting requirements

1   No, the facility did not treat, dispose, or recycle during the report year and had no plans during the report year to develop an on-site RCRA permitted treatment, disposal, or recycling system.

2   No, the facility did not treat, dispose, or recycle during the report year but is planning to develop an on-site RCRA permitted treatment, disposal, or recycling system.

3   Yes, the facility treated, disposed, or recycled on site in a unit subject to RCRA permitting requirements.

NOTE:	If you selected code 3, you should fill out one or more Forms PS to describe the existing on-site RCRA-permitted hazardous waste treatment, disposal, or recycling system.

FORM IC

Box C: RCRA-exempt treatment, disposal, or recycling
During 1995, was treatment, disposal, or recycling of RCRA hazardous waste conducted on site in units exempt from RCRA permitting requirements? Select one code from the list below and record in the response space in Box C.

Code	RCRA-exempt treatment, disposal, or recycling
1	No, the facility did not treat, dispose, or recycle during the report year and had no plans to develop during the report year an on-site RCRA-exempt treatment, disposal, or recycling system.
2	No, the facility did not treat, dispose, or recycle during the report year but is planning to develop an on-site RCRA-exempt treatment, disposal, or recycling system.
3	Yes, the facility treated, disposed, or recycled on site in a unit exempt from RCRA permitting requirements pursuant to §264.1(g) or §265.1(c).

> **NOTE:** If you selected code 3, you should fill out one or more Forms PS to describe the existing on-site RCRA-exempt hazardous waste treatment, disposal, or recycling system.

Section VII: Waste Minimization Activity during 1994 or 1995
Blocks A, B, and C are required fields.

Waste minimization means reduction, to the extent feasible, of hazardous waste generated or subsequently treated, stored, or disposed. It includes any source reduction or recycling activity undertaken by a generator resulting in: (1) the reduction of total volume or quantity of hazardous waste; (2) the reduction of toxicity of hazardous waste; or (3) both, as long as the reduction is consistent with the goal of minimizing present and future threats to human health and the environment.

> **NOTE:** Treatment (including burning and incineration) of the waste after it has exited the process is not considered waste minimization activity. The following are examples of activities that should <u>not</u> be reported here as waste minimization.
>
> - Sending waste off site for management (other than recycling).
> - Treatment to reduce volume (after the waste exits the process in which it was generated).
> - Treatment to reduce toxicity (after the waste exits the process in which it was generated).
> - Installation of filter press to reduce water content and volume.
> - Installation of equipment to comply with Clean Water Act.
>
> Bankruptcy or reduction in production volume due to economic factors are <u>not</u> waste minimization activities.

Box A: Did this site begin or expand a source reduction activity during 1994 or 1995?

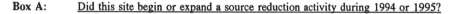

Check "Yes" or "No" in Box A.

☞	NOTE:	**Source reduction** means any practice which: (1) reduces the amount of any hazardous substance, pollutant, or contaminant entering any waste or otherwise released into the environment (including fugitive emissions) prior to recycling, treatment, or disposal; and (2) reduces impact on public health and the environment associated with the release of such substances, pollutants, or contaminants. The term includes equipment or technology modifications; process or procedure modifications; reformulation or redesign of products; substitution of raw materials; and improvements in housekeeping, maintenance, training, or inventory control. Source reduction does not include any practice that alters the physical, chemical, or biological characteristics or the volume of a hazardous substance, pollutant, or contaminant through a process or activity which itself is not integral to and necessary for the production of a product or the provision of a service.

Box B: Did this site begin or expand a recycling activity during 1994 or 1995?

Check "Yes" or "No" in Box B.

☞	NOTE:	**Recycling** means the use or reuse of waste as an effective substitute for a commercial product, or as an ingredient or feedstock in an industrial process. It also refers to the reclamation of useful constituent fractions within a waste material or removal of contaminants from a waste to allow it to be reused. As used in this report, recycling implies use, reuse, or reclamation of a waste, either on site or off site, after it has been generated. See 40 CFR, Section 261.1 (c) (4) , (5), and (7).

Box C: Did this site systematically investigate opportunities for source reduction or recycling during 1994 or 1995?

Check "Yes" or "No" in box C.

☞	NOTE:	The Pollution Prevention Research Branch of EPA's Office of Research and Development is publishing a series of industry-specific pollution prevention waste minimization guidance materials. The manuals supplement EPA's waste reduction manual issued in July 1988 titled: "Waste Minimization Opportunity Assessment Manual." The identification number for this manual is EPA/625/7-8'3/003. For copies, call the RCRA/Superfund Hotline at 1-800-424-9346 or (703) 412-9810.

Box D: Did any of the factors listed below delay or limit this site's ability to initiate new or additional source reduction activities during 1994 or 1995?

Check "Yes" or "No" for each item.

Box E: Did any of the factors listed below delay or limit this site's ability to initiate new or additional on-site or off-site recycling activities during 1994 or 1995?

Check "Yes" or "No" for each item.

FORM OI

BEFORE COPYING FORM, ATTACH SITE IDENTIFICATION LABEL OR ENTER:

SITE NAME: _____

EPA ID NO: |_|_|_|_| |_|_|_|_| |_|_|_|_| |_|_|_|

**U.S. ENVIRONMENTAL
PROTECTION AGENCY**

1995 Hazardous Waste Report

**FORM
OI**

**OFF-SITE
IDENTIFICATION**

INSTRUCTIONS: Read the detailed instructions on the reverse side before completing this form.

Site 1	A. EPA ID No. of off-site installation or transporter

C. Handler type	(CHECK ALL THAT APPLY)	B. Name of off-site installation or transporter
	□ Generator	D. Address of off-site installation
	□ Transporter	Street ___ City ___
	□ TSDR	State |_|_|_|_| Zip |_|_|_|_|_| - |_|_|_|_|

Site 1
A. EPA ID No. of off-site installation or transporter
|_|_|_|_| |_|_|_|_| |_|_|_|_| |_|_|_|
B. Name of off-site installation or transporter

C. Handler type   (CHECK ALL THAT APPLY)
□ Generator
□ Transporter
□ TSDR
D. Address of off-site installation
Street _____
City _____
State |_|_|_|_| Zip |_|_|_|_|_|_| - |_|_|_|_|

Site 2
A. EPA ID No. of off-site installation or transporter
|_|_|_|_| |_|_|_|_| |_|_|_|_| |_|_|_|
B. Name of off-site installation or transporter

C. Handler type   (CHECK ALL THAT APPLY)
□ Generator
□ Transporter
□ TSDR
D. Address of off-site installation
Street _____
City _____
State |_|_|_|_| Zip |_|_|_|_|_|_| - |_|_|_|_|

Site 3
A. EPA ID No. of off-site installation or transporter
|_|_|_|_| |_|_|_|_| |_|_|_|_| |_|_|_|
B. Name of off-site installation or transporter

C. Handler type   (CHECK ALL THAT APPLY)
□ Generator
□ Transporter
□ TSDR
D. Address of off-site installation
Street _____
City _____
State |_|_|_|_| Zip |_|_|_|_|_|_| - |_|_|_|_|

Site 4
A. EPA ID No. of off-site installation or transporter
|_|_|_|_| |_|_|_|_| |_|_|_|_| |_|_|_|
B. Name of off-site installation or transporter

C. Handler type   (CHECK ALL THAT APPLY)
□ Generator
□ Transporter
□ TSDR
D. Address of off-site installation
Street _____
City _____
State |_|_|_|_| Zip |_|_|_|_|_|_| - |_|_|_|_|

Site 5
A. EPA ID No. of off-site installation or transporter
|_|_|_|_| |_|_|_|_| |_|_|_|_| |_|_|_|
B. Name of off-site installation or transporter

C. Handler type   (CHECK ALL THAT APPLY)
□ Generator
□ Transporter
□ TSDR
D. Address of off-site installation
Street _____
City _____
State |_|_|_|_| Zip |_|_|_|_|_|_| - |_|_|_|_|

Comments:

INSTRUCTIONS FOR FILLING OUT

FORM OI - OFF-SITE IDENTIFICATION

WHO MUST COMPLETE THIS FORM?

Sites required to file the 1995 Hazardous Waste Report must submit Form OI if:

- Form OI is required by your State AND

- The site received hazardous waste from off site or sent hazardous waste off site during 1995.

PURPOSE OF THIS FORM

Form OI documents the names and addresses of off-site installations and transporters.

HOW TO COMPLETE THIS FORM

Form OI is divided into five identical parts. You must fill out one part for each off-site installation to which you shipped hazardous waste, each off-site installation from which you received hazardous waste, and each transporter you used during 1995. If these off-site installations and transporters total more than five, you must photocopy and complete additional copies of the form. You do not need to report the address, Box D, for transporters.

Throughout the form, enter "NA" if the information requested is not applicable. Use the Comments section at the bottom of the form to clarify or continue any entry. Reference the comment by entering the site number and box letter.

ITEM-BY-ITEM INSTRUCTIONS

Complete Boxes A through D for each off-site installation to which you shipped hazardous waste and each off-site installation from which you received hazardous waste during 1995.

Complete Boxes A through C for each transporter you used during the year. (The transporter address is not required in Box D).

Box A: EPA ID No. of off-site installation or transporter
Enter the 12-digit EPA ID number of the off-site installation to which you shipped hazardous waste or from which you received hazardous waste. Or, enter the EPA ID number of the transporter who shipped hazardous waste to or from your site. Each EPA ID should appear only once. If the off-site installation or transporter did not have an EPA ID number during 1995, enter "NA" in Box A.

Box B: Name of off-site installation or transporter
Enter the name of the off-site installation or transporter reported in Box A.

Box C: Handler Type
Check all boxes that apply to describe the handler type of the off-site installation or transporter reported in Box A.

Box D: Address of off-site installation
Enter the address of the off-site installation reported in Box A. If the EPA ID number reported in Box A refers to a transporter, enter "NA" in Box D.

FORM WR

BEFORE COPYING FORM, ATTACH SITE IDENTIFICATION LABEL OR ENTER:

SITE NAME: _____

EPA ID NO: ⌴_⌴_⌴_⌴ ⌴_⌴_⌴_⌴ ⌴_⌴_⌴_⌴ ⌴_⌴_⌴

U.S. ENVIRONMENTAL PROTECTION AGENCY

1995 Hazardous Waste Report

FORM WR

WASTE RECEIVED FROM OFF-SITE

INSTRUCTIONS: Read the detailed instructions beginning on page 30 of the 1995 Hazardous Waste Report booklet before completing this form.

Waste 1	A. Description of hazardous waste Instruction page 30.	B. EPA hazardous waste code Page 31.	C. State hazardous waste code Page 31.

D. Off-site source EPA ID number
Page 31.

E. Quantity received in 1995
Page 31.

F. UOM   Density
Page 31.
☐ 1 lbs/gal   ☐ 2 sg

G. Waste form code
Page 32.   ⌴B_⌴_⌴_⌴

H. RCRA-radioactive mixed
Page 32.

I. System type
Page 32.   ⌴M_⌴_⌴_⌴

Waste 2	A. Description of hazardous waste Instruction page 30.	B. EPA hazardous waste code Page 31.	C. State hazardous waste code Page 31.

D. Off-site source EPA ID number
Page 31.
☐ Check if ID same as in Waste 1

E. Quantity received in 1995
Page 31.

F. UOM   Density
Page 31.
☐ 1 lbs/gal   ☐ 2 sg

G. Waste form code
Page 32.   ⌴B_⌴_⌴_⌴

H. RCRA-radioactive mixed
Page 32.

I. System type
Page 32.   ⌴M_⌴_⌴_⌴

Waste 3	A. Description of hazardous waste Instruction page 30.	B. EPA hazardous waste code Page 31.	C. State hazardous waste code Page 31.

D. Off-site source EPA ID number
Page 31.
☐ Check if ID same as in Waste 2

E. Quantity received in 1995
Page 31.

F. UOM   Density
Page 31.
☐ 1 lbs/gal   ☐ 2 sg

G. Waste form code
Page 32.   ⌴B_⌴_⌴_⌴

H. RCRA-radioactive mixed
Page 32.

I. System type
Page 32.   ⌴M_⌴_⌴_⌴

Comments:

FORM GM

BEFORE COPYING FORM, ATTACH SITE IDENTIFICATION LABEL OR ENTER:

SITE NAME: _____

_____

EPA ID NO: |__|__|__| |__|__|__|__| |__|__|__|__| |__|__|__|

**U.S. ENVIRONMENTAL
PROTECTION AGENCY**

1995 Hazardous Waste Report

**FORM
GM**

**WASTE GENERATION
AND MANAGEMENT**

INSTRUCTIONS: Read the detailed instructions beginning on page 16 of the 1995 Hazardous Waste Report booklet before completing this form.

Sec. I   A. Waste description · Instruction page 18.

B. EPA hazardous waste code Page 19.

|__|__|__|__| |__|__|__|

|__|__|__|__| |__|__|__|__| |__|__|__|__|

C. State hazardous waste code Page 19.

|__|__|__|__|__|__| |__|__|__|__|__|__|

D. SIC code Page 19.

|__|__|__|__|

E. Origin code |__| Page 19
System
Type |M|__|__|__|

F. Source code Page 20.

|A|__|__|

G. Point of measurement
Page 20.

|__|__|

H. Form code
Page 20.

|B|__|__|__|

I. RCRA - radioactive mixed Page 20.

|__|__|

Sec. II   A. Quantity generated in 1994
Instruction Page 21.

|__|__|__|__|__|__|__|__|__| · |__|

B. Quantity generated in 1995
Page 21.

|__|__|__|__|__|__|__|__|__|__| · |__|

C. UOM   Density
Page 21.

|__|__| |__|__|__| · |__|__|__|

☐ 1 lbs/gal ☐ 2 sg

D. Did this site do any of the following to this waste: treat on site, dispose on site, recycle on site, or discharge to a sewer/POTW? Page 21.

☐ 1 Yes (CONTINUE TO SYSTEM 1)
☐ 2 No (SKIP TO SEC. III)

ON-SITE PROCESS SYSTEM 1

On-site process system type
Page 22.

|M|__|__|__|

Quantity treated, disposed, or recycled on site
in 1995

|__|__|__|__|__|__|__|__|__|__| · |__|

ON-SITE PROCESS SYSTEM 2

On-site process system type
Page 22.

|M|__|__|__|

Quantity treated, disposed, or recycled on site
in 1995

|__|__|__|__|__|__|__|__|__|__| · |__|

Sec.III   A. Was any of this waste shipped off-site in 1995   ☐ 1 Yes (CONTINUE TO BOX B)
Instruction page 22.   ☐ 2 No (SKIP TO SEC IV)

Site 1

B. EPA ID No. of facility waste was shipped to
Page 23.

|__|__|__| |__|__|__|__| |__|__|__|__| |__|__|__|

C. System type shipped to
Page 23.

|M|__|__|__|

D. Off-site
availability code
Page 23. |__|

E. Total quantity shipped in 1995
Page 23.

|__|__|__|__|__|__|__|__|__| · |__|

Site 2

B. EPA ID No. of facility waste was shipped to
Page 23.

|__|__|__| |__|__|__|__| |__|__|__|__| |__|__|__|

C. System type shipped to
Page 23.

|M|__|__|__|

D. Off-site
availability code
Page 23. |__|

E. Total quantity shipped in 1995
Page 23.

|__|__|__|__|__|__|__|__|__| · |__|

Sec. IV   A. Did new activities in 1995 result in minimization of this waste? ☐ 1 Yes (CONTINUE TO BOX B)
Instruction page 24.   ☐ 2 No (THIS FORM IS COMPLETE)

B. Activity Page 24.

|W|__|__|__| |W|__|__|__|

|W|__|__|__| |W|__|__|__|

C. Other effects Page 25.

☐ 1 Yes

☐ 2 No

D. Quantity recycled in 1995 due to new activities
Page 25.

|__|__|__|__|__|__|__|__|__| · |__|

E. Activity/production
index Page 25.

|__|__| · |__|

F. 1995 source reduction quantity Page 26.

|__|__|__|__|__|__|__|__|__| · |__|

Comments:

FORM PS

BEFORE COPYING FORM, ATTACH SITE IDENTIFICATION LABEL OR ENTER:

SITE NAME: _____

EPA ID NO: |__|__|__| |__|__|__|__| |__|__|__|__| |__|__|__|

FORM PS

U.S. ENVIRONMENTAL PROTECTION AGENCY

1995 Hazardous Waste Report

WASTE TREATMENT, DISPOSAL, OR RECYCLING PROCESS SYSTEMS

INSTRUCTIONS: Read the detailed instructions beginning on page 33 of the 1995 Hazardous Waste Report booklet before completing this form.

Sec. I   A. Waste treatment, disposal, or recycling system description
Instruction Page 38.

B. System type Page 38.	C. Regulatory status Page 39.	D. Operational status Page 39.	E. Unit types Page 39.
|__M__|__|__|	|__|__|	|__|__|	|__|__|__| |__|__|__|

Sec. II   A. 1995 influent quantity
Instruction page 40.

	B. Maximum operational capacity Page 41.

UOM   Density

Total |__|__|__|__|__|__|__|__|__|__|__| • |__| |__|__| |__|__|__| • |__|__| Total |__|__|__|__|__|__|__|__|__|__|__| • |__|

RCRA |__|__|__|__|__|__|__|__|__|__|__| • |__| □ 1 lbs/gal □ 2 sg   RCRA |__|__|__|__|__|__|__|__|__|__|__| • |__|

C. 1995 liquid effluent quantity
Instruction page 42.

D. 1995 solid/sludge residual quantity
Page 43.

UOM   Density   UOM   Density

Total |__|__|__|__|__|__|__|__|__|__|__| • |__| |__|__| |__|__|__| • |__|__| Total |__|__|__|__|__|__|__|__|__|__|__|__| • |__| |__|__| |__|__|__| • |__|__|

RCRA |__|__|__|__|__|__|__|__|__|__|__| • |__| 1 lbs/gal □ 2 sg   RCRA |__|__|__|__|__|__|__|__|__|__|__| • |__| □ 1 lbs/gal □ 2 sg

---

E. Limitation on maximum operational capacity Page 43.	F. Commercial capacity availability code Page 43.	G. Percent capacity commercially available Page 43.
1. |__|__| 2. |__|__| 3. |__|__|	|__|__|	|__|__|__| %

Comments:

Appendix F

Contingency Plan and Emergency Procedures

7.11.1. APPLICABILITY

The regulations in this section apply to the owners and operators of all hazardous waste management facilities, except as provided otherwise in 7.8.1.

7.11.2. PURPOSE AND IMPLEMENTATION OF CONTINGENCY PLAN

(a) Each owner or operator shall have a contingency plan for his facility designed to minimize hazards to human health or the environment from fires, explosions, or any unplanned sudden or non-sudden discharge of hazardous waste or hazardous waste constituents to air, soil, or surface water.

(b) The provisions of the plan shall be carried out immediately whenever there is a fire, explosion, or release of hazardous waste or hazardous waste constituents which could threaten the environment or human health.

7.11.3. CONTENT OF CONTINGENCY PLAN

(a) The contingency plan shall describe the actions facility personnel shall take to comply with 7.11.2. and 7.11.7. in response to fires, explosions, or any unplanned sudden or non-sudden discharge of hazardous waste or hazardous waste constituents to air, soil, or surface water at the facility.

(b) If a facility owner or operator already has prepared a Spill Prevention, Control and Countermeasures (SPCC) Plan in accordance with 40 CFR 112 or 40 CFR 151, or some other emergency or contingency plan, he need only amend that plan to incorporate hazardous waste management provisions sufficient to comply with the requirements of this Subpart B.

281

(c) The plan shall describe arrangements agreed to by local law enforcement agencies, fire departments, hospitals, contractors, and state and local emergency response teams to coordinate emergency services, in accordance with 7.10.7.

(d) The plan shall list names, addresses, phone numbers (office and home) of all persons qualified to act as facility emergency coordinator (see 7.11.6) and this list shall be kept up-to-date. Where more than one person is listed, one shall be named as primary emergency coordinator and others shall be listed in the order in which they will assume responsibility as alternates.

(e) The plan shall include a list of all emergency equipment at the facility (such as fire extinguishing systems, discharge control equipment, communications and alarm systems (internal and external), and decontamination equipment), where this equipment is required. This list shall be kept up-to-date. In addition, the plan shall include the location and physical description of each item on the list, and a brief outline of its capabilities.

(f) The plan shall include an evacuation plan for facility personnel where there is a possibility that evacuation could be necessary. This plan shall describe signal(s) to be used to begin evacuation, evacuation routes, and alternate evacuation routes (in cases where the primary routes could be blocked by discharges of hazardous waste or fires.).

7.11.4.   COPIES OF CONTINGENCY PLAN

A copy of the contingency plan and all revisions to the plan shall be:

(a) Maintained at the facility;

(b) Made available to the committee or its duly appointed representative upon request; and

(c) Submitted to all local law enforcement agencies, fire departments, hospitals, and state and local emergency response teams that may be called upon to provide emergency services.

7.11.5.   AMENDMENT OF CONTINGENCY PLAN

The contingency plan shall be reviewed, and immediately amended, if necessary, under any of the following circumstances:

(a) Revisions to applicable regulations;

(b) Failure of the plan in an emergency;

(c) Changes in the facility design, construction, operation, maintenance, or other circumstances that materially increase the potential for discharges of hazardous waste or hazardous waste constituents, or change the response necessary in an emergency;

(d) Changes in the list of emergency coordinators; or

(e) Changes in the list of emergency equipment.

7.11.6.   EMERGENCY COORDINATOR

At all times, there shall be at least one employee either on the facility premises or on call (i.e., available to respond to an emergency by reaching the facility within a short period of time) with the responsibility for coordinating all emergency response measures. This facility emergency coordinator shall be thoroughly familiar with all aspects of the facility's contingency plan, all operations and activities at the facility, the locations of all records in the facility, and the facility layout. In addition, this person shall have the authority to commit the resources needed to carry out the contingency plan. The emergency coordinator's rsponsibilities are more fully spelled out in 7.11.7. Applicable responsibilities for the emergency coordinator vary depending on factors such as type and variety of waste(s) handled by the facility, and type and complexity of the facility.

7.11.7. EMERGENCY PROCEDURES

(a) Whenever there is an imminent or actual emergency situation, the emergency coordinator (or his designee when the emergency coodinator is on call) shall immediately:
   (1) Activate internal facility alarms or communication systems, where applicable, to notify all facility personnel; and
   (2) Notify appropriate state or local agencies with designated response roles whenever their assistance is needed.
(b) In the event of a discharge, fire, or explosion, the facility's emergency coordinator shall immediately identify the character, exact source, amount, and a real extent of any discharged materials. He may do this by observation and/or review of facility records or manifests, and, if necessary, by chemical analysis.
(c) Concurrently, the facility's emergency coordinator shall immediately assess possible hazards to the environment or human health that may result from the discharge, fire, or explosion. This assessment shall consider both direct and indirect effects of the discharge, fire, or explosion (e.g., the effects of any toxic, irritating, or asphyxiating gases that are generated, or the effects of any hazardous surface water run-offs from water or chemical agents used to control fire and heat-induced explosions).
(d) If the emergency coordinator determines that the facility has had a discharge, fire, or explosion which could threaten human health or the environment outside the facility, he shall report his findings as follows:
   (1) If his assessment indicates that evacuation of local areas may be advisable, he shall immediately notify appropriate local authorities. He shall be available to assist appropriate officials in making the decision whether local areas should be evacuated; and
   (2) He shall immediately notify both the Utah State Department of Health as specified in Part IX of these regulations and the government officials designated as the on-scene coordinator for that geographical area, (in the applicable regional contingency plan under 40 CFR 1510) or the National Response Center (800/424-8802). The report shall include:
    (i) Name and telephone number of reporter;
    (ii) Name and address of facility;
    (iii) Time and type of incident (e.g., discharge, fire);
    (iv) Name and quantity of material(s) involved, to the extent available;
    (v) The extent of injuries, if any; and
    (vi) The possible hazards to human health, or the environment, outside the facility.

(e) During an emergency, the facility's emergency coordinator shall take all reasonable measures necessary to ensure that fires, explosions, and discharges do not occur, recur, or spread to other hazardous waste at the facility. These measures shall include, where applicable, stopping processes and operations, collecting and containing discharged waste, and removing or isolating containers.

(f) If the facility stops operations in response to a discharge, fire, or explosion, the facility's emergency coordinator shall monitor for leaks, pressure buildup, gas generation, or ruptures in valves, pipes, or other equipment, wherever this is appropriate.

(g) Immediately after an emergency, the facility's emergency coordinator shall provide for treating, storing, or disposing of recovered waste, contaminated soil or surface water, or any other material that results from a discharge, fire, or explosion at the facility.

Unless the owner or operator can demonstrate, in accordance with 2.1.2. (c) or (d), that the recovered material is not a hazardous waste, the owner or operator becomes a generator of hazardous waste and shall manage it in accordance with all applicable requirements in Parts IV, V, VII (Subpart B), and VIII of these regulations.

(h) The facility's emergency coordinator shall ensure that, in the affected area(s) of the facility:

(1) No waste that may be incompatible with the discharged material is treated, stored, or disposed of until cleanup procedures are completed; and

(2) All emergency equipment listed in the contingency plan is cleaned and fit for its intended use before operations are resumed.

(i) The facility owner or operator shall notify the Committee and other appropriate state and local authorities, that the facility is in compliance with paragraph (h) above before operations are resumed in the affected area(s) of the facility.

(j) The facility owner or operator shall record in the operating record the time, date, and details of any incident that requires implementing the contingency plan. Within 15 days after the incident, he shall submit a written report on the incident to the Committee. The report shall include:

(1) Name, address, and telephone number of the owner or operator;

(2) Name, address, and telephone number of the facility;

(3) Date, time, and type of incident (e.g., fire, discharge);

(4) Name and quantity of material(s) involved;

(5) The extent of injuries, if any;

(6) An assessment of actual or potential hazards to the environment or human health, where this is applicable; and

(7) Estimated quantity and disposition of recovered material that resulted from the incident.

Appendix G

*Leak Lookout
(EPA/530/UST-88/006)*

United States
Environmental Protection
Agency

Office of
Underground Storage Tanks
Washington, D.C. 20460

EPA/530/UST-88/006
August 1988

♻EPA # Leak Lookout

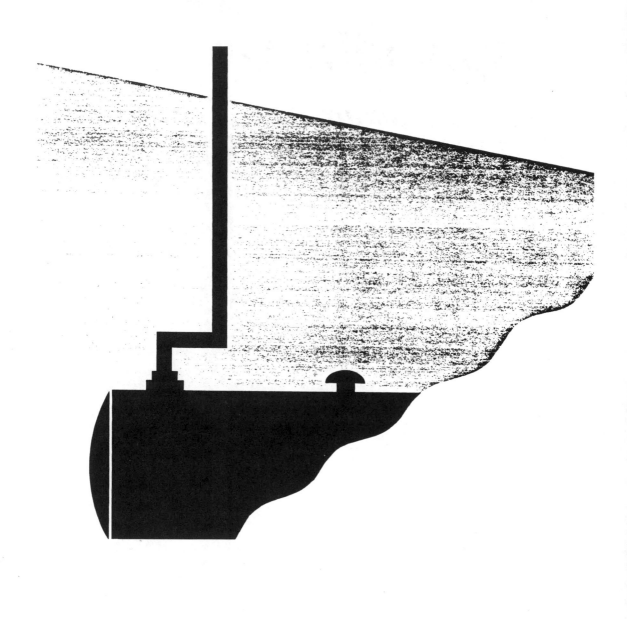

LEAK LOOKOUT

Why Worry About Leak Detection?

Because your tank or its piping may leak.
As many as 25 percent of all underground
storage tanks (USTs) may now be leaking.
Many more will leak in the near future--
possibly yours. In fact, your tank or its
piping might be leaking now, although you
may not know it. If a tank system is past
its prime (that is, over 10 years old), especially if it's not protected against corrosion,
the potential for leaking increases dramatically. Newer tank systems, particularly the
piping, can also leak. Don't let your profits
drain away.

Because it's the law. And it's the law for
a good reason. Much of our country
depends upon ground water for drinking
water, and leaked or spilled petroleum can
contaminate this vital resource. Explosions are another potential hazard of leaks.
Federal requirements for tank systems,
excluding home heating and small farm
tanks, will become effective at the end of
1988. Many State and local governments
already require specific steps to prevent,
detect, or clean up leaks; and others will
soon have similar requirements. Check
with your State and local governments to
learn what requirements apply to you.

Because it's in your best interest. Leaking UST sites can be very costly to clean
up. Imagine how much money you'd lose
if your tank could not be used for weeks
during lengthy cleanups or if local residents sued you for property damages. The
costs can run into the thousands--perhaps
as much as $500,000 and more. Detect
and clean up leaks before they hurt you
financially.

*Because it's for the good of your commu-
nity and the environment.* Petroleum
leaks can have serious and far-reaching
consequences, such as the contamination
of soil, drinking water supplies, and air.
Petroleum and its resulting poisonous
vapors can also accumulate in nearby
confined spaces, such as septic tanks and
home basements, and can cause fires or
explosions. Communities across the
nation have suffered disasters resulting
from petroleum leaks. Don't let it happen
to yours.

How Can This Brochure Help Me?

If you own or operate an underground storage tank that holds petroleum products such as gasoline, diesel fuel, or oil, you should know that there are detectors on the market that can warn you about a leak. If you choose an external leak detector, this brochure will help you select the best one for your underground storage tank. It won't tell you which product to buy or rank equipment from best to worst, but it will give you the background you need on external devices to ask the right questions when you speak with vendors about their products. You should also check with the State or local agency responsible for underground storage tanks in your area to find out what rules apply to you. (If you are not sure which agency to contact, start with your local fire department.)

This brochure will answer questions about:

- The purpose of external leak detectors;

- Equipment and installation costs;

- Compatibility with your current tank;

- The types of detectors on the market;

- How external leak detectors work;

- The pros and cons of various types of equipment; and

- What companies sell external leak detectors.

> ### What Companies Sell External Leak Detectors?
>
> Vendors who responded to a survey by the U.S. Environmental Protection Agency are listed at the end of this brochure. The U.S. Environmental Protection Agency does not endorse any specific product or company, nor does this list include all companies that sell external leak detectors.

What Kinds of Leak Detectors Are Available?

There are two basic types of leak detectors--*external leak detectors* and *internal leak detectors*. This brochure explains only *external leak detectors*, which are installed outside the underground storage tank and warn you if petroleum liquid or vapors are present in monitoring wells near the tank. Some models respond primarily to petroleum liquids while other models respond to petroleum vapors. *Internal leak detectors* monitor the fuel level inside the underground storage tank itself. There are a number of different types of internal leak detectors. Internal leak detection will be described in a future publication.

How Does an External Leak Detection System Work?

An external leak detection system has two main components: (1) the leak detector itself, and (2) one or more monitoring wells. External detectors range from simple, hand-held devices to sophisticated automatic equipment. If a leak occurs, petroleum liquid and vapor flow into nearby monitoring wells near the tank, where they are detected by the leak detection device. Some detectors use sensors located inside the monitoring wells to trigger an alarm in a central control panel; others test samples of water or air taken from the wells. (See Figure 1, below).

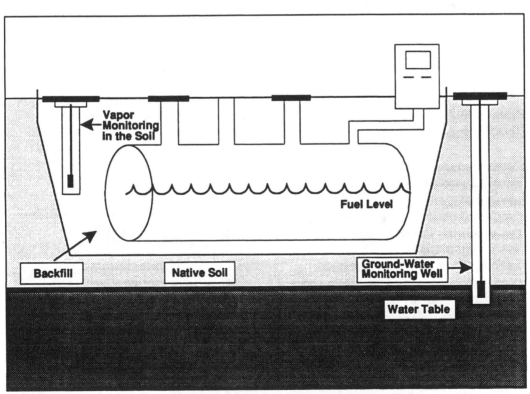

Figure 1. Underground Storage Tank System with Monitoring Wells

How Much Will It Cost?

External leak detectors range in price from about $100 for a basic manual detector to more than $14,000 for a high-end automatic system. Detectors that respond to petroleum vapors are usually more expensive than those that respond to liquid.

Installation costs, which include the cost of putting in the monitoring wells, vary considerably. But maintenance is relatively inexpensive for most models. By doing some comparison shopping, you can find an external leak detector that does the job and fits your budget.

How Do I Know I'm Getting a Good System?

Although some external leak detectors have been used with petroleum underground storage tanks for a number of years, many are new or have been used for other purposes. The newer models may work as well as those used in the past, but you'll need to make an extra effort to ensure that you're getting the right system. That's why it's important to talk with several vendors and obtain as much information as possible about their products. Find out how long their equipment has been on the market and how it has been used in the past. You'll also want to check to make sure the equipment you purchase- meets State and local requirements.

Your leak detector should come with detailed operating instructions and be backed by a warranty. You should have the detector checked every year by the company that sold you the equipment to ensure that it is working properly.

Can I Install the Leak Detector Myself?

No, it takes someone with specific technical knowledge to install monitoring wells and leak detection equipment. In addition, some States require installer certification, and others may do so in the future. Many vendors install leak detectors and monitoring wells themselves. If not, they can usually recommend an experienced contractor to do the installation. Remember, this is a lifetime investment that can save you trouble and money in the future. Don't try to cut corners on installation and end up with a system that doesn't do the job.

Do I Need Special Training to Operate the System?

If the company that sold you the equipment offers a training class, it would be worthwhile to attend. If not, ask the company's representative to walk you through the testing procedures, and see if there is a "hotline" number to call if you have a problem. Be sure to read and follow all instructions that come with the detector whether you have a manual or automatic device.

Who Should Be Responsible for the Leak Detector?

If you can't do it yourself, you should depend on a person who will follow the monitoring and testing procedures outlined in the instructions and record all readings accurately. The benefit to you is protection against ground water contamination, which can be very costly to clean up.

Can I Use an External Detector with My Current Tank?

In most cases, the answer is yes. But you have to consider two key factors: (1) the type of *backfill* in the excavation zone surrounding the tank (tank systems are buried in an "excavation zone," which extends from 1 to 3 feet beyond the tank on all sides), and (2) whether the soil around the tank has been *contaminated* by petroleum.

Backfill: Leak detectors work best when a porous material such as pea gravel or sand is used for backfill within the excavation zone. Leaking liquids and vapors flow rapidly through this type of soil to monitoring wells.

Clay soils, which are not porous, should not be used as backfill with an external detection system, because clay will delay leaking petroleum--particularly liquid petroleum--from reaching monitoring wells. However, an external detector can be used in an area with clay soil as long as a porous material is used as backfill. Most underground storage tanks installed in recent years have a pea gravel backfill, but many older tanks are surrounded by whatever type of soil happens to be at the site.

The equipment installer should check the type of backfill and the level of existing contamination before installing your leak detector. You should check with the vendor or installer to make sure that these tests are performed and that the device you plan to buy will work under the conditions at your site. Find out if your vendor has installed other leak detectors in your area and whether there have been any problems.

Existing contamination: If you install a leak detector at your existing tank site, it must be able to detect any new leaks even though there may already be contamination from a prior leak or spill. External leak detectors work best when the soil and ground water near the tank are clean.

What Kind of Equipment Do I Need?

There is a wide range of equipment to choose from. In the survey of vendors, the U.S. Environmental Protection Agency identified nearly 50 companies that sell external leak detectors, and many of them offer several models. There are other companies that sell external leak detectors that were not covered by the survey.

PAGE 6

External leak detectors come in two categories--those triggered by liquid fuel and those that react to vapor.

In addition, some leak detectors are "intermittent," which means you must perform manual tests at regular intervals, while others are "continuous"--they operate automatically. Here are some things to consider when deciding which type will best suit your needs:

 **Liquid vs.**   **Vapor**

Liquid detectors: Liquid detectors work best in areas where the ground water is close to the surface, making the soil around the tank wet. Leaking fuel seeps down through the soil to the ground water and floats on the surface of the water as it is carried to monitoring wells. Liquid detectors are suitable for use with most petroleum products.

Vapor detectors: Vapor detectors work best in places where the ground water is deep and the soil around the tank is dry. However, vapor detectors will work in shallow ground water conditions if sensors installed in the monitoring wells are not actually under water. Vapor detectors react more quickly to leaks than liquid detectors, because vapor spreads more quickly than liquid. But vapor detectors are highly sensitive and, therefore, more likely to give false readings, and they may require more maintenance than liquid devices. Vapor detectors work best with fuels that produce a large amount of fumes, such as gasoline.

Manual vs. Automatic

There is no simple rule to help you decide whether to use a manual leak detector or an automatic one. The decision depends on a number of factors including your financial resources, the soil around your tank, the number of tanks and sites, and the fuels being stored. Here are some things to consider:

Manual detectors: Manual detectors are easier to use and less expensive than automatic devices; but they require monthly tests at all tank locations, and a record must be kept of the results. Manual detectors let you know immediately if petroleum liquid or vapor is present in a monitoring well, although they only work when you are testing.

If you can take the time to do the tests and keep accurate records of the results, a manual system may be cost-effective. If not, you're probably better off with an automatic system.

Automatic detectors: Automatic detectors are usually more expensive than hand operated ones; but they can be less time consuming to operate, because you don't have to perform routine tests as you do with manual devices. Automatic detectors are working all the time and will send a warning signal if petroleum liquid or vapor enters a monitoring well. Some devices send the signal within minutes after petroleum is detected; others can take as long as 20 minutes. Except for regular maintenance, you don't have to spend much time monitoring the system.

However, just as vapor detectors are more likely to give false readings than liquid detectors, sophisticated automatic devices are more likely to malfunction than simpler manual detectors. You could think of an automatic detector as a race car that needs to be kept in tip-top condition for peak performance, while a manual detector is more like the basic sedan--it doesn't go as fast as the race car, but it can still get you where you are going.

Special Cautions

Response time: Although most automatic detectors respond rapidly once petroleum liquid or vapor enters a monitoring well, the time it takes to reach the well varies greatly depending upon soil and water conditions and the location of the wells. The more porous the soil around the tank and the closer the monitoring wells are to the tank, the more quickly leaking petroleum will reach the well.

Interference: Many automatic leak detectors, and some manual devices, react to chemicals, vapors, and other substances in the ground or air that are unrelated to your petroleum underground storage tank. When this occurs, it can cause the detector to send a false warning or miss a real tank leak. Ask your vendor what interferences may affect the leak detector you plan to buy and find out what you can do about them.

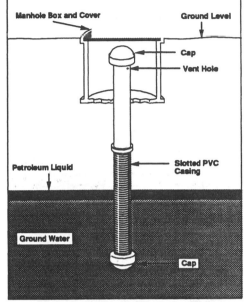

Figure 2. Details of a Monitoring Well

How Many Monitoring Wells Do I Need?

Naturally, the more you have the better your chances of finding a leak. Whatever the number of monitoring wells, however, you do not necessarily need more than one leak detector. If you are using a manual detector, you can test different wells with the same equipment. If you have an automatic detector, sensors in each well are connected to a central control panel. However, you may need additional leak detectors if you have tanks at several locations. (Figure 2 above , illustrates the components of a typical monitoring well.)

How Do External Leak Detectors Work?

Here is a brief description of the main types of external leak detectors and their strengths and weaknesses:

Manual Liquid Detectors

The two most commonly used manual liquid leak detectors are "grab samplers" and chemical-sensitive pastes. In grab sampling, a clear container with a valve on one end--known as a bailer--is lowered into a monitoring well and then removed. If fuel is present, layers of petroleum will be suspended in the water like an oil slick. In the second method, a tape measure treated with special paste is placed in a monitoring well. If fuel is present, the paste changes color. (See figure 3, next column.)

Strengths: Manual liquid detectors are widely used because they are inexpensive, accurate, and easy to operate.

Weaknesses: Manual liquid detectors provide limited information about the type of contamination found. (However, chemical analysis can identify the material collected by grab sampling.) They do not allow for continuous monitoring as do automatic devices; therefore, someone must take time to perform routine tests.

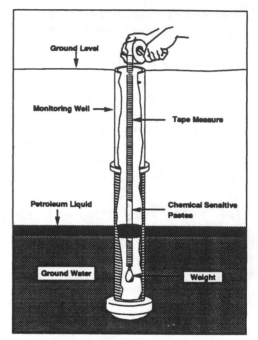

Figure 3. Chemical-Sensitive Pastes with Tape Measure

Automatic Liquid Detectors

There are three types of automatic liquid detectors. Interface probes operate by beaming light through a probe located in a shaft near the tank. If there is liquid in the shaft, the light beam is interrupted and a signal goes off in the control panel; another sensor determines whether the liquid is water or petroleum. Another type, known as a product soluble device, has a cable or hose made of material that dissolves in petroleum. The cable or hose

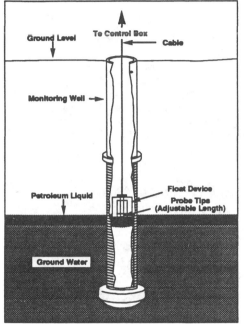

Figure 4. Thermal Conductivity Sensor

is suspended in a monitoring well, and if petroleum is present, it will dissolve the material and trigger an alarm. A third, the thermal conductor, uses a sensor with a heating element that floats in a monitoring well. If petroleum is present, a change in the rate of heat loss triggers an alarm. (See Figure 4, above.)

Strengths: Some automatic liquid detectors have been on the market for 15 years, and thousands of them are in use. The central control panels have a long service life and can accommodate multiple sensors. One type--the product soluble device--detects both petroleum liquid and vapor.

Weaknesses: Interference by ultraviolet light, water vapor, ice, and other substances can trigger false readings. The response time varies from one model to another.

Manual Vapor Detectors

There are at least six different categories of manual vapor detectors. The basic mechanism is that a sample of water or air from a monitoring well comes into contact with a flame, heating element, ultraviolet light, or other material within the detector. If petroleum is present, an electrical or chemical change occurs and can be seen by reading a meter or gauge.

Strengths: Some manual vapor detectors have been commercially available for 10 years, and many are currently in use, although not necessarily for petroleum underground storage tank leak detection. They are highly sensitive and respond within seconds. Some models can identify the exact material that has been detected. Control panels have a long service life.

Weaknesses: Manual vapor detectors require special maintenance, making them time consuming to operate. They are subject to interference, which can result in false readings. The central control panels usually support only one sensor; if you have more than one monitoring well, you will need additional devices. However some devices are portable so you only need one even if you have more than one monitoring well.

Automatic Vapor Detectors

There are three main types of automatic vapor detectors. The metal-oxide semi-conductor works on the principle that petroleum vapor will cause a change in the electrical current in a cell inside the detector. A variation on this model, called an adsistor or diffusion sensor, uses the same principle but operates somewhat differently. Another variety, known as a product permeable device, uses materials that allow vapor but not water to penetrate into the detector. Once the vapor is inside, the detector operates the same way as a metal-oxide semiconductor. A third type, the catalytic sensor, works by bringing an air sample into contact with heated filaments. If vapor is present, the temperature inside the detector will rise and trigger an alarm.

Strengths: Automatic vapor detectors are new to the underground storage tank market, but thousands of these devices have been used during the past decade for other applications. Some models can tell you what type of material has been detected. They are highly sensitive and have a rapid response time. The control panels have a long service life and can accommodate multiple sensors, which last from 1 to 10 years.

Weaknesses: Automatic vapor detectors are subject to interference from a wide range of substances, which can cause false readings.

Does the Federal Government Endorse These Products?

To repeat, the U.S. Environmental Protection Agency does not endorse companies or products. The vendor list included in this brochure is solely for your information, and it does not include all companies that sell external leak detectors.

How Can I Get More Information?

A more detailed description of the devices discussed in this brochure plus additional information on external leak detection are available. You may order the U.S. Environmental Protection Agency's complete report, "Survey of Vendors of External Petroleum Leak Detection Devices for Use with Underground Storage Tanks," from the Superintendent of Documents, Government Printing Office, Washington, DC 20402, (202) 783-3238. Please request stock number 055-000-00277-1. The cost is $4.25.

For specific requirements that apply to your tank, please call your State or local agency responsible for underground storage tanks. If you are not sure which agency to contact, start with your local fire department.

For general information on the national UST program, you can call EPA's toll-free Hotline number, 1-800-424-9346.

LIST OF VENDORS

Detector Type	Company	Address	Phone	Product

🜄 Liquid Sensors

MANUAL

Detector Type	Company	Address	Phone	Product
Grab Samplers	NEPCCO	29 Wall St. Foxboro, MA 02035	617-543-8458	Liquid Samplers and Bailers
	Norton Chemplast	150 Dey Rd. Wayne, NJ 07470	201-696-4700	Bailers
	(many others)			
Chemical-Sensitive Paste	Kolor Kut Products	P.O. Box 5415 Houston, TX 77262	713-926-4780	Water & Gasoline Finding Pastes
	J.H. McCabe Co., Inc.	P.O. Box 822 Short Hills, NJ 07078	201-635-0963	Water & Gasoline Indicator Pastes

AUTOMATIC

Detector Type	Company	Address	Phone	Product
Interface Probe Monitor Monitor	Comar, Inc.	P.O. Box 832676 Richardson, TX 75083	214-238-7691	*Models 807, 808, & 809 Tank Monitors *PLD-17 Pipeline
Gauging	EMTEK, Inc.	27 Harvey Rd. Bedford, NH 03210	603-627-3131	Electronic Well Light (EWGL-12)
	Marine Moisture Control	60 Inip Dr. Inwood, NY 11696	718-327-3430 800-645-7339	Sonic Ullage Interface Probe
	Groundwater Technology, Inc.	220 Norwood Park S. Norwood, MA 02062	617-769-7600	Interface Probe
Product Soluble Devices	EMTEK, Inc.	27 Harvey Rd. Bedford, NH 03102	603-627-3131	Detectron
	IFP Enterprise	680 Fifth Ave. New York, NY 10019	212-265-3800	Oil Fuse
	In-Situ, Inc.	210 South 3rd St. Laramie, WY 82070	307-742-8213	Petrochemical Release Monitors
	K&E Associates	3312 Industry Dr. Long Beach, CA 90806	213-424-1517	PMS-800

Detector Type	Company	Address	Phone	Product
Product Soluble Devices (Cont.)	Pump Engineer Associates	921 National Ave. Addison, IL 60101	312-543-2214	Sentinel
	Technology 2000, Inc.	265 Ballardvale St. Wilmington, MA 01887	617-658-2900	TOLTECH Hydrocarbon Monitor
Electrical Resistivity Sensors	Control Devices	2009-A West Detroit St. Broken Arrow, OK 74012	918-251-0387	Wik-Stik
	Total Containment	15 E. Uwchlan Ave. Exton, PA 19341	215-524-9274	Total Containment Cable-TC3000
Thermal Conductivity Devices	FCI	1755 LaCosta Meadows Dr. San Marcos, CA 92069	800-854-1993 615-744-6950	785 Leak Detection Systems
	Leak-X	560 Sylvan Ave. Englewood Cliff, NJ 07632	201-569-8989	Leak-X System
	Groundwater Technology, Inc.	220 Norwood Park S. Norwood, MA 02062	617-769-7600	CMS Variable Level System
Pollulert Systems (Mallory)		P.O. Box 706 Indianapolis, IN 46206	800-343-2126	FD102 and FD103
	Universal Sensors and Devices, Inc.	9205 Alabama Ave, Unit C Chatsworth, CA 91311	818-998-7121	Leak Alert System

Vapor Sensors

MANUAL

Detector Tubes	MSA	10770 Moss Ridge Road Houston, TX 77043	800-672-2222 713-690-6268	Samplair Pump & Test Kit
	National Draeger, Inc.	P.O. Box 120 Pittsburgh, PA 15230	412-787-8383	Gas & Vapor Detection Products
Combustible Gas Detectors	MSA	P.O. Box 426 Pittsburgh, PA 15230	800-672-2222 412-967-3000	Several models

Detector Type	Company	Address	Phone	Product
Photoionization Detectors	AID, Inc.	Rt. 41 & Newark Rd. Avondale, PA 19311	215-268-3181	Model 580
	Astro Int. Corp.	100 Park Ave. League City, TX 77573	713-332-2484	Trace Gas Analyzer-1010
	HNU	160 Charlemont St. Newton Highlands, MA 02161	617-964-6690	Model PI 101
	Photovac International	741 Park Ave. Huntington, NY 11743	516-351-5809	*TIP *Underground Tank Monitor
Portable GCs	AID, Inc.	Rt. 41 & Newark Rd. Avondale, PA 19311	215-268-3181	Model 590
	HNU	160 Charlemont St. Newton Highlands, MA 02161	617-964-6690	Models 201,301D, 501
	Microsensor Technology Inc.	41762 Christy St. Fremont, CA 94538	415-490-0900	Michromonitor
	Photovac, Inc.	739B Park Ave. Long Island, NY 11743	516-351-5809	Photovac 10A10
	Sentex Sensing Tech., Inc.	553 Broad Ave. Ridgefield, NJ 07657	201-945-3694	Scentor
	XON Tech	6862 Hayvenhurst Ave. Van Nuys, CA 91406	818-787-7380	GC-810
FIDs	Foxboro Analytical	330 Neponset Ave. Foxboro, MA 02035	617-543-8750	Century OVA
Infrared	Foxboro Analytical	330 Neponset Ave. Foxboro, MA 02035	617-543-8750	Miran
	Horiba Instrumental	121 Duryea Ave. Irvine, CA 92714	800-446-7422	IR Hydrocarbon Gas Analyzer
AUTOMATIC Catalytic Sensor Devices	Bacharach Instrument Co.	625 Alpha Dr. Pittsburgh, PA 15238	412-963-2235	*TLV Sniffer *Model 303 *Model H
Gaspointer	Gas Tech, Inc.	8445 Central Ave. Newark, CA 94560	415-794-6200	Model 1238

Detector Type	Company	Address	Phone	Product
Catalytic Sensor Devices (Cont.)	Industrial Scientific Devices Corp.	355 Steubenville Pike Oakdale, PA 15071	412-788-4353	LD-222
	Intek Corp.	P.O. Box 42821/606 Houston, TX 77042	713-498-5855	IGD
	Lumidor Safety Products	5364 NW 167th St. Miami, Florida 33014	305-625-6511	Model CRP-1
	Detector Electronics	207 East Java Dr. P.O. Box 3566 Sunnyvale, CA 94088	408-734-1221	Combustible Gas Detection System
Metal Oxide Semiconductors	API/Ronan	12410 Benedict Ave. Downey, CA 90242	213-803-1497	TRS 76
	Armstrong Monitoring	215 Colonade Rd. S. Nepean, Ontario Canada K2E 7K3	613-225-9531	4200 Sensor
	Azonic Technology Corp.	1671 Mabury Rd. San Jose, CA 95133	408-729-4900	Enviro-Ranger
	Calibrated Instrument, Inc.	731 Saw Mill River Rd. Ardsley, NY 10502	914-693-9232	Pure Air Monitor
	Enmet Corp.	2308 S. Industrial Hwy. Ann Arbor, MI 48104	313-761-1270	Several models
	Genelco, Inc.	11649 Chairman Dr. Dallas, TX 75243	214-341-8410	Soil Sentry
	Harco Technologies Corp.	1216 E. Tower Rd. Schaumberg, IL 60195	312-882-3777	Multi Ram 12
	International Sensor Technology	17771 Fitch St. Irvine, CA 92714	714-863-9999	AG5000 & AG5100
	MSA	600 Penn Center Blvd. Pittsburgh, PA 15235	412-776-8802	Tankgard
	Groundwater Technology, Inc.	220 Norwood Park S. Norwood, MA 02062	617-769-7600	Vapor CMS
	Sierra Monitor	1991 Tarob Court Milpitas, CA 95035	408-262-6611	Model 201
Metal Oxide	Universal	9205 Alabama Ave.	818-998-7121	Leak Alert

Detector Type	Company	Address	Phone	Product
Semiconductors (Cont.)	Sensors and Devices, Inc.	Unit C Chatsworth, CA 91311		
	U.S. Industrial Products Co.	13564 Pumice St. Norwalk, CA 90650	213-921-4342	Tank Monitor
Product Permeable Devices	Teledyne Geotech	3401 Shiloh Rd. Garland, TX 75041	214-271-2561	LASP System
	W.L. Gore	1505 N. 4th St. Flagstaff, AZ 86002	602-526-1290	LEAKLEARN
Diffusion Sensors	Adsistor Technology	11300 N.E. 25th St. P.O. Box 98115 Seattle, WA 98125	206-523-6468	Adsistor Sensor
	Emco Wheaton	Chamberlain Rd. Conneaut, OH 44030	216-599-8151	Leak Sensor II Vapor Probe
	EMMCO, Inc.	2525 Lehigh Pl. Costa Mesa, CA 92626	714-545-6030	Env. Control Safety Monitoring System
	Spearhead Tech., Inc.	P.O. Box 51160 Seattle, WA 98115	604-688-8245	STI 2X12 Inground Tank Monitor

Appendix H

Straight Talk on Tanks
(EPA/530/UST-90/012)

United States
Environmental Protection
Agency

Solid Waste And
Emergency Response
(OS-420)

EPA/530/UST-90/012
August 1990

♻EPA   Straight Talk On Tanks

As An Owner Or Operator Of Underground Storage Tanks . . .

- Do you understand the basic leak detection requirements for underground storage tanks (USTs)?

- Are you confused about choosing the most appropriate leak detection method for your UST?

These are important questions, because your USTs must have leak detection when they are installed or by compliance dates based on the age of the tank. This booklet contains information you can use to answer questions about UST leak detection requirements and methods.

STRAIGHT TALK ON TANKS begins with an overview of the regulatory requirements for leak detection. Each following section focuses on one leak detection method or the special requirements for piping. You will find answers to many basic questions about how leak detection methods work and which methods are most appropriate for your UST site. These sections provide information you can use in questioning vendors and making your selection of leak detection.

An Overview: Leak Detection Requirements For Underground Storage Tank Systems

Federal UST regulations require all UST systems to have leak detection. This section explains when you need leak detection and what your basic leak detection choices are.

When Do You Have To Have Leak Detection?

Tanks and piping installed after December 22, 1988 must have leak detection when they are installed. For USTs installed before that date, the leak detection requirements are phased in over 5 years, depending on the age of the tank. By 1993 all USTs must meet the Federal leak detection requirements. *State and local regulations may be more stringent than Federal requirements, so you should always check to see which requirements you need to meet.* (You will find information on contacting your State on pages 28 through 30 of this booklet.) The chart below shows how the Federal requirements are phased-in over 5 years for existing USTs:

If your UST was installed . . .	It must have leak detection by December of . . .
before 1965 or unknown	1989
1965 -- 1969	1990
1970 -- 1974	1991
1975 -- 1979	1992
1980 -- Dec. 1988	1993

There is a special deadline for existing pressurized piping systems because their pressurized contents pose an especially great threat to the environment. *Existing pressurized piping systems must meet leak detection requirements by December 22, 1990.* Other kinds of existing piping must comply by the dates shown in the phase-in schedule above.

What Leak Detection Methods Are OK?

You need to remember that State or local regulations may differ from the Federal requirements, so be sure to check and see which requirements apply to your UST. Rather than requiring specific technologies, the Environmental Protection Agency has identified a variety of general leak detection methods that owners and operators can use to meet the Federal requirements. You can use:

• Ground-Water Monitoring

• Vapor Monitoring

• Secondary Containment with Interstitial Monitoring

• Automatic Tank Gauging Systems

These are all *monthly* monitoring methods and eventually everyone must use at least one of them. However, as a temporary method (for 10 years after new tank installation and for up to 10 years for existing tanks), you can combine tank tightness testing and manual monthly inventory control (or manual tank gauging if you have a very small tank).

Not all of these leak detection methods can be used for both tanks and piping. Leak detection methods for piping include ground-water monitoring, vapor monitoring, secondary containment with interstitial monitoring, and tightness testing. Pressurized piping must also have an automatic line leak detector. See later sections on suction and pressurized piping for full discussions of the requirements for piping.

A brief description of leak detection methods appears on the next two pages. More complete descriptions appear in the following sections.

Ground-Water Monitoring

Ground-water monitoring senses the presence of liquid product floating on the ground water. This method requires installation of monitoring wells at strategic locations in the ground near the tank and along the piping runs. To discover if leaked product has reached ground water, these wells can be checked periodically by hand or continuously with permanently installed equipment. This method cannot be used at sites where ground water is more than 20 feet below the surface.

Vapor Monitoring

Vapor monitoring senses and measures product "fumes" in the soil around the tank and piping to determine the presence of a leak. This method requires installation of carefully placed monitoring wells. Vapor monitoring can be performed manually on a periodic basis or continuously using permanently installed equipment.

Secondary Containment with Interstitial Monitoring

Secondary containment consists of placing a barrier -- by using a vault, liner, or double-walled structure -- around the UST. Leaked product from the inner tank or piping is directed towards an "interstitial" monitor located between the inner tank or piping and the outer barrier. Interstitial monitoring methods range from a simple dip stick to a continuous automated vapor or liquid sensor permanently installed in the system.

Automatic Tank Gauging Systems

Monitors permanently installed in the tank are linked electronically to a nearby control device to provide information on product level and temperature. During a test period of several hours when nothing is put into or taken from the tank, these monitors are used to automatically calculate the changes in product volume that can indicate a leaking tank.

Leak Detection Methods for Tanks and Piping

1- Ground-Water Monitoring

2- Vapor Monitoring

3- Secondary Containment with Interstitial Monitoring

4- Automatic Tank Gauging Systems

5- Tank Tightness Testing and Inventory Control

6- Manual Tank Gauging

7- Leak Detection for Underground Suction Piping

8- Leak Detection for Pressurized Underground Piping

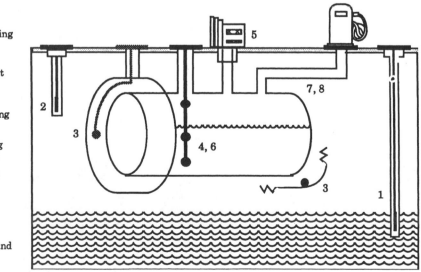

Tank Tightness Testing and Inventory Control

This is strictly a combination method using periodic tank tightness testing and monthly inventory control.

Tightness tests require temporarily installing equipment in the tank. There are two types of tightness tests: volumetric and non-volumetric. A volumetric test involves filling the tank to a specified level and precisely measuring the change in level and temperature over several hours. Non-volumetric test methods include ultrasound techniques and tracer gas detectors. These are sophisticated tests and must be performed by trained, experienced professionals.

In addition to tightness testing, you must use monthly inventory control. Inventory control is basically like balancing a checking account. Every *month* the product volume is balanced between what is delivered and sold from the tank (this is what the "bank" says you have) with *daily* measurements of tank volume taken with a gauge stick (these measurements indicate what you actually have). If your "account" doesn't balance, you may have a leak.

Remember, this combined method can be used only during the first 10 years following new tank installation or upgrade of your existing UST. After that, you must use monthly monitoring methods.

Manual Tank Gauging

One additional method, manual tank gauging, can be used for smaller tanks, but it has several restrictions. In order to meet the Federal leak detection requirements, this method can be used by itself *only for small tanks up to 1,000 gallons*. It requires keeping the tank undisturbed for at least

36 hours, during which no product can be added or removed. During that period, you measure the contents of the tank twice at the beginning and twice at the end of the test period *every week*. At the end of the month, you average your weekly tests and compare the volume lost, if any, to the permissible standards shown on page 18. For tanks over 1,000 gallons but no more than 2,000 gallons this method is allowed only in *combination with tank tightness testing*. This combined method, however, can be used only during the first 10 years following tank installation or upgrade.

Which Method Is Best For You?

Choosing leak detection is not a cut-and-dried process. There is no one leak detection system that is best for all sites, nor is there a particular type of leak detection that is consistently the least expensive.

Each of the leak detection methods has advantages and disadvantages. For example, vapor detection devices work rapidly and most effectively in dry soils, while liquid detectors are most appropriate for areas with a high water table. Identifying the correct option or combination of options depends on a number of factors including cost, tank type, ground-water depth, soil type, and other variables.

The various factors that influence the selection and use of leak detection options are discussed in the Federal regulations and in the following sections.

You will want to find the best fit between what you need and what is available. The next page also contains a table listing a few of the factors that could influence your selection of the leak detection method that is best for your site.

Some Factors To Consider in Selecting Leak Detection
(For a full discussion of many more factors, see the following sections of this booklet.)

Detection Option	Site-Specific Factors	Tank-Related Factors	Cost Factors
Ground-Water Monitoring	Do not use if ground-water level is greater than 20 ft, if clay soil is present, or if existing product is already on the ground water.	Product must be able to float on water and not mix easily with water.	Well installation: $15 - $70/ft depth Equipment: $200 - $5,000 per tank.
Vapor Monitoring	Do not use at sites where soil is saturated with water, the backfill is clay, or soil vapor levels are too high.	Product must evaporate easily or substance that evaporates easily must be added to the tank.	$1,200 - $6,000 per tank for equipment and installation.
Secondary Containment with Interstitial Monitoring	Site conditions (such as too much water) may require use of containment that completely surrounds tank or piping.	A double-walled system must be able to detect a release through the inner wall.	Total installed cost of $5,000 - $12,000 per tank.
Automatic Tank Gauging System (ATGS)	If water collects in excavation, ATGS must have a water sensor.	To date, used primarily at sites with gasoline and diesel in tanks under 15,000 gallons.	Cost per tank: Equipment = $2,300 - $3,900 Installation = $500 - $3,000.
Tank Tightness Testing	Volumetric methods must account for presence of ground water and product temperature.	To date, used primarily at sites with gasoline and diesel tanks under 15,000 gallons.	$250 - $1,000 per test per tank for problem-free test. If problems occur, costs may be much higher.
Inventory Control	None	None	Under $200, but *must be combined with tank tightness testing.*
Manual Tank Gauging	None	Limited to tanks under 1,001 gallons when used alone or under 2,000 gallons when combined with tightness testing.	Under $200, but *may also require tightness testing.*
Automatic Line Leak Detectors	None	Used only for pressurized lines.	Total installed cost of $400 - $2,000 per line.
Line Tightness Testing	None	Used only for piping.	$50 - $100 per test per line if conducted with tank test. May be more expensive if conducted alone. Must do test every 3 years.

Ground-Water Monitoring

NOTE: Ground-water monitoring cannot be used at sites where ground water is more than 20 feet below the surface.

Will I be in compliance?

When installed and operated according to manufacturer's instructions, a ground-water monitoring system meets the *Federal* leak detection requirements for new and existing USTs. Operation of a ground-water monitoring system at least once each month fulfills the requirements for the life of the tank. Ground-water monitoring can also be used to detect leaks in piping (see the later sections on leak detection for piping). You should find out if there are *State or local* limitations on the use of ground-water monitoring or requirements that are different from those presented below.

How does it work?

Operation

- Ground-water monitoring involves the use of one or more permanent monitoring wells placed close to the UST. The wells are checked at least monthly for the presence of product that has leaked from the UST and is floating on the ground-water surface.

- The two main components of a ground-water monitoring system are the monitoring well (typically a well of 2-4 inches in diameter) and the monitoring device.

Installation

- The number of wells and their placement is very important. Many State and local agencies have developed regulations for this, usually requiring somewhere between one and four monitoring wells per UST (additional ones may be required for piping).

- Before installation, a site assessment is necessary to determine the soil type, ground-water depth and flow direction, and the general geology of the site.

Variations

- Detection devices may be permanently installed in the well for automatic, continuous measurements of leaked product.

- Detection devices are also available in manual form. Manual devices range from a bailer (used to collect a liquid sample for visual inspection) to a device that can be inserted into the well to electronically indicate the presence of leaked product. Manual devices must be operated at least once a month.

What are the regulatory requirements?

- Ground-water monitoring can only be used if the stored substance does not easily mix with water and floats on top of water.

- If ground-water monitoring is to be the sole method of leak detection, the ground water must not be more than 20 feet below the surface, and the soil between the well and the UST must be sand, gravel or other coarse materials.

- Monitoring wells must be properly designed and sealed to keep them from becoming contaminated from outside sources. The wells must also be clearly marked and locked.

- Wells should be placed in, or very near to, the UST backfill so that they can detect a leak as quickly as possible.

- Product detection devices must be able to detect one-eighth inch or less of leaked product on top of the ground water.

Will it work at my site?

- In general, ground-water monitoring works best at UST sites where:

    - The ground-water surface extends beneath the tank;

    - Monitoring wells are installed in the tank backfill;

    - Ground water is between 2 and 10 feet from the surface; and

    - There are no previous releases of product that would falsely indicate a current release.

    A site assessment is critical for determining these site-specific conditions.

What other information do I need?

- The proper design and construction of a monitoring well system is crucial to effective detection of leaked product and should be performed by an experienced contractor. Before construction begins, any specific State or local construction requirements should be identified.

- Purchasing a ground-water monitoring system is similar to any other major purchase. You should "shop around," ask questions, get recommendations, and select a company that meets the needs of your UST site.

How much does it cost?

- The capital costs for ground-water monitoring are generally much greater than the annual operating costs.

- The following cases illustrate the effect that different factors have on the cost of ground-water monitoring:

    Case #1
    One tank — well in backfill — short piping runs — manual monitoring — two wells installed:

    Equipment Cost = $200-250
    Installation Cost = $15-25/ft well depth
    Annual Operating Cost = Under $100

    Case #2
    One tank — well not in backfill — long piping runs — automated monitoring — five wells installed:

    Equipment Cost = $2,200-5,000
    Installation Cost = $50-70/ft well depth; conduit to the central console = $500-2,000
    Annual Operating Cost = Under $200

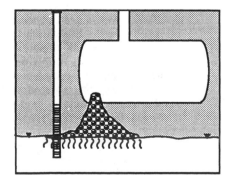

Vapor Monitoring

Will I be in compliance?

When installed and operated according to manufacturer's instructions, vapor monitoring meets the *Federal* leak detection requirements for new and existing USTs. Operation of a vapor monitoring system at least once each month fulfills the requirements for the life of the tank. Vapor monitoring can also be installed to detect leaks from piping (see the later sections on leak detection for piping). You should find out if there are *State or local* limitations on the use of vapor monitoring or requirements that are different from those presented below.

How does it work?

Operation

- Vapor monitoring measures "fumes" from leaked product in the soil around the tank to determine if the tank is leaking.

Variations

- Fully automated vapor monitoring systems have permanently installed equipment to continuously gather and analyze vapor samples and respond to a release with a visual or audible alarm.

- Manually operated vapor monitoring systems range from equipment that immediately analyzes a gathered vapor sample, to devices that gather a sample that must be sent to a laboratory for analysis. Monitoring results from manual systems are generally less accurate than those from automated systems. Manual systems must be used at least once a month to monitor a site.

Calibration and Maintenance

- All vapor monitoring devices should be calibrated annually to a gas standard to ensure that they are properly responding to vapor.

- Maintenance items vary depending upon the system. Manual systems usually require more maintenance than automated systems.

Installation

- Vapor monitoring requires the installation of monitoring wells within the tank backfill.

- Usually one well per 20-40 feet surrounding tanks and piping is sufficient (the proper number depends upon the site conditions).

What are the regulatory requirements?

- The UST backfill must be sand, gravel or another material that will allow the vapors to easily move to the monitor.

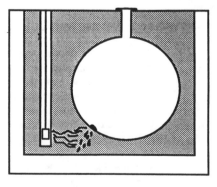

- The backfill should be clean enough that previous contamination does not interfere with the detection of a current leak.

- The substance stored in the UST must vaporize easily so that the vapor monitor can detect a release.

- High ground water, excessive rain, or other sources of moisture must not interfere with the operation of vapor monitoring for more than 30 consecutive days.

- Monitoring wells must be locked and clearly marked.

Will it work at my site?

- Before installing a vapor monitoring system, a site assessment should determine whether vapor monitoring is appropriate at the site. A site assessment usually includes at least a determination of the ground-water level, background contamination, stored product type, and soil type.

- Some vapor monitoring systems can overcome site problems, such as clay backfill. You should discuss any problems that may apply to your site with the equipment salesman and your contractor to ensure they have considered the problems and will compensate for them, if necessary, when installing your vapor monitoring system.

What other information do I need?

- Purchasing a vapor monitoring system is similar to any other major purchase. You should "shop around," ask ques-

tions, get recommendations, and select a method and a company that can meet the needs of your site.

How much does it cost?

- The cost of a vapor monitoring system is influenced by the UST site condition, the required number of monitoring wells, their depth, whether an automated or manual system is chosen, the complexity of the chosen system, and its maintenance. If a site needs to be cleaned up before a system can be installed, costs would increase. However, vapor monitoring has very low annual operating costs (unless a manual system requires laboratory analysis). Here are two possible cases:

Case #1
One tank — 20 feet of piping — manual monitoring using laboratory analysis — two wells installed:

Equipment Cost = $ 200-400
Installation Cost = $ 1,000-2,000
Annual Operating Cost = $ 1,200

Case #2
One tank — 70 feet of piping — automated monitoring and results — five wells installed:

Equipment Cost = $2,800-$3,000
Installation Cost = $2,000-4,000
Annual Operating Cost = Under $25

Secondary Containment With Interstitial Monitoring

NOTE: Secondary containment with interstitial monitoring is required for hazardous substance USTs, and the requirements are different from those for petroleum USTs. Consult your State or local agency for regulations on hazardous substance USTs.

Will I be in compliance?

When installed and operated according to manufacturer's specifications, secondary containment with interstitial monitoring meets the *Federal* leak detection requirements for new and existing USTs. Operation of the monitoring device at least once each month fulfills the requirements for the life of the tank. Secondary containment with interstitial monitoring can also be used to detect leaks from piping (see the later sections on leak detection for piping).

You should find out if *State or local* requirements allow all of the types of secondary containment and interstitial monitoring or have other restrictions that are different from those described below. In some jurisdictions, secondary containment is required for all USTs.

How does it work?

Secondary containment

- Secondary containment provides a barrier between the tank and the environment.

- The barrier holds the leak between the tank and the barrier long enough for the leak to be detected.

- The barrier is shaped so that a leak will be directed towards the monitor.

- Barriers include:

  - Double-walled tanks, in which an outer tank partially or completely surrounds the primary tank;

  - Leakproof excavation liners that partially or completely surround the tank;

  - Leakproof liners that closely surround the tank (also known as "jackets"); and

  - Concrete vaults, with or without lining.

- Clay and other earth materials *cannot* be used as barriers.

Interstitial monitors

- Monitors are used to check the area between the tank and the barrier for leaks and alert the operator if a leak is suspected.

- Some monitors indicate the physical presence of the leaked product, either liquid or gaseous. Other monitors check for a change in condition that indicates a hole in the tank, such as a loss of pressure or a change in the level of water between the walls of a double-walled tank.

- Monitors can be as simple as a dipstick used at the lowest point of the containment to see if liquid product has leaked and pooled there. Monitors can also be sophisticated automated systems that continuously check for leaks.

What are the regulatory requirements?

- The barrier must be immediately around or beneath the tank.

- The interstitial monitor must be checked at least once every 30 days.

- A double-walled system must be able to detect a release through the inner wall.

- An excavation liner must:

  - Direct a leak towards the monitor;

  - Not allow the specific product being stored to pass through it any faster than 10^{-6} cm/sec;

  - Be compatible with the product stored in the tank;

  - Not interfere with the UST's cathodic protection;

  - Not be disabled by moisture;

  - Always be above the ground water and the 25-year flood plain; and

  - Have clearly marked and locked monitoring wells, if they are used.

Will it work at my site?

- In areas with high ground water or a lot of rainfall, it may be necessary to select a secondary containment system that completely surrounds the tank to prevent moisture from interfering with the monitor.

What other information do I need?

- Correct installation is fairly difficult yet is crucial both for the barrier and the interstitial monitor. Therefore, trained and experienced installers are necessary.

- The purchase of secondary containment with interstitial monitoring is similar to any other major purchase. You should "shop around," ask questions, get recommendations, and select a method and company that can meet the needs of your UST site.

How much does it cost?

- The costs of the secondary containment depend on the size and number of the tanks, how much of the tank is surrounded by the barrier, the product, the type of containment, and the site conditions. The cost of the interstitial monitor depends on the number of tanks, degree of automation, and type of monitor.

- Here are possible costs for containment and monitoring at a typical station with three 10,000-gallon tanks:

  - Monitored double-walled tanks cost $15,000 - $30,000 above the cost of single-walled, unmonitored tanks; or

  - An excavation liner (not covering the top of the tank) and monitoring could cost $10,000 - $16,000.

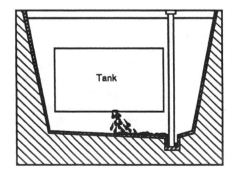

Automatic Tank Gauging Systems

Will I be in compliance?

When installed and operated according to manufacturer's specifications, automatic tank gauging systems (ATGS) meet the *Federal* leak detection requirements for new and existing USTs. A test performed each month fulfills the requirements for the life of the tank. (For additional leak detection requirements for piping, see the later sections on leak detection for piping.) You should find out if *State or local* requirements allow ATGS or have other requirements that are different from those presented below.

How does it work?

- The product level and temperature in a tank are measured *continuously* and automatically analyzed and recorded by a computer.

- In the "inventory mode," the ATGS replaces the use of the gauge stick to measure product level and perform inventory control. This mode records the activities of an in-service tank, including deliveries.

- In the "test mode," the tank is taken out of service and the product level and temperature are measured for at least one hour.

What are the regulatory requirements?

- The ATGS must be able to detect a leak at least as small as 0.2 gallons per hour. By December 1990, the ATGS must also be able to meet the Federal regulatory requirements regarding probabilities of detection and false alarm.

Will it work at my site?

- ATGS have been used primarily on tanks containing gasoline or diesel, with a capacity of less than 15,000 gallons.

- If considering using an ATGS for larger tanks or products other than gasoline or diesel, discuss its applicability with the manufacturer's representative.

- Water around a tank may hide a leak by temporarily preventing the product from leaving the tank. To detect a leak in this situation, the ATGS should be capable of detecting water in the bottom of a tank.

What other information do I need?

- The ATGS probe is permanently installed through a pipe (not the fill pipe) on the top of the tank. Each tank at a site must be equipped with a separate probe.

- The ATGS probe is connected to a monitor that displays ongoing product level information and the results of the

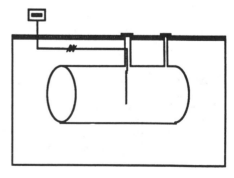

monthly test. Printers can be connected to the monitor to record this information.

- For most ATGS, up to 8 tanks can be connected to a single monitor.

- ATGS usually are equipped with alarms for high and low product level, high water level, and theft.

- ATGS can be linked with computers at other locations, from which the system can be programmed or read.

- No product should be delivered to the tank or withdrawn from it for at least 6 hours before the monthly test or during the test (which generally takes 1 to 6 hours).

- An ATGS can be programmed to perform a test more often than once per month, if so desired.

- Purchasing an ATGS is similar to any other major purchase. You should "shop around," ask questions, get recommendations, and select a method and company that can meet the needs of your site.

How much does it cost?

- Equipment costs —

  Monitor: $1,700-2,700; varies with manufacturer and whether a printer is included.
  Probes: $500-1,100/probe; varies with manufacturer.
  Cables: $0.15-1.00/foot; varies with the contractor and the part of the country.

- Installation costs for a typical 3-tank system —

  For a site that already has conduits for cables: $500-1,500.

  For a site in which conduit must be laid an average distance: $2,500-3,000.

  For a very complex site with many conduits running long distances and where rewiring is necessary: up to $10,000.

- Annual operating costs for a typical 3-tank system —

  About $50-100 for electricity, printer paper, and maintenance.

- For a typical 3-tank system the total equipment cost ranges from about $5,000 for a basic system to $10,000 for a top-of-the-line system. The installation cost averages about $2,000-3,000.

Tank Tightness Testing and Inventory Control

Will I be in compliance?

When performed according to manufacturer's specifications, periodic tank tightness testing combined with monthly inventory control can temporarily (as described below) meet the *Federal* leak detection requirements for new and existing USTs. In addition, you should determine if *State or local* requirements have limitations on the use of these methods or requirements different from those presented below.

These two leak detection methods must be used together, because neither method alone meets the Federal requirements for leak detection for tanks. Tightness testing is also an option for underground piping, as described in the later sections on leak detection for piping.

Because they must be used together, both tank tightness testing and inventory control are discussed in this section. Tank tightness testing is discussed first, followed by inventory control.

Tank Tightness Testing

How does it work?

Tightness tests include a wide variety of methods. Other terms used for these methods include "precision testing" and "volumetric testing."

Operation

- There are a few methods that do not measure the level or volume of the product. Instead, these methods use a principle such as acoustics to determine the physical presence of a hole in the tank. With such methods, all of the factors in the following bullets may not apply.

- Most tightness test methods are "volumetric" methods in which the change in product level or volume in a tank over several hours is measured very precisely (in milliliters or thousandths of an inch).

- For most methods, changes in product temperature also must be measured very precisely (thousandths of a degree) concurrently with level measurements because temperature changes cause volume changes that interfere with finding a leak.

- For most methods, a net decrease in product volume (subtracting out temperature-induced volume changes) over the time of the test indicates a leak.

- The testing equipment is temporarily installed in the tank, usually through the fill pipe.

- The tank must be taken out of service for the test, generally 6 to 12 hours, depending on the method.

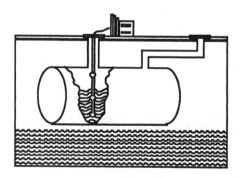

- Many test methods require that the product in the tank be at a certain level before testing, which often requires adding product from another tank on-site or purchasing additional product.

Varieties of methods

- Some tightness test methods require all of the measurements and calculations to be made by hand by the tester. Other tightness test methods are highly auto-mated. After the tester sets up the equipment, a computer controls the measurements and analysis.

- There are several different acceptable ways to measure product temperature: mixing the product so it is all one tem-perature; using a sensor that calculates an average temperature by measuring temperature throughout the depth of the product; and using at least 3 tem-perature sensors at different product levels to calculate an average tempera-ture.

- A few methods measure properties of the product that are independent of temperature, such as the mass of the product, and so do not need to measure product temperature.

What are the regulatory requirements?

- The tightness test method must be able to detect a leak at least as small as 0.1 gallon per hour. By December 1990, the tightness test method must also be able to meet the Federal regulatory require-ments regarding probabilities of detec-tion and false alarm.

- Tightness tests must be performed peri-odically as shown in the following table:

MINIMUM TESTING FREQUENCY	
New tanks	Every 5 years for 10 years following installation
Existing tanks, upgraded	Every 5 years for 10 years following upgrade
Existing tanks, not upgraded	Every year until 1998

("Upgraded" tanks have corrosion protection and spill/overfill prevention devices.)

- After the applicable time period listed above, you must have a monitoring method that can be performed at least once per month. See the other sections of this booklet for allowable monthly monitoring options.

Will it work at my site?

- Tank tightness testing has been used primarily on tanks less than 15,000 gal-lons in capacity containing gasoline and diesel.

- If you are considering using tightness testing for larger tanks or products other than gasoline or diesel, discuss the method's applicability with the manu-facturer's representative.

What other information do I need?

- For most methods, the test is performed by a testing company. You just observe the test.

- Manifolded tanks generally should be disconnected and tested separately.

- Depending on the method, up to 4 tanks can be tested at one time. Generally, an automated system is necessary to test 3 or 4 tanks at a time.

- Procedure and personnel, not equipment, are usually the most important factors in a successful tightness test. Therefore, well-trained and experienced testers are very important. Some States and local authorities may have tester certification programs.

- Purchasing a tightness test is similar to any other major purchase. You should "shop around," ask questions, get recommendations, and select a method and a company that can meet the needs of your site.

What does it cost?

- There are no capital costs for test equipment.

- The *total* cost per test is highly variable. The prices quoted by testing companies range from about $250 to $1,000 per tank, with most between $500 and $800. These prices are for a simple test with no problems.

- The final cost for a tank tightness test can be significantly higher. Some factors that would add to the cost of a test that you should ask about are:

  - Product to fill the tank to the minimum testing level, if it is product that you would not buy otherwise.

  - Lost business from shutting down the tank during normal business hours.

  - Replacing or repairing parts of the tank system before a test can be performed.

  - Uncovering part of the tank system and then recovering it, to correct problems such as vapor pockets or piping that must be valved off.

- Costs can be reduced if a large number of tanks are to be tested and if you check the tank and do the necessary repairs and replacements before the test crew arrives.

Inventory Control

How does it work?

- Inventory control is basically like balancing a checking account. Every *month* the product volume is balanced between what is delivered and what is sold from the tank (this is what the "bank" says you have) with *daily* measurements of tank volume taken with a gauge stick (these measurements indi-

cate what you actually have). If your "account" doesn't balance, you may have a leak.

- UST inventories are determined in the morning and in the evening or after each shift by using a gauge stick and the data is recorded on a ledger form. The level on the gauge stick can be translated to a volume of product in the tank using a calibration chart, which is often furnished by the UST manufacturer.

- The amounts of product delivered to and withdrawn from the UST each day are also recorded.

- At least once each month, the gauge stick data and the sales and delivery data are reconciled and the month's overage or shortage is determined. If the overage or shortage is greater than or equal to 1.0 percent of the tank's flow-through volume plus 130 gallons of product, the UST may be leaking.

What are the regulatory requirements?

- Inventory control must be used in conjunction with periodic tank tightness tests.

- The gauge stick should be long enough to reach the bottom of the tank and marked so that the product level can be determined to the nearest one-eighth of an inch.

- A monthly measurement must be taken to identify any water at the bottom of the tank.

- Deliveries must be made through a drop tube that extends to within one foot of the tank bottom.

- Product dispensers must be calibrated to the local weights and measures standards.

Will it work at my site?

- If your tank is not level, inventory control may need to be modified. You will need to create your own tank chart by adding increments of product and reading the corresponding level.

What other information do I need?

- Inventory control is a practical, commonly used management tool that does not require closing down the tank operation.

- You can perform inventory control yourself.

- Gauge accuracy can be significantly improved by the use of product-finding paste.

- Your product supplier, jobber or contractor may be able to teach you the proper technique for inventory control, and may be able to supply you with recording forms and a gauge stick.

How much does it cost?

- The main costs are the price of a gauge stick and perhaps some product-finding paste. The cost of these items is less than $200.

Manual Tank Gauging

NOTE: Manual tank gauging can only be used for smaller tanks. Tanks 1,000 gallons or less can use this method alone, but tanks from 1,001 - 2,000 gallons can only use manual tank gauging when it is combined with tank tightness testing. Manual tank gauging cannot be used for tanks over 2,000 gallons.

If you are considering using manual tank gauging to meet the Federal UST leak detection requirements, this section provides information about its appropriate use. This method should not be confused with inventory control, which is described in the previous section of this booklet.

Will I be in compliance?

As described below, when performed according to recommended practices, manual tank gauging meets the *Federal* leak detection requirements for USTs with a capacity of 1,000 gallons or less for the life of the tank. (For additional leak detection requirements for piping, see the following sections on leak detection for piping.) You should find out if *State or local* requirements have limitations on the use of manual tank gauging or have requirements that are different from those presented below.

How does it work?

- Four liquid level measurements must be taken weekly, two at the beginning and two at the end of **at least a 36-hour period** during which nothing is added to or removed from the tank.

- The average of the two consecutive ending measurements are subtracted from the average of the two beginning measurements to indicate the change in product volume.

- Every week, the calculated change in tank volume is compared to the standards *shown in the table below.* If the calculated change exceeds the weekly standard, the UST may be leaking. Also, monthly averages of the four weekly test results must be compared to the monthly standard in the same way.

Tank Capacity	Weekly Standard (one test)	Monthly Standard (4-test avg.)	Minimum Duration of Test
If Manual Tank Gauging is the ONLY leak detection method used:			
up to 550 gallons	10 gallons	5 gallons	36 hours
551 - 1,000 gal. (when largest tank is 64" x 73")	9 gallons	4 gallons	44 hours
1,000 gal. (if tank is 48" x 128")	12 gallons	6 gallons	58 hours
If Manual Tank Gauging is combined with Tank Tightness Testing:			
1,001 - 2,000 gal.	26 gallons	13 gallons	36 hours

What are the regulatory requirements?

- Liquid level measurements must be taken with a gauge stick that is marked to measure the liquid to the nearest one-eighth of an inch.

- Manual tank gauging may be used as the sole method of leak detection for tanks with a capacity of 1,000 gallons or less for the life of the tank.

- For tanks with a capacity of 1,001 - 2,000 gallons, manual tank gauging **must be combined with tightness testing according to the schedule below.** See the earlier section on tank tightness testing for details on this method.

MINIMUM TIGHTNESS TESTING FREQUENCY	
New tanks	Every 5 years for 10 years following installation
Existing tanks, upgraded	Every 5 years for 10 years following upgrade
Existing tanks, not upgraded	Every year until 1998

("Upgraded" tanks have corrosion protection and spill/overfill prevention devices.)

- Please note that this combined method will meet the Federal requirements only temporarily. After the applicable time period listed above, you must have a monitoring method that can be performed at least once a month. See the other sections of this booklet for allowable monthly monitoring options.

- Tanks greater than 2,000 gallons in capacity *may not use* this method of leak detection to meet these regulatory requirements.

Will it work at my site?

- Manual tank gauging is inexpensive and can be an effective leak detection method when used as described above with tanks of the appropriate size.

What other information do I need?

- You can perform manual tank gauging yourself. Correct gauging, recording and interpretation are the most important factors for successful tank gauging.

- Your product supplier, jobber or contractor can probably teach you the proper technique for manual tank gauging, and may be able to supply you with recording forms and a gauge stick.

- The accuracy of tank gauging can be greatly increased by spreading product-finding paste on the gauge stick before taking a measurement.

How much does it cost?

- For tanks less than 1,001 gallons, the only costs are the price of a gauge stick and perhaps some product-finding paste. These costs are nominal (less than $200).

- For tanks between 1,001-2,000 gallons, there is the additional cost of periodic tank tightness testing. These costs are highly variable (from about $250 to over $1,000 for each test).

Leak Detection for Underground Suction Piping

Will I be in compliance?

When installed and operated according to manufacturer's specifications, the leak detection methods discussed here meet the *Federal* regulatory requirements for the life of new and existing underground suction piping systems. You should find out if *State or local* requirements allow all these methods or have other requirements that are different from those described below.

What are the regulatory requirements?

- No leak detection is required if the suction piping has (1) enough slope so that the product in the pipe can drain back into the tank when suction is released and (2) has only one check valve, which is as close as possible beneath the pump in the dispensing unit. If a suction line is to be considered exempt based on these design elements, there must be some way to check that the line was actually installed according to these plans.

- If a suction line does not meet all of these design criteria, one of the following leak detection methods must be used:

  - A line tightness test at least every 3 years; or

  - Monthly vapor monitoring; or

  - Monthly ground-water monitoring; or

  - Monthly interstitial monitoring.

The line tightness test must be able to detect a leak at least as small as 0.1 gallon per hour. By December 1990, the test must also meet the Federal regulatory requirements regarding probabilities of detection and false alarm.

- Ground-water, vapor, and interstitial monitoring have the same regulatory requirements for piping as they do for tanks. See the earlier sections of this booklet on those methods.

How do the methods work?

Line tightness testing

- The line is taken out of service and pressurized. A drop in pressure over time, preferably one hour, suggests a possible leak.

- Suction lines are not pressurized very much during a tightness test (less than 15 pounds per square inch).

- Tightness tests must be conducted at least every three years.

- Most line tightness tests are performed by a testing company. You just observe the test.

- Some *tank* tightness test methods can be performed to include a tightness test of the connected piping.

- For most line tightness tests, no permanent equipment is installed.

- The line must be taken out of service for the test, ideally for several hours to allow the line to stabilize before the test.

- In the event of trapped vapor pockets, it may not be possible to conduct a valid line tightness test. There is no way to tell definitely before the test begins if this will be a problem, but longer complicated piping runs with a lot of risers and dead ends are more likely to have vapor pockets.

Ground-water or vapor monitoring

- Ground-water monitoring checks for leaked product floating on the ground water near the piping.

- Vapor monitoring detects product that leaks into the soil and evaporates there.

- A monitoring well should be installed every 20 to 40 feet.

- UST systems using ground-water or vapor monitoring for the tanks are well suited to use the same monitoring method for the piping.

- See the earlier sections on ground-water and vapor monitoring. Use of these methods with piping is similar to that for tanks.

Secondary containment with interstitial monitoring

- A barrier is placed between the piping and the environment. Barriers such as double-walled piping or a leakproof liner in the piping trench can be used.

- A monitor is placed between the piping and the barrier to sense a leak if it occurs. Monitors range from a simple stick that can be put in a sump to see if a liquid is present, to continuous automated systems, such as those that monitor for the presence of evaporated product.

- Proper installation of secondary containment is the most important and the most difficult aspect of this leak detection method. Trained and experienced installers are necessary.

- See the section on secondary containment for additional information. Secondary containment for piping is similar to that for tanks.

What other information do I need?

- Purchasing piping leak detection is similar to any other major purchase. You should "shop around," ask questions, get recommendations, and select a method and company with experience and one that can meet the needs of your site.

How much does it cost?

Line tightness tests

- When performed at the same time as a tank tightness test, a typical line test costs about $50-100. The price varies with the length and complexity of the piping.

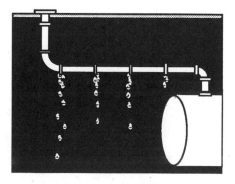

If a testing company comes on-site to perform only a line tightness test, the cost will probably be much higher unless you can negotiate for a package deal for a larger number of tests. Not all tightness testing companies will do independent line tests.

Ground-water or vapor monitoring

- If you have already selected ground-water or vapor monitoring for your tanks, the additional cost to include piping in the monitoring network may be relatively small if only wells need to be added. If an underground cable or an extra control panel is needed for an automated system, the cost will be higher. See the sections on ground-water and vapor monitoring for costs of tank monitoring.

Secondary containment with interstitial monitoring

- The total installed cost for double-walled piping for a typical 3-tank station is $10,000, not including digging the trenches. Costs vary with size of pipe, length of run, site conditions, and contractor. In general, double-walled piping systems cost about 3 times as much as single-walled systems.

- For a typical station, trench liners cost about $25 to $40 per linear foot for 2-inch pipes, depending on the number of pipes. Installation is about $800 to $1,500, depending on site conditions.

- The costs of a monitor range from essentially nothing for a dipstick to a total installed cost of about $1,000 for an electronic sensor (not including control panel).

Leak Detection for Pressurized Underground Piping

Will I be in compliance?

When installed and operated according to manufacturer's specifications, the leak detection methods discussed here meet the *Federal* regulatory requirements for the life of new and existing pressurized underground piping systems. You should find out if *State or local* requirements allow all of these methods or have other requirements that are different from those described below.

When do I have to start?

New pressurized piping must have leak detection when it is installed. **Existing** pressurized piping must meet the leak detection regulatory requirements by **December 22, 1990.**

What are the regulatory requirements?

- Each pressurized piping run must have *one* leak detection method from *each* set below:

  An Automatic Line Leak Detector:

  — Automatic flow restrictor *or*
  — Automatic flow shutoff *or*
  — Continuous alarm system

  And One Other Method:

  — Monthly ground-water monitoring *or*
  — Monthly vapor monitoring *or*
  — Monthly interstitial monitoring *or*
  — Annual tightness test

- The automatic line leak detector (LLD) must be able to detect a leak at least as small as 3 gallons per hour at a line pressure of 10 pounds per square inch within 1 hour by shutting off the product flow, restricting the product flow, or triggering an audible or visual alarm.

- The line tightness test must be able to detect a leak at least as small as 0.1 gallon per hour when the line pressure is one and one-half times its normal operating pressure.

- By December 1990, automatic LLDs and line tightness tests must also be able to meet the Federal regulatory requirements regarding probabilities of detection and false alarm.

- Ground-water, vapor, and interstitial monitoring have the same regulatory requirements for piping as they do for tanks. See the earlier sections of this booklet on those methods.

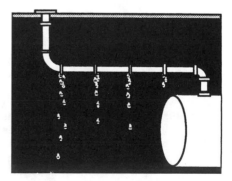

How do the methods work?

Automatic line leak detectors (LLDs)

- Flow restrictors and flow shutoffs can monitor the pressure within the line in a variety of ways: whether the pressure decreases over time; how long it takes for a line to reach operating pressure; and combinations of increases and decreases in pressure.

- If a possible leak is detected, a *flow restrictor* keeps the product flow through the line at 3 gallons per hour, well below the usual flow rate.

- If a possible leak is detected, a *flow shutoff* completely cuts off product flow in the line or shuts down the pump.

- A *continuous alarm system* constantly monitors line conditions and immediately triggers an audible or visual alarm if a leak is suspected. Automated vapor or interstitial line monitoring systems can also be set up to operate continuously and sound an alarm, flash a signal on the console, or even ring a telephone in a manager's office when a leak is detected.

- Both automatic flow restrictors and shutoffs are permanently installed directly into the pipe or the pump housing.

- Vapor and interstitial monitoring systems can be combined with automatic shutoff systems so that whenever the monitor detects a possible release the piping system is shut down. This would qualify as a continuous alarm system. Such a setup would meet the monthly monitoring requirement as well as the LLD requirement.

Line tightness testing

- The line is taken out of service and pressurized, usually above the normal operating pressure. A drop in pressure over time, preferably one hour, suggests a possible leak.

- Tightness tests must be conducted annually.

- Most line tightness tests are performed by a testing company. You just observe the test.

- Some *tank* tightness test methods can be performed to include a tightness test of the connected piping.

- For most line tightness tests, no permanent equipment is installed.

- The line must be taken out of service for the test, ideally for several hours to allow the line to stabilize the test.

- In the event of trapped vapor pockets, it may not be possible to conduct a valid line tightness test. There is no way to tell definitely before the test begins if this will be a problem, but longer complicated piping runs with a lot of risers and dead ends are more likely to have vapor pockets.

Secondary containment with interstitial monitoring

- A barrier is placed between the piping and the environment. Double-walled piping or a leakproof liner in the piping trench can be used. A monitor is placed between the piping and the barrier that senses a leak if it occurs. Monitors range from a simple stick that can be put into a sump to see if a liquid is present, to continuous automated systems, such as those that monitor for the presence of evaporated product.

- Proper installation of secondary containment is the most important and the most difficult aspect of this release detection method. Trained and experienced installers are necessary.

- See the section on secondary containment for additional information. Secondary containment for piping is similar to that for tanks.

Ground-water or vapor monitoring

- Ground-water monitoring checks for leaked product floating on the ground water near the piping.

- Vapor monitoring detects product that leaks into the soil and evaporates there.

- A monitoring well should be installed every 20 to 40 feet.

- UST systems using ground-water or vapor monitoring for the tanks are well suited to use the same monitoring method for the piping.

- See the sections on ground-water and vapor monitoring for additional information. Use of these methods with piping is similar to that for tanks.

What other information do I need?

- Purchasing piping release detection is similar to any other major purchase. You should "shop around," ask questions, get recommendations, and select a method and company that can meet the needs of your UST site.

How much does it cost?

Automatic LLDs

- Automatic flow restrictors: total installed cost about $300-$400 per line.

- Automatic shutoff devices: total installed cost for one line is about $2,000. There may be cost savings for multiple lines or when you include tank monitoring in the same system.

- The annual operating costs are negligible.

Line tightness tests

- When performed at the same time as a tank tightness test, a typical line test costs about $50-100. The price varies with the length and complexity of the piping.

- If a testing company comes on-site to perform only a line tightness test, the cost will probably be much higher unless you can negotiate a package deal for a large number of tests. Not all tightness testing companies will do independent line tests.

Secondary containment with interstitial monitoring

- The total installed cost for double-walled piping at a typical 3-tank station is about $10,000, not including digging the trenches. Costs vary with the size of pipe, length of run, site conditions, and contractor. In general, double-walled piping systems cost about 3 times as much as single-walled systems.

- For a typical station, trench liners cost about $25 to $40 per linear foot for 2-inch pipes, depending on the number of pipes. Installation is about $800 to $1,500, depending on site conditions.

- Costs for interstitial monitoring devices range from essentially nothing for a dipstick to a total installed cost of about $1,000 per line for an electronic sensor (not including the control panel).

Ground-water or vapor monitoring

- If you have already selected ground-water or vapor monitoring for your tanks, the additional cost to include the piping in the monitoring network may be relatively small if only wells need to be added. If an underground cable or an extra control panel is needed for an automated system, the costs will be higher. See the sections on ground-water and vapor monitoring for costs of tank monitoring.

Need More Information?

If this booklet does not answer all your questions, contact your State UST Program Office for additional information. Contact information for these State UST Program Offices is listed on pages 28 through 30.

As you seriously explore leak detection options, you may want to take advantage of the following sources of information before you make your final selection:

- Local agencies, such as your fire department, that regulate USTs

- Trade association representatives

- State petroleum marketers association

- Contractor, jobber, or equipment supplier

- Leak Detection Technology Association (202) 835-2355/828-1000

- Petroleum Equipment Institute (918) 494-9696

For additional information about Federal UST requirements, contact the Environmental Protection Agency's RCRA/Superfund Hotline, by calling (800) 424-9346 or (202) 382-3000, Monday — Friday, 8:30 a.m. — 7:30 p.m. EST.

State UST Program Offices

AK UST CONTACT
Dept. of Environmental Conservation
P.O. Box 0
3220 Hospital Drive
Juneau, AK 99811-1800　907-465-2630

AL UST CONTACT
Dept. of Environmental Management
Ground-Water Branch/Water Division
1751 Congressman W. L. Dickerson Dr.
Montgomery, AL 36130　205-271-7986

AR UST CONTACT
Dept. of Pollution Control & Ecology
P.O. Box 9583
8001 National Drive
Little Rock, AR 72219　501-562-7444

AZ UST CONTACT
Department of Environmental Quality
2005 North Central Avenue, Room 300
Phoenix, AZ 85004　602-257-6984

CA UST CONTACT
State Water Resources Control Board
Division of Loans and Grants
P.O. Box 944212
2014 T Street
Sacramento, CA 94244-2120　916-739-4324

CO UST CONTACT
CO Department of Health
Hazardous Materials and Waste
    Management Program
Underground Tank Program
4210 East 11th Avenue
Denver, CO 80220　303-331-4830

CT UST CONTACT
CT Dept. of Environmental Protection
Underground Storage Tank Program
State Office Building
165 Capitol Avenue
Hartford, CT 06106　203-566-4630

DC UST CONTACT
DC Dept. of Consumer and Reg. Affairs
614 H Street, NW Rm. 516
Washington, D.C. 20013-7200　202-783-3205

DE UST CONTACT
DE Dept. of Natural Resources &
    Environmental Control
Underground Storage Tank Branch
715 Grantham Lane
New Castle, DE 19720　302-323-4588

FL UST CONTACT
Dept. of Environmental Regulation
Tank Section
Twin Towers Office Building - Rm 403
2600 Blair Stone Road
Tallahassee, FL 32399-2400　904-488-3936

GA UST CONTACT
Environmental Protection Division
Underground Storage Tank Unit
3420 Norman Berry Drive - 7th Floor
Hapeville, GA 30354　404-669-3927

HI UST CONTACT
Department of Health
500 Ala Moana Blvd.
Honolulu, HI 96813　808-543-8226

IA UST CONTACT
IA Department of Natural Resources
Henry A. Wallace Building
900 East Grand
Des Moines, IA 50319　515-281-8692

ID UST CONTACT
ID Department of Health & Welfare
450 West State Street
Boise, ID 83710　208-334-5847

IL UST CONTACT
IL Office of State Fire Marshal
3150 Executive Park Drive
Springfield, IL 62703-4259　217-785-5878

IN UST CONTACT
5500 West Bradbury Avenue
Indianapolis, IN 46241　317-243-5055

KS UST CONTACT
KS Department of Health & Environment
Bureau of Environmental Remediation
Underground Storage Tank Section
Forbes Field, Building 740
Topeka, KS 66620　913-296-1678

KY UST CONTACT
Division of Waste Management
Underground Storage Tank Section
18 Reilly Road
Frankfort, KY 40601　502-564-6716

LA UST CONTACT
Dept. of Environmental Quality
Underground Storage Tank Division
P.O. Box 44274, 438 Main Street
Baton Rouge, LA 70804　504-342-7808

MA UST CONTACT
MA Department of Public Safety
Underground Storage Tank Program
P.O. Box 490, East Street, Bldg. #5
Tewksbury, MA 01876   508-851-9813

MD UST CONTACT
MD Department of Environment
Hazardous & Solid Waste Mgmt. Admin.
Underground Storage Tank Section
2500 Broening Highway
Baltimore, MD 21234   301-631-3442

ME UST CONTACT
ME Dept. of Environmental Protection
State House - Station 17
Hospital Street, Ray Building
Augusta, ME 04333   207-289-2651

MI UST CONTACT
MI Department of State Police
Fire Marshal Division
7150 Harris Drive
Lansing, MI 48913   517-334-7090

MN UST CONTACT
MN Pollution Control Agency
Underground Storage Tank Program
520 Lafayette Road North
St. Paul, MN 55155   612-296-7743

MO UST CONTACT
MO Department of Natural Resources
P.O. Box 176
205 Jefferson Street
Jefferson City, MO 65102   314-751-7428

MS UST CONTACT
Department of Environmental Quality
Bureau of Pollution Control
Underground Storage Tank Section
P.O. Box 10385, 2380 Hwy 80 West
Jackson, MS 39289-0385   601-961-5171

MT UST CONTACT
MT Dept. of Health & Environmental Sci.
Solid & Hazardous Waste Bureau
111 North Last Chance Gulch
Arcade Building, Basement
Helena, MT 59701   406-444-5970

NC UST CONTACT
Pollution Control Branch
Division of Environmental Management
Dept. of Env., Health and Natural Res.
P.O. Box 27687
Raleigh, NC 27611-7687   919-733-8486

ND UST CONTACT
ND Department of Health
Division of Waste Management
Box 5520, 1200 Missouri Avenue
Bismarck, ND 58502-5520   701-224-2366

NE UST CONTACT
NE State Fire Marshal's Office
Underground Storage Tank Division
P.O. Box 94677
246 South 14th Street
Lincoln, NE 68509   402-471-9465

NH UST CONTACT
NH Dept. of Environmental Services
Underground Storage Tank Program
6 Hazen Drive, P.O. Box 95
Concord, NH 03301   603-271-3444

NJ UST CONTACT
Dept. of Environmental Protection
Div. of Water Resources (CN-029)
401 East State Street
Trenton, NJ 08625   609-984-3156

NM UST CONTACT
Environmental Improvement Division
Underground Storage Tank Bureau
1190 St. Francis Drive
Harold Runnels Building, Room N2150
Santa Fe, NM 87503   505-827-0188

NV UST CONTACT
Dept. of Conservation & Natural Res.
Division of Environmental Protection
Capitol Complex
201 S. Fall Street
Carson City, NV 89710   702-687-5872

NY UST CONTACT
Dept. of Environmental Conservation
Bulk Storage Section, Div. of Water
50 Wolf Road, Room 326
Albany, NY 12233-3520   518-457-4351

OH UST CONTACT
OH Department of Commerce
7510 East Main Street
P.O. Box 525
Reynoldsburg, OH 43068   614-752-7938

OK UST CONTACT
Corporation Commission
Underground Storage Tank Program
Jim Thorpe Building
2101 North Lincoln Blvd.
Oklahoma City, OK 73105   405-521-3107

OR UST CONTACT
OR Dept. of Environmental Quality
811 SW Sixth Avenue
Portland, OR 97204   503-229-6652

PA UST CONTACT
PA Dept. of Environmental Resources
Non-point Source & Storage Tank Section
P.O. Box 2063, Fulton Building, 12th Floor
Harrisburg, PA 17120   717-657-4080

RI UST CONTACT
RI Dept. of Environmental Management
Underground Storage Tank Section
291 Promenade St.
Providence, RI 02908   401-277-2234

SC UST CONTACT
Dept. of Health and Environ. Control
Ground-Water Protection Division
2600 Bull Street
Columbia, SC 29201   803-734-5332

SD UST CONTACT
Dept. of Water & Natural Resources
Office of Water Quality
523 East Capitol
Joe Foss Building
Pierre, SD 57501-3181   605-773-3351

TN UST CONTACT
Dept. of Health & Environment
200 Doctors Building
706 Church Street
Nashville, TN 37247   615-741-4081

TX UST CONTACT
Texas Water Commission
Underground Storage Tank Section
P.O.Box 13087, 1700 North Congress
Austin, TX 78711   512-463-7786

UT UST CONTACT
UT Department of Health
Bureau of Solid & Hazardous Waste
Solid and Hazardous Waste Section
P.O. Box 16700
Salt Lake City, UT 84116-0700   801-538-6752

VA UST CONTACT
VA State Water Control Board
P.O. Box 11143
2111 North Hamilton Street
Richmond, VA 23230-1143   804-367-6685

VT UST CONTACT
VT Dept. of Natural Resources
Underground Storage Tank Program
103 South Main Street, West Building
Waterbury, VT 05676   802-244-8702

WA UST CONTACT
WA Department of Ecology
Solid & Hazardous Waste Program/
   Underground Storage Tank Unit
4224 Sixth Avenue
Rowesix, Bldg. 4, Mail Stop PV-11
Olympia, WA 98504-8711   206-459-6272

WI UST CONTACT
WI Dept. of Industry, Labor & Human
   Relations
P.O. Box 7969
201 East Washington Avenue
Madison, WI 53707-7969   608-267-9725

WV UST CONTACT
WV Waste Management Division
WV Department of Natural Resources
Underground Storage Tank Section
1260 Greenbrier Street
Charleston, WV 25311   304-348-5935

WY UST CONTACT
WY Dept. of Environmental Quality
Water Quality Division
Herschler Building, 4th Floor
122 West 25th Street
Cheyenne, WY 82002   307-777-7081

AS UST CONTACT
Environmental Protection Agency
Office of the Governor
American Samoa Government
ATTN: UST Program
Pago Pago, American Samoa 96799
     684-633-2682

CNMI UST CONTACT
Division of Environmental Quality
P.O. Box 1304
Commonwealth of Northern Mariana Ids
Saipan, CM 96950   607-234-6984

GU UST CONTACT
Environmental Protection Agency
IT&E
Harmon Plaza, Complex Unit D-107
130 Rojas Street
Harmon, Guam 96911   671-646-8863

PR UST CONTACT
Water Quality Control
Environmental Quality Board
P.O. Box 11488
Commonwealth of Puerto Rico
Santurce, Puerto Rico 00910   809-725-8410

VI UST CONTACT
Environmental Protection Division
Dept. of Planning and National Res.
Suite 213, Nisky Center
Charlotte Amalie
St. Thomas, Virgin Islands 00802   809-774-3320

Glossary

Acid gas A by-product of incomplete combustion of solid waste and fossil fuels with a pH value of less than 6.5.

Acid rain A form of pollution created by the release of acid gases from incinerators, factories, or fossil-fuel burning power plants; any acidifying atmospheric deposition (e.g., acid snow, acid fog, acid fallout) commonly has a low pH because of the presence of sulfuric or nitric acid.

Administrative order An order issued by the EPA administrator (or designee) to a violator of the Resource Conservation and Recovery Act (RCRA) provisions that impose enforceable legal duties (e.g., forcing a facility to comply with specific regulations); four types of RCRA orders are compliance orders, corrective action orders, monitoring and analysis orders, and imminent hazard orders.

Aeration The process of exposing compost material to air.

Aerobic Requiring the presence of free oxygen.

Agricultural solid waste Manure, plant stalks, hulls, and leaves produced from farming.

Air classification A process of separating light and heavy shredded solid waste by injecting an air stream into a controlled chamber.

Air classifier A mechanical device that separates solid waste into light and heavy components by using a high-speed air stream.

Air knife A blower device that employs an air steam to push selected material(s) off a conveyor.

Alloy Metal produced by combining a basic metal with other metals or nonmetals to attain certain properties.

Anaerobic In the absence of free oxygen.

Animal bedding An agricultural product, occasionally made from waste paper, for use in livestock quarters.

Antiscavenge ordinance A governmental regulation prohibiting the unauthorized collection of secondary materials set out for pickup by a designated collector.

Aquifer Rock or sediment in a formation or group of formations, or part of a formation that is saturated and sufficiently permeable to transmit significant amounts of water to wells and springs.

Ash The residue that remains after a solid waste or fossil fuel has been incinerated.

Association of State and Territorial Solid Waste Management Officials (ASTSWMO) An organization based in Washington, D.C., composed of state agencies charged with regulating landfill disposal and incineration.

At-the-desk-separation The sorting and storage of recyclable office papers on or beside an employee's desk.

Auto wrecking Dismantling scrap automobiles to recover reusable and recyclable materials, followed by crushing and/or shredding the automobiles.

Avoided costs Solid waste management costs saved by a recycling program. One such saving comes from avoiding disposal fees. Another avoided cost is the saving in garbage collection fees through rerouting and extending truck life.

Back-end system   The portion of a resource recovery facility where reclaimed materials are extracted from the residue of incinerated municipal solid waste.

Backyard composting   The controlled biodegradation of leaves, grass clippings, and/or other yard wastes on the site where they were generated.

BACT   Acronym for best available control technology for control of pollution, as defined by the EPA.

Bacteria   Single-cell, microscopic organisms, some of which are important in the treatment of solid waste. Bacteria can be aerobic (requiring oxygen to live), anaerobic (able to live without oxygen), or facultative (able to live with or without oxygen).

Baghouse   An air pollution control device that uses a large filter bag to trap particulate emissions before they are released into the atmosphere.

Basic oxygen furnace (BOF)   A steel production furnace that oxidizes molten pig iron. Pure oxygen enters the furnace at high speeds by a lance immersed in the charge.

Beneficiation   In recycling, the mechanical process to remove contaminants and clean scrap glass containers; originally a mining industry term for the treatment of a material to improve its form or properties, such as crushing ore to remove impurities.

Best engineering judgment (BEJ)   The specific requirements that an owner or operator of a hazardous waste facility must comply and incorporate into the facility's operating permit.

Beverage Industry Recycling Program (BIRP)   A coalition of state beverage producers, packagers, wholesalers, and retailers that undertakes activities in support of recycling, particularly buy-back centers; operate in a handful of states.

Biennial Report   A report (EPA Form 8700-13A) submitted by generators of hazardous waste to the regional administrator due March 1 of each even-numbered year. The report includes information on the generator's activities during the previous calendar year. The owner or operator of a treatment, storage, and disposal facility must also prepare and submit a biennial report using EPA Form 8700-1313.

Bill of lading   A form that accompanies the shipment of a load of secondary materials and acknowledges its receipt.

Bioaccumulation   The accumulation of potentially toxic substances in organisms at successively higher levels on the food chain; also called *biomagnification.*

Bioconversion   The conversion of organic waste by means of biological decomposition caused by bacteria and/or fungi to produce usable gases or compost products.

Biodegradable   A material that can be consumed by organisms and broken down to simple substances such as carbon dioxide and water.

Bottom ash   The residual solids left after incinerating waste; also called *residue.*

Broker   A firm that purchases but typically does not take physical possession of secondary materials from processors for resale to consumers, acting as an intermediary in the marketplace

Brown goods   Obsolete electronic products, such as radios, stereos, and televisions.

Bulky waste   Large waste such as furniture, appliances, tires, branches, stumps, and trees.

Buy-back   The repurchase of recyclable products from the public.

Buy-back center   A facility that purchases reclaimed waste material from the public or waste brokers to reintroduce on the commodity market in a raw or processed form.

Buy-back recycling center   A commercially located, staffed recycling facility that purchases small amounts of postconsumer secondary materials (e.g., aluminum cans, glass containers, and newspapers) from the public but does little processing of them.

By-product   A material produced without separate commercial intent during the manufacture of processing of other materials.

Cap/cover   A permanent layer of impervious material (e.g., clay, polyethylene liner, PVC liner) added to the cover upon closure of a landfill.

Chlorinated hydrocarbon   Synthetic organic molecules in which one or more of the hydrogen atoms are replaced by chlorine atoms, resulting in nonbiodegradeable compounds that therefore pose a potential health hazard. Many have been shown to produce cancer in laboratory animals. Also referred to as *organochlorides.*

Civil action   A lawsuit regarding hazardous waste filed in court against a person or business that has failed to comply with statutory or regulatory requirements or an administrative order or has contributed to a release of hazardous wastes or constituents, categorized as compliance, corrective, monitoring and analysis, and imminent hazard.

Classification The act of separating waste materials manually by screening, or by air classification into categories of size, weight, and/or color.

Clean Air Act (CAA) Regulations to prevent discharges of substances that may harm public health or natural resources into the air by both stationary sources of pollution (e.g., factories) and mobile sources (e.g., cars, trucks, aircraft). (*See* 40 CFR 50-80.)

Clean Water Act (CWA) The legislation that regulates the discharge of nonhazardous waste into surface waters by municipal, industrial, and other specific and nonspecific sources and ultimately to eliminate all discharges into surface waters. Its interim goal is to make all surface waters usable for fishing and swimming. (*See* 40 CFR 100-140, 40 CFR 400-470.)

Closure The act of the owner or operator of a waste facility to secure it or a unit in it pursuant to the requirements of 40 CFR Part 264 to minimize the need for further maintenance, and controls or eliminate, to the extent necessary threats to human health and the environment and to eliminate postclosure escape of waste. A written closure plan must be submitted and approved.

Co-composting The composting of municipal solid waste and wastewater treatment plant sludge.

Code of Federal Regulations (CFR) A document containing all finalized federal regulations.

Cogeneration The production of both steam and electricity by a power generation facility (e.g., a waste-to-energy facility).

Collection The process of picking up wastes at homes, businesses, institutions, and other locations, loading them into an enclosed collection vehicle, and hauling them away for processing.

Combustion The ignition of oxygen with an organic substance that results in the production of energy.

Combustion air Air blown into a furnace to provide oxygen to a fire that burns refuse; commonly is blown from under the grate (under fire air) and directly into and over the flame (overfire air).

Co-mingle (or commingle) To blend together similar recycled materials such as mixed brown, green, and clear glass but to separate from disposable materials in the waste stream.

Commercial solid waste Solid waste generated by wholesale, retail, or service businesses and multiunit residential structures; one form of municipal solid wastes.

Common carrier A firm licensed to ship materials for a fee.

Compactor A power-driven device used to compress and reduce the volume of wastes or secondary materials.

Compatibility The ability of materials to exist together without adverse environmental effects or health risks; primarily applied to waste fluid combinations and liner materials.

Compliance order/action An order or action issued under Section 30008-(a) of RCRA, requiring any person not complying with a requirement of RCRA to take steps to come into compliance.

Component separation The separation of wastes into classes including paper, cardboard, plastics, food waste, glass, metals, yard waste, leather, rubber, and miscellaneous other materials.

Composition The components of solid waste, with the amount of each component expressed as a percentage of the total waste.

Compost As a noun, a mixture of organic wastes aerobically decomposed to an intermediate, relatively stable state that is environmentally inert; can be used as a soil conditioner; as a verb, to decay.

Composting The controlled disposal of solid organic wastes by means of biological decomposition to a state where the product, compost, is environmentally inert. Compost is used as a soil conditioner. Composting can be conducted out doors in windrows or in mechanical aeration tanks within a resource recovery facility.

Comprehensive Environmental Response Compensation Liability Act (CERCLA) Passed in 1980, gives the federal government the power to respond to releases or threatened releases of any hazardous substance into the environment and of a pollutant or contaminant that may present an imminent and substantial danger to public health or welfare; established the Hazardous Substance Trust Fund (SuperFund), which is available to finance responses taken by the federal government. Spills or other releases of significant quantities of a substance listed as hazardous must be reported at once to the National Response Center at (800) 424-8802. (*See* 40 CFR 300.)

Conservation The preservation and wise use of natural resources.

Construction and Demolition Waste Waste material produced in the construction, remodeling, repair, or demolition of buildings, homes, industrial plants, pavements, and structures.

Consumption The amount of any resource or energy used in a given time by a given number of people.

Containment Any method or technology that prevents migration of waste into the environment.

Contaminant (1) Any solute that enters the hydrologic cycle through human behavior, or (2) a material that is harmful to the recycling process when included with a recyclable material; called *contraries* in some countries.

Contamination Impurity.

Contingency Plan A document setting out an organized, planned, and coordinated course of action to be followed in case of a fire or explosion or a release of hazardous waste or hazardous constituents from a treatment, storage, or disposal facility that could threaten human health or the environment.

Corrective Action/ Order An order the EPA issues that requires corrective action under RCRA Section 3008(h) at a facility when there has been a release of hazardous waste or constituents into the environment. Corrective action may be required beyond the facility boundary and regardless of when the waste was placed at the facility.

Cover The soil applied over waste at a landfill at the end of each work day.

Criminal action A prosecuting action taken by the U.S. government or a state toward any person(s) who has knowingly and willfully not complied with the law. Such an action can result in the imposition of fines or imprisonment.

Criteria pollutants Those airborne chemical pollutants used to define the National Ambient Air Quality standard. In accordance with the Clean Air Act, these include carbon monoxide, ozone, lead, sulfur dioxide, nitrogen oxides, and hydrocarbons.

Decomposition The breakdown of organic wastes by various means. Complete chemical oxidation leaves only carbon dioxide, water, and inorganic solids.

Deep-well injection The subsurface implacement of fluids through a bored, drilled, or driven well, or through a dug well whose depth is greater than the largest surface dimension.

Demolition wastes Bulky wastes produced from the destruction of buildings, roads, sidewalks, etc. These wastes usually include large, broken pieces of concrete; miscellaneous construction metals; bricks; and glass.

Designated facility A hazardous waste treatment, storage, or disposal facility that has received an EPA or state permit (or has interim status) and has been designated on the manifest by the generator as the facility to which the generator's waste should be delivered.

Destruction removal efficiency (DRE) A specific mathematical formula used to determine how efficiently an incinerator works. By law, incinerators must destroy and remove 99.99:99.9999% of each POHC. (*See* 40 CFR Part 264.343).

Detection level The minimum concentration of a substance that analytical techniques can detect with some degree of accuracy in various environmental samples, such as groundwater or soil.

Discharge The accidental or intentional spilling, leaking, pumping, pouring, emitting, or dumping of waste onto any land or into any water.

Disposal All activities associated with the long-term handling of both collected solid wastes and residual wastes that occur after solid waste is processed or recovered for conversion products. Ultimate disposal of wastes is usually accomplished by means of sanitary landfilling.

Disposal facility A facility or part of a facility at which solid and/or hazardous waste is intentionally placed into or on any land or water, and at which waste will remain after closure.

Domestic waste *See* residential solid waste.

Dry scrubber An air pollution control device that removes particulate and gaseous pollutants from exhaust emissions by injecting a reagent into the exhaust flue of a combustion chamber.

Eddy current separation The separation of nonferrous scrap metals from mixed materials. The mixed materials pass through a varying magnetic field, creating eddy currents in the nonferrous scrap. These currents counteract with the magnetic field and exert a repelling force on the metals, thus separating the metals from the other materials.

Electric-arc furnace A metal recycling furnace generating heat by using an electric arc between carbon electrodes and the furnace charge. In electric arc steel furnaces the charge is almost entirely ferrous scrap.

Electrodynamic separator Equipment used to induce a magnetic field in the waste stream, usually on a conveyor belt, that causes a field of opposite magnetic polarity to push normally nonmagnetic white materials (aluminum cans) out of the waste stream for recycling; also referred to as an *eddy-current separator*.

Electrostatic precipitator (ESP) An air pollution control device that traps electrically charged particles out of an air emissions exhaust stream by attracting them to plates with an opposite electrical charge.

Elutriation The separation of small, light-density particles (usually resulting from shredding and screening operations) from coarser particles by using an upward flow of air passing through a screen.

End user Mills and other industrial facilities (e.g., paper mills, steel mills, and glass container production plants) where secondary materials are converted into new materials.

Endangered and threatened species Species whose populations are so reduced in number or whose range is so limited in geographic extent that further reduction in numbers or in size of available habitat could inalterably reduce the breeding success of the species and lead to subsequent extinction. These species are listed in Section 4 of the Endangered Species Act.

Energy recovery A form of resource recovery in which the organic part of the waste is converted to usable energy. Energy recovery from processed or raw refuse is achieved through combustion of the waste to produce high pressure steam used in an electric generation facility, through pyrolysis to produce an oil or gas product, and through anaerobic digestion to produce methane gas.

EPA Acronym for the U.S. Environmental Protection Agency.

EPA characteristics Any of four characteristics—ignitability, corrosivity, recativity, and EPA oxicity—that cause waste to be classified as hazardous according to RCRA.

EPA identification number The unique number assigned by the EPA to each generator or transporter of hazardous waste, and each treatment, storage, or disposal facility.

EP toxicity A test, called the extraction procedure (EP), that is designed to identify wastes likely to leach hazardous concentrations of a particular toxic constituent into the groundwater as a result of improper management. It is a characteristic of hazardous waste.

Exception report A report that generators who transport hazardous waste off-site must submit to the regional administrator if they do not receive a copy of the manifest signed and dated by the owner or operator of the designated facility to which their waste was shipped within 45 days from the date on which the initial transporter accepted the waste.

Existing facility A hazardous waste management facility that was in operation or for which construction began on or before November 19, 1980.

Feasibility study A planning document used as a first step to determine what types of waste management should be practiced, as well as their capital and operating costs, environmental effects, and recommended schedule of implementation; also referred to as a *planning study.*

Federal Hazardous Substances Act (FHSA) Allows the Consumer Product Safety Commission to ban or regulate hazardous materials produced for use by consumers. Under the act, the commission has labeling authority over consumer products that are toxic, corrosive, flammable, irritant, or radioactive. (*See* 16 CFR 1500-1512.)

Federal Insecticide, Fungicide, and Rodenticide Act (FIFRA) Provides regulatory authority for registration and use of pesticides and similar products intended to kill or control insects, rodents, weeds, and other living organisms. (*See* 40 CFR 162-180.)

Federal Register A document published daily by the federal government that contains either proposed or final regulations.

Final status A hazardous waste management facility that has interim status acquires final status when final administrative disposition has been made of its permit application.

Garbage Discarded food wastes and wastes likely to decompose.

Generator Defined under RCRA as any person who first creates a hazardous waste or any person who first makes the waste subject to the Subtitle C regulations (e.g., imports a hazardous waste, initiates a shipment of a hazardous waste from a treatment, storage or disposal facility [TAD] or mixes hazardous wastes of different Department of Transportation [DOT] shipping descriptions by placing them into a single container). In more general usage, a generator is an individual, company, organization, or activity that produces wastes or secondary materials.

Grab sample A single sample of a secondary material taken at no set time for evaluation or testing.

Grade A class of secondary material that is distinguished from similar classes on the basis of quality, color, use, content, appearance, density, or other factors. Grades of recyclable materials can be determined informally as part of common industry practices or officially as part of a trade association or governmental effort.

Ground Cover Material used to cover the soil surface to control erosion and leaching, shade the ground, and offer protection from excessive heaving and freezing; some are produced from yard-waste compost.

Groundwater Fresh water deposits, accumulating in subterranean aquifers in a zone of saturation. These deposits are recharged by infiltrating rain and snow melt and serve as reservoirs for wells and springs.

Guidance Documents issued mainly to elaborate and provide direction on the implementation of regulations.

Hazard ranking system The model used to determine inclusion of a waste site on the EPA's National Priorities List for CERCLA (SuperFund) cleanup.

Hazardous Materials Transportation Act (HMTA) Provides authority for regulating the transportation of hazardous material by road, air, and rail. The Department of Transportation's Materials Transportation Bureau (MTB) identified particular quantities and forms of materials as hazardous and specifies packaging, labeling, and shipping requirements for the materials that pose a risk to health, safety or property. (*See* 49 CFR 106, 107, 171-179.)

Hazardous and Solid Waste Amendments (HSWA) The 1984 act (Public Law 98-616) that significantly expanded both the scope and the coverage of RCRA.

Hazardous waste As defined in RCRA, a solid waste, or combination of solid wastes, that because of its quantity, concentration, or physical, chemical, or infectious characteristics may (1) cause or significantly contribute to an increase in mortality or an increase in serious irreversible, or incapacitating reversible, illness or (2) pose a substantial present or potential hazard to human health or the environment when improperly treated, stored, transported, or disposed of, or otherwise managed. As defined in the regulations, a solid waste is hazardous if it meets one of four conditions: (1) exhibits a characteristic of a hazardous waste (40 CFR Sections 261.20 through 262.24), (2) has been listed as hazardous (40 CFR Sections 261.31 through 261.33), (3) is a mixture containing a listed hazardous waste and a non-solid waste (unless the mixture is specifically excluded or no longer exhibits any of the characteristics of hazardous waste), or (4) is not excluded from regulation as a hazardous waste.

Hazardous waste handling facility A large-quantity generator that becomes involved in the on-site handling and treatment of hazardous waste and is subject to many of the same requirements of a TSDF, but without

being permitted for permanent storage or disposal.

Household waste *See* residential solid waste.

Incineration The process by which solid, liquid, or gaseous combustible wastes are burned and changed into gaseous byproducts and residue (referred to as *ash*).

Incinerator Any enclosed device in which controlled flame combustion is such that it cannot be classified as a boiler or as an industrial furnace.

Incompatible Two or more substances that cause fire, explosion, generation of flammable or toxic gases, or other violent reactions when mixed.

Industrial wastes Solid and liquid wastes generated by industry. Often this is in the form of slags, sludges, cakes, fines, and dusts. Only a few communities consider industrial waste a subset of municipal solid wastes.

Injection well *See* deep-well injection.

Inner liner A continuous layer of material placed inside a tank or container to protect the construction materials of the tank or container from the contained waste or reagents used to treat the wastes.

Institutional solid wastes Solid wastes generated by schools, hospitals, universities, museums, governments, and other institutions. Some communities define institutional solid wastes as *commercial solid wastes*. Institutional solid wastes are one portion of municipal solid wastes.

Institutional wastes Wastes from schools, hospitals, universities, museums, government buildings containing a high percentage of paper, medical waste, and food waste.

Integrated waste management A solid waste management strategy that ranks the preferred alternatives in the following order source reduction and reuse, recycling, resource recover (e.g., front-end recovery, waste-to-energy incineration), and landfill disposal.

Interim status Allows owners and operators of hazardous waste facilities (specifically TSDs) that were in existence, or for which construction had commenced, prior to November 19, 1980, to continue to operate without a permit after this date. Owners and operators of TSDs are eligible for interim status on an ongoing basis if the TSD is in existence on the effective date of regulatory changes under RCRA that cause the facility to be subject to Subtitle C regulation. Owners and operators in interim status are subject to Subtitle C regulation. Owners and operators in interim status are subject to and must comply with the applicable standards in 40 CFR Part 265. Interim status is gained through the notification process and by submitting Part A of the permit application.

Intermodal shipping The linking of two forms of transportation, such as trucks and railroads, to ship materials. For example, one might use intermodal shipping by loading secondary materials in a truck trailer, having it trucked to a railroad yard, putting the trailer on a rail car, moving the trailer by rail, and unloading it for truck delivery to the receiving mill.

Junkyard A common term for an auto dismantling or scrap metal processing facility.

Land treatment A facility or part of a facility at which waste is applied onto or incorporated into the soil surface. Such facilities are *disposal facilities* if the waste remains after closure.

Landfill A disposal facility or part of a facility where waste is placed in or on land and which is not a land treatment facility, a surface impoundment, or an injection well. (*Also see* sanitary landfill.)

Leachate Any liquid (often rainwater or other precipitation), including any suspended components in the liquid, that has moved through or drained from waste. Often contaminated with decomposed wastes, bacteria, and other materials; not to be confused with a leak.

Liability The state of being legally responsible for property damage or bodily injury caused during operation, closure, or postclosure phases of a waste management facility.

Liner A continuous layer of natural (e.g., clay) or synthetic material (e.g., HDPE, PVC), beneath or on the sides of a surface impoundment, landfill, or landfill cell that restricts the downward or lateral escape of waste, waste constituents, or leachate. Sophisticated liner systems today often combine layers of welded HDPE plastic sheets, sands, and clay. Leachate collection pipes are laid above the liner to collect leachate and drain it to a treatment plant.

Listed wastes Hazardous wastes that have been placed on one of three lists developed by the EPA nonspecific source wastes; specific source wastes; and commercial chemical products. These lists were developed by examining different types of waste and chemical products to see if they exhibit one of four characteristics, meet the statutory definition of hazardous waste, are acutely toxic or acutely hazardous, or are otherwise toxic.

Manifest The shipping document (EPS Form 8700-22) used for identifying the quantity, composition, origin, routing, and destination of hazardous waste during its transportation from the point of generation to the point of treatment, storage, or disposal.

Marsh A wetland dominated by emergent vegetation.

Mass burn Combustion of solid waste without preprocessing, as in a mass-burn incinerator.

Materials balance An accounting of the mass of solid waste entering and leaving a process such as an incinerator.

Materials recovery The extraction from the waste stream of economically reusable substances or byproducts by manual and/or mechanical separation and recovery. (*See* Resource Recovery.)

Migration route The environmental medium, that is, air, water, or land, through which waste can be released.

Mixed waste Randomly associated solid waste materials.

Monitoring Methods used to inspect and collect data on a facility's operational parameters or on contiguous air, groundwater, surface water, or soil quality.

Municipal wastes The combined residential and selected commercial solid wastes generated in a given municipality.

National Priorities List The list of hazardous waste sites targeted for clean-up action under CERCLA (SuperFund).

National Response Center Under CERCLA, all spills or discharges into the environment of certain amounts of materials designated as hazardous must be reported immediately to the National Response Center. The center maintains a 24-hour-a-day telephone line for reporting spills (800/424-8802). When the center is called, responsibility for dealing with the release is immediately assigned to either the Coast Guard or the EPA, depending on the location and type of emergency.

Natural resources The supply of materials, not created by humans, that are used for making goods; also called *raw, primary,* or *virgin materials.*

New facility A hazardous waste TSD facility that began operation or for which construction was begun after November 19, 1980.

NIMBY (Not-in-My-Back-Yard) An assertive response from a community or individual to the potential introduction of a waste management facility in their vicinity. NIMBY is often perceived by waste management experts as an unreasonable, reactionary response to their rational proposal. NIMBY has also been explained as a reasonable response to a perceived fear of health risk, loss of property value, or diminished quality of life.

Occupational Safety and Health Administration (OSHA) This agency provides the regulatory vehicle for assuring the safety and health of workers in firms generally employing more than ten people. Its goal is to set standards of safety that will prevent injury and/or illness among workers. Safety, chiefly encompassing the physical work place, and health, which governs exposure to settings that could induce acute or chronic health effects, are covered by the act. (*See* 29 CFR 1910, 1915, 1918, 1926.)

On-site Means on the same or geographically contiguous property that may be divided by public or private right(s)-of-ways, provided the entrance and exit between the properties is at a crossroads, intersection, and access is by crossing as opposed to going along the right(s)-of-way. Noncontiguous properties owned by the same person but connected by a right-of-way which the person controls and to which the public does not have access are also considered on-site property.

On-site handling, storage, and processing All activities associated with the handling, storage, and processing of solid wastes before being collected and taken to the disposal area.

One-hundred-year flood plain Areas adjacent to streams where the probability of flooding in any given years is one in a hundred.

Partial closure The closure of a discrete part of a hazardous waste facility in accordance with the applicable closure requirements of 40 CFR Parts 264 and 265. For example, partial closure may include the closure of a trench, a unit operation, a landfill cell, or a pit, while other parts of the same facility continue in operation. (A proposed redefinition was published on March 19, 1985; *see* 40 CFR 11068.)

Particulate matter Emissions from a waste-to-energy facility in the form of solid particles or condensable vapors (vapors that can become liquids under altered temperature or pressure conditions). Particulates are emitted as a result of incomplete combustion of fuel as well as the entrainment of noncombustible inert matter in the flue gases.

Permit An authorization, license, or equivalent control document issued by the EPA or an authorized state to implement the regulatory requirements of RCRA Subtitle C Parts 264 and 265 for hazardous waste TSD facilities.

Permit-by-rule A provision of Subtitle C whereby a hazardous waste management facility is deemed to have a RCRA permit if it is permitted under the Safe Drinking Water Act, the Clean Water Act, or the Marine Protection, Research, and Sanctuaries Act, and also meets a few additional Subtitle C requirements as specified at 40 CFR Section 270.60. The term *permit* does not include RCRA interim status, nor refer to any permit that has not yet been the subject of final agency action, such as a draft permit or a proposed permit.

Permit requirements Requirements in a RCRA permit, including ambient, performance, design, and/or operating standards contained in the regulations that the owner or operator must meet in perpetuity in constructing, operating, closing, and caring for the facility.

PM-10 Particulate matter below 10 microns in size. This smaller particulate size has been of particular concern to the regulatory agencies, who have recently promulgated stricter emissions standards for PM-10.

Point source Any discernible, confined, and discrete conveyance, including, but not limited to any pipe, ditch, channel, tunnel, conduit, well, discrete fissure, container, rolling stock, concentrated animal feeding operation, vessel, or other floating craft, from which pollutants are or may be discharged. This term does not include return flows from irrigated agriculture.

Pollution The contamination of the environment by the exposure of wastes or other harmful materials. A result of human activity.

Postclosure requirements Monitoring and maintenance requirements for closed hazardous waste management units throughout the postclosure care period (i.e., 30 years); these are specified as part of facility specific permit conditions. A specific requirement of hazardous waste management units.

Principal organic hazardous constituents (POHC) Constituents specified in a hazardous waste incinerator's facility permit from among those constituents listed in 40 CFR Part 261, Appendix VIII, for each waste feed to be burned. The specification is based on degree of difficulty of incineration, of organic constituents in the waste, and on their concentration or mass in the waste feed.

Process waste Discards from an industrial process.

Pyrolysis A waste treatment process that involves the chemical decomposition of material by heat, in the absence of oxygen, yielding a gaseous or liquid product that can be used as a fuel.

Qualifying Facility Status Certificate A certificate issued by the Federal Energy Regulatory Commission to the effect that the facility qualifies as a small power production facility (as defined in section 210 of PURPA).

Reclamation The restoration of air, land, or water to a better or more useful state, such as reclamation of strip-mined land by sanitary landfilling or reclamation of abandoned landfills by recreational park land.

Recoverable resource Material that can be reused or recycled after servicing a specific purpose because it still has useful physical or chemical properties.

Recovered material Material and byproducts that have been recovered or diverted from solid waste. This term does not include those materials and byproducts generated from, and commonly reused within, an original manufacturing process (e.g., mill broke and home scrap).

Recyclable When commonly used, refers to the technical ability of a material to be reused in manufacture. A more precise definition incorporates requirements that a recycling collection, processing and market system be in place and economically functioning for a material to be recyclable. Using this definition, many materials are technically capable of being recycled but are not considered recyclable due to the lack of a viable recovery system.

Reduction To diminish the amount of solid waste destined for disposal.

Regional administrator The highest ranking official in each of the ten EPA regions.

Regulation he legal mechanism that spells out how a statute's broad policy directives are to be carried out. Regulations are published in the *Federal Register* and then codified in the Code of Federal Regulation (CFR).

Regulatory compliance Meeting the requirements of federal or state regulations regarding facility design, construction, operation, performance, closure, and postclosure care.

Remedial action Those clean-up measures consistent with a permanent remedy taken instead of or in addition to removal.

Remedial investigation/feasibility study A comprehensive study of hazardous waste site conditions, threats posed, the effectiveness of alternative remedial actions and costs associated with each alternative. Recommends activities to be carried out to address site-contamination problems and pollutant sources.

Removal action The cleanup or removal of released hazardous substance from the environment to prevent, minimize, or totally mitigate damage.

Representative sample A sample of a whole (e.g., waste pile, lagoon, groundwater, or waste stream) that can be expected to have the average properties of the whole.

Residential wastes Wastes generated in homes generally consisting of consumer goods wastes, including newspaper, cardboard, beverage and food cans, plastics, glass, and food, garden, and lawn wastes.

Residue The solid matter remaining after completion of a physical or chemical process, such as incineration.

Resource conservation Reduction of the amounts of solid waste that are generated, reduction of overall resource consumption, and utilization of recovered resources. Accomplished through waste reduction, recycling, composting, and resource recovery techniques.

Resource Conservation and Recovery Act (RCRA) A 1976 federal law on which much of the U.S. Environmental Protection Agency's solid and hazardous waste program is based. Commonly referred to as RCRA, this act is an amendment to the first piece of federal policy on solid waste management called the Solid Waste Disposal Act of 1965. RCRA was amended in 1980 and again on November 8, 1984, by HSWA. Although RCRA was passed to control all varieties of solid waste disposal, both hazardous and nonhazardous, and to encourage recycling and alternative energy sources, its major emphasis during the 1970s and 1980s was the control of hazardous waste disposal. (*See* 40 CFR 240-271.)

Resource recovery The recovery of materials or energy from waste often via a high technology, physical/chemical conversion facility.

Reuse To use a material more than one time.

Safe Drinking Water Act (SDWA) This act requires the establishment of uniform federal standards for drinking water quality and development of a system to regulate underground injection of wastes and other substances that could potentially contaminate undergroundwater sources. Note that surface water is protected under the CWA. Among other things, the act bans the underground injection of certain materials in or near an undergroundwater source and requires issuing of permits, monitoring, and recordkeeping for underground injection that is permitted. (*See* 40 CFR 140-149.)

Sanitary landfill As defined by the American Society of Civil Engineers, a method of disposing of refuse on land without creating nuisances or hazards to public health or safety by utilizing the principles of engineering to confine the refuse to the smallest practical volume and to cover it with a layer of earth at the end of each day's operation, or such more frequent intervals as may be necessary.

Scrubber   A piece of equipment used to control acid gas emissions at a waste-to-energy facility.

Secondary containment   This applies to containers and tanks. In container systems, secondary containment consists of a base (concrete or other impervious material) with the capacity to contain 10 percent of the volume of the container or a volume of the largest container, whichever is greater. In tank systems, secondary containment includes one or more of the following: an external liner, a vault, a double-walled tank, or equivalent device.

Site   The land or water area where any facility or activity is physically located or conducted, including adjacent land used in connection with the facility or activity.

Slag   The material that forms from the melting and subsequent cooling of ash and solid byproducts within the combustion chamber of an incinerator.

Sludge   Any solid, semisolid, or liquid waste generated from a municipal, commercial, or industrial wastewater treatment plant, water supply treatment plant, storm water control facility, or air pollution control facility, exclusive of the treated effluent from a wastewater treatment plant.

Slurry wall   An underground vertical wall made of relatively impermeable material that significantly retards leachate and groundwater migration.

Small-quantity generator   A generator who produces less than 100 kilograms of hazardous waste per month (or accumulates less than 100 kilograms at any one time) or one who produces less that 1 kilogram of acutely hazardous waste per month (or accumulates less than 1 kilogram of acutely hazardous waste at any one time).

Solid Waste   As defined in RCRA, any garbage, refuse, sludge from a waste treatment plant, water supply treatment plant, or air pollution control facility and other discarded material, including solid, liquid, semisolid, or contained gaseous material resulting from industrial, commercial, mining, and agricultural operations, and from community activities, but does not include solid or dissolved material in domestic sewage, or solid or dissolved materials in irrigation return flows or industrial discharges that are point sources subject to permits under the Clean Water Act, or special nuclear or byproduct material as defined by the Atomic Energy Act of 1954. In more common usage this term means any waste materials produced by residents, businesses, institutions, and industry, but not liquids, hazardous wastes, and other nonsolid materials.

Solid waste management   A directed approach to the systematic management of solid wasted from the point of generation to the point of final disposal. This includes the elements of generation, on-site storage, collection, transfer and transport, recycling, reduction, processing, recovery, and final disposal.

Solid waste management plan   A plan developed to define the roles and objectives of managing solid wastes at any level: city, country, regional, state, or federal.

Solid waste management program   Incorporating solid waste management elements into a program to find a solution to existing or potential solid waste problems. The program can include engineering, master planning, financing, and addressing legal, institutional, and social concerns.

Source reduction   An action that reduces the generation of waste at the source. This often refers to the decreased generation of household solid waste. This is accomplished by reduced consumer consumption, increased product durability, reparability, or reusability, and reduced packaging. (*Also see* waste reduction and waste minimization.)

Special wastes   Any waste requiring special handling, such as scrap tires, used motor oil, hospital wastes, or household hazardous wastes.

Stabilization   The process by which wastes are rendered relatively inert, uniform, biologically inactive, nuisance-free, or harmless.

State hazardous waste plan   A scheme generated at the state level to deal with the management of hazardous waste generated, treated, stored, or disposed of within the state of transported outside the sate.

Statute   The law as passed by Congress and signed by the president.

Storage   The holding of hazardous waste for a temporary period, at the end of which the hazardous waste is treated, disposed of, or stored elsewhere.

SuperFund   *See* CERCLA.

Surface impoundment   A facility or part of a facility that is a natural topographic depression, excavation, or diked area formed primarily of earthen materials (although it may be lined with synthetic materials), which is designed to hold an accumulation of liquid wastes or wastes containing free liquids and which is not an injection well. Examples of surface impoundments are holding, storage, settling, and aeration pits, ponds, and lagoons.

Surface water Any body of water that is found above ground (e.g., canals, creeks, lakes, ponds, rivers, streams).

Thermal treatment The treatment of waste in a device using high temperatures as the primary means to change the chemical, physical, or biological character or composition of the waste. Incineration is an example of thermal treatment.

Topographic map A map indicating surface elevations of an area through the use of contour lines. It also shows population centers and other cultural and land-use features, surface water drainage patterns, and forests. These maps enable quick identification of areas of slopes that are more suitable for sanitary landfills.

Total dissolved solids (TDS) A measurement of all solids that are dissolved in water, waste water, or leachate, usually stated in milligrams per liter or parts per million.

Totally enclosed treatment facility A facility for the treatment of hazardous waste that is directly connected to an industrial production process and that is constructed and operated in a manner that prevents the release of any hazardous waste or any constituent of hazardous waste into the environment during treatment. An example is a pipe in which waste acid is neutralized.

Toxic Capable of producing injury, illness, or damage to living organisms through ingestion, inhalation, or absorption through any body surface. The United States Academy of Sciences defines the toxicity of a given material using four parameters (1) rate of release to the environment, (2) residence time in the environment, (3) potential for bioaccumulation, and (4) adverse effects on health.

Toxic Substances Control Act (TSCA) Provides the regulatory vehicle for controlling exposure and use of raw industrial chemicals that fall outside the jurisdiction of other environmental laws. Where other environmental laws control chemicals during use, transport, and disposal, the TSCA was passed to assure chemicals would be evaluated before use to make sure they pose no unnecessary risk to human health or the environment. (*See* 40 CFR 700-799.)

Transporter Any person engaged in the off-site transportation of hazardous waste within the United States, by air, rail, highway, or water, if such transportation requires a manifest under 40 CFR Part 262.

Trash Commonly used term for solid wastes, including yard wastes but not food wastes.

Treatment Any method, technique, or process including neutralization designed to change the physical, chemical, or biological character or composition of any hazardous waste so as to neutralize it or render it nonhazardous or less hazardous or to recover it, make it safe to transport, store, or dispose of, or amenable for recovery, storage, or volume reduction.

Trial Burn A trial incineration of hazardous waste that tests an incinerator's destruction removal efficiency.

TSDF Acronym for hazardous waste treatment, storage, or disposal facility.

Underground Storage Tank (UST) Under RCRA, an underground storage tank is defined as any tank with at least 10 percent of its volume buried below the ground, including any pipes attached to the tank. Thus, above ground tanks with extensive underground piping may be regulated under Subtitle I of RCRA.

Unsaturated zone The geological layer below the earth's surface and above the uppermost seasonal level of the water table.

Waste diversion credit A financial incentive provided to municipalities or private recycling operations based on the tonnage diverted from the waste stream.

Waste exchange An action leading to the reduction of waste generated by another firm.

Waste management practices Refers to aspects of a facility's design, operation, and closure that ensure protection of human health and the environment while treating, storing, or disposing of waste.

Waste minimization An action leading to the reduction of waste generation, particularly by industrial firms. (*Also see* waste reduction and source reduction.)

Waste reduction The decreased generation of solid waste. This is accomplished by changing or reducing consumer consumption, increasing product durability, reparability, or reusability, changing packaging practices or reducing packaging, and introducing new production technologies that are less wasteful. (*Also see* waste minimization and source reduction.)

Waste sources Agricultural, residential, commercial, and industrial areas that generate solid wastes. Can also include treatment plants that generate sludge through processing.

Waste stream The waste material output of a community, region, or facility.

Waste-to-energy (WTE) A term used for municipal solid waste incinerators that recover the heat of combustion and use it to generate steam for heating or conversion to electricity.

Water table Either (1) the upper limit of the part of the soil or underlying rock material that is wholly saturated with water or (2) the upper surface of the zone of saturation in groundwaters in which the hydrostatic pressure is equal to atmospheric pressure.

Water-wall furnace Field-erected incineration equipment, the side of which consists of water-carrying tubes. As water circulates through the tubes, it extracts the energy produced by waste combustion and turns into steam.

Wet scrubber Fragile environments such as bogs, bayous, swamps, marshes, tidal flats, or other areas that are regularly inundated or saturated by groundwater or surface water with a frequency sufficient to support a prevalence of vegetative or aquatic life that requires saturated soil conditions for growth and reproduction.

Yard waste Leaves, grass clippings, branches, and other organic wastes produced as part of yard and garden development and maintenance.

Readings

GENERAL

Freeze, R. A., and Cherry, J. A. *Groundwater.* Upper Saddle River, NJ: Prentice Hall, 1991.

Hall, R. M., and Case, D. R. *All About Environmental Auditing.* Washington, DC: Federal Publications, Inc., 1994.

Hasan, S. E. *Geology and Hazardous Waste Management.* Upper Saddle River, NJ: Prentice Hall, 1996.

Masters, G. M. *Basic Environmental Technology.* Upper Saddle River, NJ: Prentice Hall, 1991.

Nathanson, J. *Basic Environmental Technology.* Upper Saddle River, NJ: Prentice Hall, 1997.

POLLUTION PREVENTION

Dennison, M. S. *Pollution Prevention, Strategies, and Technologies.* Rockville, MD: Government Institutes, 1996.

Freeman, H. *Hazardous Waste Minimization.* New York: McGraw-Hill, 1990.

Freeman, H., ed. *Industrial Pollution Prevention Handbook.* New York: McGraw-Hill, 1995.

Theodore, L., and McGuinn, Y. *Pollution Prevention.* New York: Van Nostrand Reinhold, 1992.

Theodore, L., Dupont, R., and Reynolds, J., eds. *Pollution Prevention Problems and Solutions.* Lausanne, Switzerland: Gordon and Breach Science Publishers.

U.S. EPA. *Innovative Treatment Technologies,* EPA/540/9-91/002. Washington, DC, 1991.

U.S. EPA, Office of Solid Waste and Emergence Response. *Let's Reduce and Recycle: Curriculum for Solid Waste Awareness,* EPA/330-SW-90-005. Washington, DC, 1990.

Vadja, G.F., and Stouch, J.C. *An Integrate Approach to Waste Minimization.* Pittsburgh, PA: Air and Waste Management, 1990.

SOLID AND HAZARDOUS WASTE

Blackman, W. D., Jr. *Basic Hazardous Waste Management.* Boca Raton, FL: Lewis Publishers/CRC Press, 1993.

Dominquez, G. S., and Bartlett, K. G. (eds.). *Hazardous Waste Management: The Law of Toxics and Toxic Substances.* Boca Raton, FL: CRC Press, 1986.

Powers, P. W. *How to Dispose of Toxic Substances and Industrial Wastes.* Park Ridge, NJ: Noyes Data Corp., 1976.

Shields, E. "Guidelines for Selecting a TSD Facility." *Hazardous Materials and Waste Management Magazine,* January 1985.

Tchobanoglouse, G., et al. *Integrated Solid Waste Management.* New York: McGraw-Hill, 1993.

U.S. EPA. *Does Your Business Produce Hazardous Waste? Many Small Businesses Do,* EPA/530-SW-90-027. Washington, DC.

U.S. EPA. *RCRA Handbook,* 2nd ed. Concord, MA: ERT, 1986.

RADIOLOGICAL WASTE

Berlin, R. E., and Stanton, C.C. *Radioactive Waste Management.* New York: John Wiley & Sons, 1989.

Craig, R. *Land Ban: Its Impact on Mixed Waste.* Proceedings of U.S. EPA Mixed Waste Workshop. Denver, CO, July 19–20, 1988.

Hild, W. "The Role of Incineration in the Management of Radwaste." *Incineration of Radioactive Waste,* ed. C. Eid and A.J. Van Loons. London: Graham & Trotman Ltd, 1985.

Kocher, D. C. "Classification and Disposal of Radioactive Wastes—Testing and Legal and Regulatory Requirements," *Rad Prot Man,* July 1990, 58–78.

League of Women Voters. *The Nuclear Waste Primer: A Handbook for Citizens.* New York: Nick Lyons Books, 1985.

Lindenfeld, P. *Radioactive Radiations and Their Biological Effects.* College Park, MD: American Association of Physics Teachers, 1984.

Murray, R. *Understanding Nuclear Waste,* DOE/RW-0262. Columbus, OH: Battelle Press, 1989.

National Council on Radiation Protection and Measurements. "Ionizing Radiation Exposures of the Population of the United States." *NCRP Report 93.* Bethesda, MD, 1987.

New York State Energy Research and Development Authority. *Handbook of Disposal Technologies for Low-Level Radioactive Waste.* New York, June 1987.

U.S. Department of Energy. *Transporting Radioactive Material... Answers to Your Questions,* DOE/DP-0064. Washington, DC, 1989.

U.S. Department of Energy. *Yucca Mountain Studies,* DOE/RW/0293P. Washington, DC, 1989.

U.S. NRC-EPA. *Combined Siting Guidelines for Disposal of Mixed-Low Level Radioactive and Hazardous Waste.* Washington, DC, March 13 1987.

U.S. NRC-EPA. *Guidance on the Definition and Identification of Commercial Mixed-Low Level Radioactive and Hazardous Waste.* Washington DC, January 8, 1987.

U.S. NRC-EPA. *Joint NRC-EPA Guidance on a Conceptual Design Approach for Commercial Mixed-Low Level Radioactive and Hazardous Waste Disposal Facilities.* Washington, DC, August 3, 1987.

MEDICAL WASTE

Center for Environment. *Model Guidelines for State Medical Waste Management.* Lexington, KY: Council of State Governments, 1992.

Freeman, D. S., and Siskind, G. H. *Medical Waste Handbook.* Deerfield, IL: Clark Boardman Callaghan.

REMEDIAL TECHNOLOGIES

Brunner, C. R. *Site Cleanup by Incineration.* Silver Spring, MD: Hazardous Materials Control Research Institute, 1988.

U.S. EPA. *Innovative and Alternative Technology Assessment Manual,* EPA/430/9-78/009. Washington, DC: Office of Water Program Operations (WH-547), 1978.

U.S. EPA. *Review of In-Place Treatment Techniques for Contaminated Surface Soils,* vols. 1 and 2, EPA/540/2-84-003a and b. Cincinnati, OH: Office of Solid Waste and Emergency Response, MERL, 1984.

U.S. EPA. *Permit Guidance Manual on Hazardous Waste Land Treatment Demonstrations,* final version, EPA/530/SW86-032. Washington, DC: U.S. EPA, Office of Solid Waste, 1986.

U.S. EPA. *RCRA Facility Investigation (RFI) Guidance,* vols. I–IV, EPA/530/SW-87-001 to 004; OSWER Directive 9502.00-6C. Washington, DC: Office of Solid Waste, Management Division, 1987.

U.S. EPA. *A Compendium of Technologies Used in the Treatment of Hazardous Waste,* EPA/625/8-87/014. Cincinnati, OH: Center for Environmental Research Information, 1987.

U.S. EPA. *Guide to Technical Resources for the Design of Land Disposal Facilities,* EPA/625/6-88/018. Cincinnati, OH: Center for Environmental Research Information, 1988.

U.S. EPA. *Guidance for Conducting Remedial Investigations and Feasibility Studies under CERCLA,* interim final, EPA/540/G-89/004. Washington, DC: Office of Emergency and Remedial Response, 1988.

U.S. EPA. *RCRA Corrective Action: Technologies and Applications,* seminar publication, EPA/625/4-80/020. Cincinnati, OH: Center for Environmental Research Information, 1989.

U.S. EPA. *Design and Construction of RCRA/CERCLA Final Covers,* seminar publication, EPA/625/4-91/025. Washington, DC: Office of Research and Development, 1991.

U.S. EPA. *RCRA Corrective Action Stabilization Technologies,* proceedings, EPA/625/R-92/041. Washington, DC: Office of Research and Development, 1992.

U.S. EPA. *Bioremediation of Hazardous Waste,* EPA/60/R-92/126. Washington, DC: Office of Research and Development, 1992.

U.S. EPA. *A Technology Assessment of Soil Vapor Extraction and Air Sparging,* EPA/600/SR-91/173. Cincinnati, OH: Center of Research and Development, RREL, 1992.

U.S. EPA. *Bioremediation Using the Land Treatment Concept,* EPA/600/R-93/164. Washington, DC: Office of Research and Development, 1993.

U.S. EPA. *Innovative Treatment Technologies: Annual Status Report,* 6th ed., EPA 542-R-94-005, no. 6. Washington, DC: Office of Solid Waste and Emergency Response (5102W), 1994.

U.S. EPA. *Superfund Innovative Technology Evaluation Program, Technology Profiles,* 7th ed., EPA/540/R-94/526. Washington, DC: Office of Research and Development, 19914.

Water Pollution Control Federation. *Hazardous Waste Site Remediation Management,* special publication. Alexandria, VA: WPCF, 1988.

Water Pollution Control Federation. "Hazardous Waste Treatment Processes, Including Environmental Audits and Waste Reduction." *Manual of Practice,* FD-18. Alexandria, VA: WPCF, 1990.

Index

ISBN 0-02-389545-4

90000

9 780023 895456